Non-Thermal Processing Technologies for the Fruit and Vegetable Industry

Fruits and vegetables rapidly spoil due to growth of microorganisms, which further render them unsafe for human consumption. The traditional methods of food preservation, which involve drying, canning, salting, curing, and chemical preservation, can significantly affect food quality by diminishing nutrients during heat processing. This can alter the texture of the products, leave chemical residues in the final processed products, which in turn has greater impact over consumers' safety and health concerns. To combat this problem, various current non-thermal food processing techniques can be employed in fruit and vegetable processing industries to enhance consumer satisfaction for delivering wholesome food products to the market, thus increasing demand.

Non-Thermal Processing Technologies for the Fruit and Vegetable Industry introduces the various non-thermal food processing techniques especially employed in the fruit and vegetable processing industries; it deals with the effect of several non-thermal processing techniques on quality aspects of processed fruit and vegetable products and keeping quality and consumer acceptability.

Key Features:

- Describes the high-pressure processing techniques employed for processing fruit- and vegetable-based beverages.
- Discusses the safety aspects of using various innovative non-thermal-based technologies for the fruit and vegetable processing industries.
- Explains ozone application, cold plasma, ultrasound and UV irradiation for fruits and vegetables with their advantages, disadvantages, process operations, mechanism for microbes in activation etc.
- Presents the commercially viable and economically feasible non-thermal processing technologies for the fruit and vegetable industry.

This book addresses professors, scientists, food engineers, research scholars, students, and industrial personnel for the stability enhancement of fruit- and vegetable-based food products by using novel non-thermal food processing techniques. Readers will come to know the current and emerging trends in use of non-thermal processing techniques for their application in several fruit- and vegetable-based food processing industries.

Non-Thermal Processing Technologies for the Fruit and Vegetable Industry

Edited by
M. Selvamuthukumaran

CRC Press
Taylor & Francis Group
Boca Raton London New York

CRC Press is an imprint of the
Taylor & Francis Group, an **Informa** business

First edition published 2023
by CRC Press
6000 Broken Sound Parkway NW, Suite 300, Boca Raton, FL 33487-2742

and by CRC Press
4 Park Square, Milton Park, Abingdon, Oxon, OX14 4RN

CRC Press is an imprint of Taylor & Francis Group, LLC

Library of Congress Cataloging-in-Publication Data

Names: Selvamuthukumaran, M., editor.
Title: Non-thermal processing technologies for the fruit and vegetable
   industry / edited by M. Selvamuthukumaran.
Description: First edition. | Boca Raton, FL : CRC Press, 2022. | Includes
   bibliographical references and index.
Identifiers: LCCN 2022020793 (print) | LCCN 2022020794 (ebook) | ISBN
   9781032080291 (hbk) | ISBN 9781032119274 (pbk) | ISBN 9781003222170
   (ebk)
Subjects: LCSH: Fruit--Processing. | Vegetables--Processing.
Classification: LCC TP440 .N66 2022 (print) | LCC TP440 (ebook) | DDC
   664/.8--dc23/eng/20220725
LC record available at https://lccn.loc.gov/2022020793
LC ebook record available at https://lccn.loc.gov/2022020794

ISBN: 9781032080291 (hbk)
ISBN: 9781032119274 (pbk)
ISBN: 9781003222170 (ebk)

DOI: 10.1201/9781003222170

Typeset in Kepler Std
by Deanta Global Publishing Services, Chennai, India

I profoundly thank

the Almighty,

my family,

my friends,

and

everybody

who have constantly encouraged and helped me to complete this book successfully

M. Selvamuthukumaran

Contents

Preface

Fruits and vegetables rapidly spoil due to the growth of microorganisms, which further renders them unsafe for human consumption. The traditional methods of food preservation—drying, canning, salting, curing, chemical preservation—which can significantly affect food quality by diminishing nutrients during heat processing, alter the texture of the products, and leave chemical residues in the final processed products, which in turn has a greater impact on consumer safety and health concerns. To combat this problem, nowadays various non-thermal food processing techniques can be employed in the fruit and vegetable processing industries to enhance consumers' satisfaction for delivering wholesome food products to the market. The available non-thermal processing techniques to preserve such fruits and vegetables include high pressure processing, cold plasma, ozone processing, supercritical CO_2, and UV irradiation.

This book introduces the various non-thermal food processing techniques especially employed in the fruit and vegetable processing industries. It describes the high pressure processing techniques employed for processing fruit- and vegetable-based beverages, and their effect on keeping quality acceptable for the consumer. It deals with the effect of several non-thermal processing techniques on quality aspects of processed fruit and vegetable products. It deals with the safety aspects of using various innovative non-thermal-based technologies for the fruit and vegetable processing industries. It also explains ultrasound, ozone application, cold plasma, and UV irradiation for fruits and vegetables along with their advantages, disadvantages, process operations, mechanism for microbes in activation etc. It narrates the commercially viable and economically feasible non-thermal processing technologies for the fruit and vegetable industry. In a nutshell, this book will mostly benefit food scientists, food process engineers, academicians, students, and food industrial persons by providing in-depth knowledge of non-thermal processing of foods for fruit and vegetable quality retention and also for effectively providing consumer acceptability.

I would like to express my sincere thanks to all the contributors, without whose continuous support this book would not have seen daylight. We would like also to express our gratitude to Mr. Steve Zollo and all other CRC Press personnel, who have made every cooperative effort to make this book a great standard publication at a global level.

M. Selvamuthukumaran

About the Editor

M. Selvamuthukumaran is a professor at the Department of Food Science and Technology, Hamelmalo Agricultural College, Eritrea. He was a visiting professor at Haramaya University, School of Food Science and Postharvest Technology, Institute of Technology, Dire Dawa, Ethiopia. He earned his Ph.D. in Food Science from the Defence Food Research Laboratory affiliated with the University of Mysore, India. His core area of research is the processing of underutilized fruits for the development of antioxidant rich functional food products. He has transferred several technologies to Indian firms as an outcome of his research work, and thus received several awards and citations for his endeavors. He has published several international papers and book chapters in the area of antioxidants and functional foods, and has guided several national and international postgraduate students in the area of food science and technology.

Contributors

Pervin Basaran
Istanbul Technical University
Istanbul, Turkey

Ayşenur Betül Bilgin
Istanbul Technical University
Istanbul, Turkey

Ayse Merve Cellat
Istanbul Technical University
Istanbul, Turkey

Dilara Devecioglu
Istanbul Technical University
Istanbul, Turkey

Dilara Nur Dikmetas
Istanbul Technical University
Istanbul, Turkey

Esra Dogu-Baykut
Istanbul Medeniyet University
Istanbul, Turkey

Gurbuz Gunes
İstanbul Teknik Üniversitesi
Istanbul, Turkey

Funda Karbancioglu-Guler
Istanbul Technical University
Istanbul, Turkey

Celale Kirkin
Istanbul Technical University
Istanbul, Turkey

Amanda Malik
Sant Longowal Institute of Engineering and
 Technology
Longowal, India

Sergio I. Martínez-Monteagudo
New Mexico State University
Las Cruces, New Mexico, USA

Vikas Nanda
Sant Longowal Institute of Engineering and
 Technology
Longowal, India

Sebahat Oztekin
Bayburt University
Bayburt, Turkey

Kirty Pant
Sant Longowal Institute of Engineering and
 Technology
Longowal, India

Luis Sabillón
New Mexico State University
Las Cruces, New Mexico, USA

Feride Sonverdi
Istanbul Technical University
Istanbul, Turkey

Mamta Thakur
ITM University
Gwalior, India

Chapter 1

Introduction to Non-Thermal Processing Applications in the Fruit and Vegetable Processing Industry

M. Selvamuthukumaran

CONTENTS

1.1 INTRODUCTION

Thermal processing of fruits and vegetables leads to the destruction of several nutritive components. Therefore it is essential to adopt non-thermal processing methods in order to retain valuable antioxidants and nutrients. The various non-thermal processing methods specifically adopted for the fruit and vegetable processing industry are represented in Figure 1.1 and briefly described below:

1.2 NON-THERMAL PROCESSING TECHNIQUES

1.2.1 Cold Plasma

This is one of the non-thermal processing techniques that is widely employed in the fruit and vegetable processing industry. The temperature used for processing various food products by using this technique can range from 25 to 65°C (Niemira, 2012). Ionization of gas leads to the formation of free radicals, plasma composition is typically based on gas composition, which are ionized

DOI: 10.1201/9781003222170-1

1

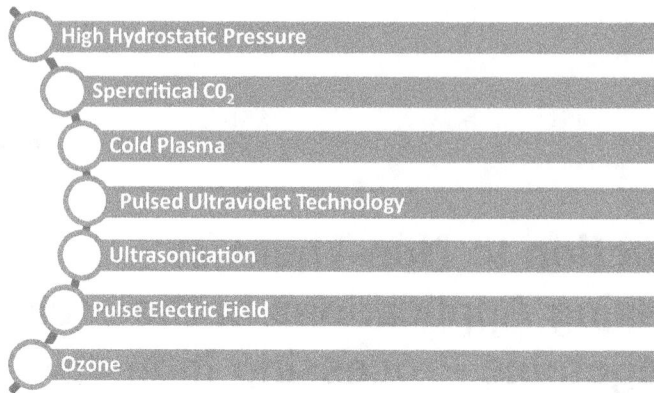

Figure 1.1 Various non-thermal processing methods adopted for the fruit and vegetable processing industry.

(Alves Filho et al., 2019), and for the creation of plasma various gases are used, including oxygen, nitrogen, helium, argon etc. (Keener and Misra, 2016). These gases can be subjected to either electrical energy or thermal energy or a magnetic field, so that plasma can be created, which may contain either a negative or positive ion or a singlet oxygen or ozone (Misra and Roopesh, 2019). This technique is highly useful for destroying enzymes, killing microbes on food surfaces, and treating packing material exclusively used for a variety of foods (Chizoba Ekezie et al., 2017). This can be even generated at room temperature, which prevents the nutrient degradation and thermal shock of foods, which are sensitive to heat especially when processing takes place at higher temperatures (Thirumdas et al., 2014).

The mechanism involved in microbes' destruction is ascribed due to the reactive species' effect on microbial cells. DNA is destroyed by these species, which can enhance protein oxidation, destroy microbial cellular components, and which ultimately lead to cell death (Phan et al., 2017). Bang et al. (2020) successfully enhanced the stability of mandarin oranges by adopting this technique. They used both antimicrobial washing cum in-package cold plasma treatment technique in order to destroy the microbes. By this technique, they successfully destroyed *Penicillium digitatum* by using kV energy of 26–27 for a treatment time period of a minimum of one minute to a maximum of four minutes. Treated orange fruits had a shortened ripening period compared to control. As a result of cold plasma treatment, they found that sensory, textural, and nutritional qualities were not further affected. Gan et al. (2021) applied this technique in chokeberry juice and studied the effect of *E. coli* and *S.cerevisiae*. They found that a four-minute treatment time period had successfully diminished the loads of *E. coli* and *S. cerevisiae* by 2.27 and 1.23 log CFU/ml. Hou et al. (2019) found that treatment of cold plasma to blueberry juice reduced the Bacillus species in juice to 7.2 log at six minutes. The color and bioactive constituents were retained as a result of cold plasma treatment. Fruit-based products like cloudy apple juice (Illera et al., 2019), tomato juice (Starek et al., 2019), whey incorporated grape juice (Amaral et al., 2018), apple, orange, and sour cherry nectar (Dasan and Boyaci, 2018) were treated successfully with these cold plasma treatment techniques.

1.2.2 Supercritical Fluid Processing

In this technique, supercritical fluids were used and replaced with organic solvents (Temelli et al., 2012). These organic solvents can leave traces of chemical residue after processing, or cause safety issues. Heating of fluid beyond critical pressure and critical temperature leads to a state known as "supercritical state" or "supercritical fluid". Such fluids possess properties of both gas i.e. diffusivity and viscosity, and liquid properties such as density. This technique is successfully employed in the fruit and vegetable processing industry to extract various bioactive constituents of plant origin. Beyond extraction, it can also be used to inactivate microorganisms. Fluid properties can be modified with a subsequent change in pressure and temperature. CO_2 is extensively used for extraction of bioactive constituents. A supercritical state can be achieved at a temperature of 31.1°C and pressure of 7.4 MPa compared to organic solvents. These supercritical fluids are non-toxic and can be easily separated from food products that are being used for extraction (Deotale et al., 2021).

Lefebvre et al. (2020) successfully extracted antioxidants from rosemary by using this technique. They found that the use of a temperature of 25°C and a CO_2 pressure of 20 MPa does not affect the extracted products' purity. Santos et al. (2020a) extracted the bioactive constituents from leaves of feijoa using this technique. They used a temperature of 55°C and pressure of 30 MPa, which led to further extraction of antioxidants and antibacterial components, which successfully reduced the growth of pathogenic bacteria especially *E. coli*. By using this technique, several researchers had extracted oil from fruit seed (Pavlic et al., 2020; Priyanka and Khanam, 2020; Santos et al., 2020b; Ferrentino et al., 2020), from olive fruit (Al-Otoom et al., 2014), from ginger (Salea et al., 2017), and also bioactive extraction i.e. lycopene, quercetin, carotenoids, anthocyanins, polyphenols were also successfully extracted by using this technique. (Pinto et al., 2020a; Gallego et al., 2019; Torres-Ossandón et al., 2018), and these extractants can be used as an ingredient in nutraceutical product formulation. The mechanism involved in microbial inactivation by using this technique reduces the pH of bacterial cells and results in cell rupturing and bursting. It also leads to enzyme inactivation; thereby the load is reduced (Spilimbergo and Bertucco, 2003). It can be used for enhancing the stability of fresh produce like fruits and vegetables and even juices (Silva et al., 2020). Supercritical CO_2 effect against microbes was studied in pomegranate juice by Bertolini et al. (2020). They observed that bacterial growth seems to be below the detection limit for stored product after a time period of 28 days. It also enhanced the phenolic content by 22%, they also further reported that antioxidant level of treated juice was higher, when compared to the traditional pasteurization processing method.

1.2.3 High Hydrostatic Pressure (HHP)

In this technique, water is used as a medium for generating applications to food products. This technique enhances the stability of the food products to a greater extent by significantly reducing the microbial population viz. gram positive and negative bacteria, as well as yeast and mold. Temperature and pressure used significantly decide the load of microbes on food products. Pressure applied for foods varied from 200 to 700 MPa (Van Loey et al., 2003). Therefore, subjecting foods to high pressure for shorter duration results in excellent food quality (Huang et al., 2020). The major advantage of using this technique is that it doesn't require any sophisticated equipment facilities. The treating foods need to be kept in a compartment, where pressure is generated by loading water into the chamber, and then food is subjected to specific pressure. Daher et al. (2017) reported that for destroying gram negative bacteria pressure in range of 350–450 MPa is being used and for gram positive bacteria, which may require use of higher pressure i.e. 1100 MPa. The mechanism involved in microbial reduction is that subjecting foods to higher pressure may result in microbial cell membrane damage, further leading to breaking of protein structure, thereby altering the metabolic pathway, so that microbial population is reduced (Ferrentino et al., 2020).

Bulut and Karatzas (2021) applied HHP on orange fruit juice at 250 MPa for a time period of 900s. They found that the microbial load was diminished to 4.88, 4.15, and 4.61 log CFU/ml under juice pH of 3.2, 4.5, and 5.8. In addition to microbial load reduction, they can be employed for extraction of antioxidants. The antioxidants were successfully extracted by using this technique in tomato waste (Nincevi et al., 2020), grape pomace (Cascaes et al., 2020), and gooseberry juice (Torres-Ossandón et al., 2018). Rios-Corripio et al. (2020) reported that this technique increases the physical as well as chemical properties of fermented juice and enhances bioactive constituents of fermented juice.

1.2.4 Pulsed Ultraviolet Technology

Ultraviolet technology is adopted to decrease microbes in food material surfaces. Food material indirectly exposed to UV-A radiation in the electromagnetic spectrum range of 320–400 nm, UV-B in the range of 280–320 nm and UV-C in the range of 200–280 nm (Popovic et al., 2021). The mechanism involved in microbial reduction is when the food is exposed to UV-C the wavelength is absorbed by microbial cells, nucleic acids, and the bond will be broken as a result of photon absorption, and resulting in the generation of dimers. These dimers can block the transcription and translation process of DNA which induces genetic material malfunctioning, and which can ultimately bring cell death (Guerrero-Beltrán and Ochoa-Velasco, 2021). Similarly, photons of both UV-A and UV-B are also significantly reduced with the use of microbial cells (Koutchma et al., 2021).

Fenoglio et al. (2020) inactivated the microbes by subjecting the fruit juices to UV-C with 390 mJ/cm^2. They achieved *L. plantarum* with a log reduction of 6.3, for *E. coli*, it is 5.1 log CFU, and for *S. cerevisiae* it is around 5.5. Other researchers also confirmed similar findings of microbial reduction in orange juice (Ferreira et al., 2020), cantaloupe melon juice (Fundo et al., 2019), and apple juice (Xiang et al., 2020). Dyshlyuk et al. (2020) projected that treatment of UV-rays for freshly harvested fruits and vegetables produce significantly reduces the microbial population with enhanced antioxidant activity. It is a simple technique, and the operation is very simple, therefore the fruit and vegetable processing industry can adopt easily this technique to enhance the shelf life of their products. The UV effect can be widely achieved especially when it is merged with other non-thermal processing techniques.

1.2.5 Ozone

The chemical name of ozone is O_3 and it possesses 3 molecular oxygen. Ozone is a colorless gas with a peculiar odor. Ozone is generated when oxygen (O_2) is merged with single oxygen. It can be used to control food spoilage as the gas possesses strong antibacterial properties. The produced gas can be directly used on food materials or it can be incorporated with water, so that ozonized water can be utilized by the food industry, and this ozone can cause the microorganism to die. It changes cell permeability by causing damage to the cell membrane of microbes. It also affects the protein structure which ultimately results in microbe enzyme malfunctioning, thereby hindering the metabolic activity, which leads to the cell death of the microorganism (Oner and Demirci, 2016; Tiwari et al., 2010). Pinto et al. (2020b) studied the stability of fruits by treating the fruits with ozone. They found that ozone treatment of freshly harvested produce augmented the textural as well as physic-chemical properties with significant microbial load reduction for a time period of 15 days, when fruits were packed and stored at MAP. Similarly other researchers – Patil et al. (2010), Choi et al. (2012), Porto et al. (2020) – also reported microorganism reduction in various fruit juices. In addition to these properties, toxin levels also can be reduced in foods when subjected to ozone treatment (Tiwari et al., 2010).

1.2.6 Ultrasonication

In the food processing industry, ultrasonication is an emerging technique (Chemat et al., 2011). It is nothing but a sound wave which can possess some frequency that is greater than ordinary human hearing frequency i.e. more than 20 kHz (Mason and Cintas, 2007). Several expansions cum compression effects were obtained particularly when ultrasonic waves oscillated through medium. In the presence of air, small cavities are formed and can grow to a particular size and collapse later. During cavity collapsing, a large amount of energy is produced, which ultimately results in enhancing the mass and heat transfer rates (Bhangu and Ashokkumar, 2016). It can be used with various frequencies, starting from low to medium and high (20–100 kHz) (100 kHz–1 MHz) (1–100 MHz) (Mason et al., 2015) (Figure 1.2).

In medium frequency levels, higher shear forces are produced, particularly when low frequency ultrasonication is used and lesser shear forces are produced – a higher frequency level was typically used. This process is carried out by placing the food in an ultrasonic bath in which either unpacked or packed foods are kept. Sound waves are created in a bath that can generate an ultrasound effect, thus the desirable changes in foods were achieved (Li et al., 2021).

For food processing, a frequency between 20 kHz and 100 kHz can be used. This technique is used to extract various bioactive constituents from plant sources because the yield is significantly improved with the enhancement of physic-chemical characteristics of extracted compounds. Cheila et al. (2020) adopted this technique for bioactive extraction from velamen leaves. They found that the yield of bioactives was enhanced to 94% with time duration of 39.5 minutes. Similarly, oil is extracted from olive fruit by this technique (Cavallo et al., 2020). Iftikhar et al. (2020), Qin et al. (2021), Wu et al. (2020), Martínez-Ramos et al. (2020) reported that bioactives were extracted from various plant parts, fruits, and vegetables.

In addition to extraction, it is also used in the initial processing step before apple dehydration; by using this technique the water activity is reduced and it also further reduces drying time, and the product obtained by this technique contains a good texture. This is achieved at 25 kHz frequency with a time period of 15 minutes. It also retains the texture of foods after the rehydration process, which was demonstrated by Tao et al. (2019) in white cabbage, Ricce et al (2016) in

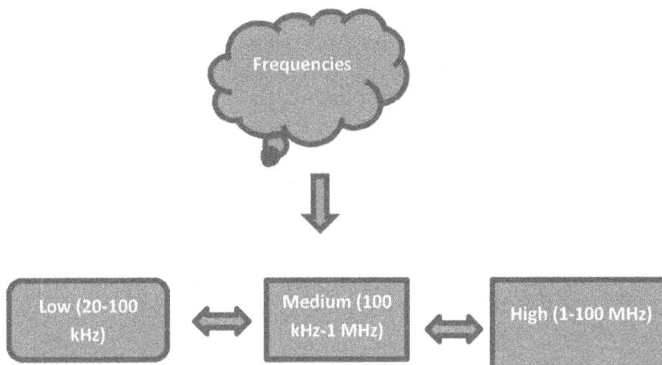

Figure 1.2   Classification of frequencies for food applications.

TABLE 1.1   APPLICATION OF PULSED ELECTRIC FIELD IN FRUIT AND VEGETABLE PRODUCTS

Products studied	Effect achieved	Reference
Pineapple and orange juice	*E. coli* inactivation	Preetha et al. (2021)
Apple and carrot juice	Inactivation of spoilage enzyme	Mannozzi et al. (2019)
Apple peel	Functional component extraction	Wang et al. (2020)
Tomato	Functional component extraction	Pataro et al. (2020)
Carrot	Reduction of drying time	Liu et al. (2020)
Potato	Starch modification	Chen et al. (2020)

carrot, Szadzinska et al. (2017) in green pepper. They noted that this technique is a substitute for the traditional pasteurization/sterilization of food products, ascribed to a significant reduction of microorganisms in food products and it can also be used for carbonated beverage degassing.

1.2.7 Pulse Electric Field

This is also one of the emerging non-thermal processing techniques used in the food processing sector (Table 1.1). In this method, a pulse with higher field intensity is applied to food for a short duration (Niu et al., 2020). Food is subjected to pulse field intensity ranging from 25 to 85 kV/cm for a duration of milliseconds to nanoseconds. It is widely used for microbial inactivation. It consists of a pulse creation unit, food treatment chamber, and control system. The treated foods need to be kept in between two electrodes which are made of stainless steel (Arshad et al., 2020). It can be used typically for liquid or semi-solid foods (Aadil et al., 2015).

The mechanism involved in microbial cell death is a cell wall that gets damaged as a result of exposure to high field intensity pulses. Metabolic enzymes are inactivated, being ascribed to the presence of hydrogen peroxide in treated products, which can allow all lipids and proteins to undergo oxidation resulting in cell death (Barba et al., 2015). Efficiency in microorganism reduction is mainly based on the intensity applied, the exposure time, and temperature.

1.3 CONCLUSION

Various non-thermal processing techniques can be successfully utilized for processing fruits and vegetables, and these techniques are still in their initial stages, therefore these methods need to be commercialized at massive level so that fruit and vegetable products processed with this technique can be readily and easily available.

REFERENCES

Aadil RM, Zeng XA, Ali A, Zeng F, Farooq MA, Han Z, et al. Influence of different pulsed electric field strengths on the quality of the grapefruit juice. *Int J Food Sci Technol* (2015) 50(10):2290–6. https://doi .org/10.1111/ijfs.12891.

Al-Otoom A, Al-Asheh S, Allawzi M, Mahshi K, Alzenati N, Banat B, et al. Extraction of oil from uncrushed olives using supercritical fluid extraction method. *J Supercrit Fluids* (2014) 95:512–8. https://doi.org /10.1016/j.supflu.2014.10.023.

Alves Filho EG, de Brito ES, Rodrigues S. Effects of cold plasma processing in food components. In: Bermudez-Aguirre D, editor. *Advances in cold plasma applications for food safety and preservation*. Washington, DC: Elsevier Inc. (2019). pp. 253–68. https://doi.org/10.1016/B978-0-12-814921-8.00008-6.

Amaral GV, Silva EK, Costa ALR, Alvarenga VO, Cavalcanti RN, Esmerino EA, et al. Whey-grape juice drink processed by supercritical carbon dioxide technology: Physical properties and sensory acceptance. *LWT* (2018) 92:80–6. https://doi.org/10.1016/j.lwt.2018.02.005.

Arshad RN, Abdul-Malek Z, Munir A, Buntat Z, Ahmad MH, Jusoh YMM, et al. Electrical systems for pulsed electric field applications in the food industry: An engineering perspective. *Trends Food Sci Technol* (2020) 104:1–13. https://doi.org/10.1016/j.tifs.2020.07.008.

Bang IH, Lee ES, Lee HS, Min SC. Microbial decontamination system combining antimicrobial solution washing and atmospheric dielectric barrier discharge cold plasma treatment for preservation of mandarins. *Postharvest Biol Technol* (2020) 162:111102. https://doi.org/10.1016/j.postharvbio.2019.111102.

Barba FJ, Parniakov O, Pereira SA, Wiktor A, Grimi N, Boussetta N, et al. Current applications and new opportunities for the use of pulsed electric fields in food science and industry. *Food Res Int* (2015) 77:773–98. https://doi.org/10.1016/j.foodres.2015.09.015.

Bertolini FM, Morbiato G, Facco P, Marszałek K, Pérez-Esteve É, Benedito J, et al. Optimization of the supercritical CO_2 pasteurization process for the preservation of high nutritional value of pomegranate juice. *J Supercrit Fluids* (2020) 164:1–11. https://doi.org/10.1016/j.supflu.2020.104914.

Bhangu SK, Ashokkumar M. Theory of sonochemistry. *Top Curr Chem* (2016) 374:1–28. https://doi.org/10.1007/978-3-319-54271-3_1.

Bulut S, Karatzas KAG. Inactivation of Escherichia coli K12 in phosphate buffer saline and orange juice by high hydrostatic pressure processing combined with freezing. *LWT* (2021) 136:110313. https://doi.org/10.1016/j.lwt.2020.110313.

Cascaes Teles AS, Hidalgo Chávez DW, Zarur Coelho MA, Rosenthal A, Fortes Gottschalk LM, Tonon RV. Combination of enzymeassisted extraction and high hydrostatic pressure for phenolic compounds recovery from grape pomace. *J Food Eng* (2020) 288:110128. https://doi.org/10.1016/j.jfoodeng.2020.110128.

Cavallo C, Carlucci D, Carfora V, Caso D, Cicia G, Clodoveo ML, et al. Innovation in traditional foods: A laboratory experiment on consumers' acceptance of extra-virgin olive oil extracted through ultrasounds. *NJAS Wageningen J Life Sci* (2020) 92(1):100336. https://doi.org/10.1016/j.njas.2020.100336.

Cheila CB, dos Anjos GL, Nóbrega RSA, da Magaton S, de Miranda FM, Dias F. Greener ultrasound-assisted extraction of bioactive phenolic compounds in Croton heliotropiifolius Kunth leaves. *Microchem J* (2020) 159:105525. https://doi.org/10.1016/j.microc.2020.105525.

Chemat F, Zill-E-Huma, Khan MK. Applications of ultrasound in food technology: Processing, preservation and extraction. *Ultrason Sonochem* (2011) 18(4):813–35. https://doi.org/10.1016/j.ultsonch.2010.11.023.

Chen BR, Wen QH, Zeng XA, Abdul R, Roobab U, Xu FY. Pulsed electric field assisted modification of octenyl succinylated potato starch and its influence on pasting properties. *Carbohydr Polym* (2020) 254:117294. https://doi.org/10.1016/j.carbpol.2020.117294.

Chizoba Ekezie FG, Sun DW, Cheng JH. A review on recent advances in cold plasma technology for the food industry: Current applications and future trends. *Trends Food Sci Technol* (2017) 69:46–58. https://doi.org/10.1016/j.tifs.2017.08.007.

Choi MR, Liu Q, Lee SY, Jin JH, Ryu S, Kang DH. Inactivation of Escherichia coli O157:H7, Salmonella typhimurium and listeria monocytogenes in apple juice with gaseous ozone. *Food Microbiol* (2012) 32(1):191–5. https://doi.org/10.1016/j.fm.2012.03.002.

Daher D, Le Gourrierec S, Pérez-Lamela C. Effect of high pressure processing on the microbial inactivation in fruit preparations and other vegetable based beverages. *Agriculturists* (2017) 7:1–18. https://doi.org/10.3390/agriculture7090072.

Dasan BG, Boyaci IH. Effect of cold atmospheric plasma on inactivation of Escherichia coli and physico-chemical properties of apple, orange, tomato juices, and sour cherry nectar. *Food Bioprocess Technol* (2018) 11(2):334–43. https://doi.org/10.1007/s11947-017-2014-0.

Deotale SM, Dutta S, Moses JA, Anandharamakrishnan C. Advances in supercritical carbon dioxide assisted sterilization of biological matrices. In: Knoerzer K, Juliano P, Smithers G, editors. *Innovative food processing technologies*. Cambridge. (2021). pp. 660–77. https://doi.org/10.1016/B978-0-08 -100596-5.22932-6.

Dyshlyuk L, Babich O, Prosekov A, Ivanova S, Pavsky V, Chaplygina T. The effect of postharvest ultraviolet irradiation on the content of antioxidant compounds and the activity of antioxidant enzymes in tomato. *Heliyon* (2020) 6(1):e03288. https://doi.org/10.1016/j.heliyon.2020.e03288.

Fenoglio D, Ferrario M, Schenk M, Guerrero S. Effect of pilotscale UV-C light treatment assisted by mild heat on E. coli, L. plantarum and S. cerevisiae inactivation in clear and turbid fruit juices. storage study of surviving populations. *Int J Food Microbiol* (2020) 332:108767. https://doi.org/10.1016/j.ijfood-micro.2020.108767.

Ferreira TV, Mizuta AG, de Menezes JL, Dutra TV, Bonin E, Castro JC, et al. Effect of ultraviolet treat-ment (UV–C) combined with nisin on industrialized orange juice in Alicyclobacillus acidoterrestris spores. *LWT* (2020) 133:109911. https://doi.org/10.1016/j.lwt.2020.109911.

Ferrentino G, Giampiccolo S, Morozova K, Haman N, Spilimbergo S, Scampicchio M. Supercritical fluid extraction of oils from apple seeds: Process optimization, chemical characterization and compari-son with a conventional solvent extraction. *Innov Food Sci Emerg Technol* (2020) 64:102428. https://doi.org/10.1016/j.ifset.2020.102428.

Fundo JF, Miller FA, Mandro GF, Tremarin A, Brandão TRS, Silva CLM. UV-C light processing of cantaloupe melon juice: Evaluation of the impact on microbiological, and some quality characteristics, during refrigerated storage. *LWT* (2019) 103:247–52. https://doi.org/10.1016/j.lwt.2019.01.025.

Gallego R, Bueno M, Herrero M. Sub- and supercritical fluid extraction of bioactive compounds from plants, food-by-products, seaweeds and microalgae – An update. *TrAC Trends Anal Chem* (2019) 116:198–213. https://doi.org/10.1016/j.trac.2019.04.030.

Gan Z, Feng X, Hou Y, Sun A, Wang R. Cold plasma jet with dielectric barrier configuration: Investigating its effect on the cell membrane of E. coli and S. cerevisiae and its impact on the quality of chokeberry juice. *LWT* (2021) 136:110223. https://doi.org/10.1016/j.lwt.2020.110223.

Guerrero-Beltrán JÁ, Ochoa-Velasco CE. Ultraviolet-C light technology and systems for preservation of fruit juices and beverages. In: Knoerzer K, Muthukumarappan K, editors. *Innovative food processing technologies*. Voctoria. (2021). pp. 210–26. https://doi.org/10.1016/B978-0-08-100596-5.22937-5.

Hou Y, Wang R, Gan Z, Shao T, Zhang X, He M, et al. Effect of cold plasma on blueberry juice quality. *Food Chem* (2019) 290:79–86. https://doi.org/10.1016/j.foodchem.2019.03.123.

Huang HW, Hsu CP, Wang CY. Healthy expectations of high hydrostatic pressure treatment in food pro-cessing industry. *J Food Drug Anal* (2020) 28(1):1–13. https://doi.org/10.1016/j.jfda.2019.10.002.

Iftikhar M, Zhang H, Iftikhar A, Raza A, Begum N, Tahamina A, et al. Study on optimization of ultrasonic assisted extraction of phenolic compounds from rye bran. *LWT* (2020) 134:110243. https://doi.org/10 .1016/j.lwt.2020.110243.

Illera AE, Chaple S, Sanz MT, Ng S, Lu P, Jones J, et al. Effect of cold plasma on polyphenol oxidase inacti-vation in cloudy apple juice and on the quality parameters of the juice during storage. *Food Chem X* (2019) 3:100049. https://doi.org/10.1016/j.fochx.2019.100049.

Keener KM, Misra NN. Future of cold plasma in food processing. In: Cullen PJ, Schluter O, editors. *Cold plasma in food and agriculture: Fundamentals and applications*. Washington, DC: Elsevier Inc. (2016). pp. 343–60. https://doi.org/10.1016/B978-0-12-801365-6.00014-7.

Koutchma T, Bissonnette S, Popovic V. An update on research, development and implementation of UV and pulsed light technologies for nonthermal preservation of milk and dairy products. In: Knoerzer K, Muthukumarappan K, editors. *Innovative food processing technologies*. Voctoria. (2021). pp. 256–76. https://doi.org/10.1016/B978-0-08-100596-5.22680-2.

Lefebvre T, Destandau E, Lesellier E. Sequential extraction of carnosic acid, rosmarinic acid and pigments (carotenoids and chlorophylls) from Rosemary by online supercritical fluid xtractionsupercritical fluid chromatography. *J Chromatogr A* (2020) 1639:461709. https://doi.org/10.1016/j.chroma.2020.461709.

Li W, Gamlath CJ, Pathak R, Martin GJO, Ashokkumar M. Ultrasound – The physical and chemical effects integral to food processing. In: Knoerzer K, Juliano P, Smithers G, editors. *Innovative food process-ing technologies*. Cambridge. (2021). pp. 329–58. https://doi.org/10.1016/B978-0-08-100596-5.22679-6.

Liu C, Pirozzi A, Ferrari G, Vorobiev E, Grimi N. Impact of pulsed electric fields on vacuum drying kinetics and physicochemical properties of carrot. *Food Res Int* (2020) 137:109658. https://doi.org/10.1016/j.foodres.2020.109658.

Mannozzi C, Rompoonpol K, Fauster T, Tylewicz U, Romani S, Rosa MD, et al. Influence of pulsed electric field and ohmic heating pretreatments on enzyme and antioxidant activity of fruit and vegetable juices. *Foods* (2019) 8(7):247. https://doi.org/10.3390/foods8070247.

Martínez-Ramos T, Benedito-Fort J, Watson NJ, Ruiz-López II, Che-Galicia G, Corona-Jiménez E. Effect of solvent composition and its interaction with ultrasonic energy on the ultrasound-assisted extraction of phenolic compounds from Mango peels (Mangifera indica L.). *Food Bioprod Process* (2020) 122:41–54. https://doi.org/10.1016/j.fbp.2020.03.011.

Mason TJ, Cintas P. Sonochemistry. *Handb Green Chem Technol* (2007) 2021:372–96. https://doi.org/10.1002/9780470988305.ch16.

Mason TJ, Chemat F, Ashokkumar M. Power ultrasonics for food processing. In: Ashokkumar M, editor. *Power ultrasonics: Applications of high-intensity ultrasound*. Cambridge: Elsevier Ltd. (2015). pp. 815–43. https://doi.org/10.1016/B978-1-78242-028-6.00027-2.

Misra NN, Roopesh MS. Cold plasma for sustainable food production and processing. In: Vorobiev E, Chemat F, editors. *Green food processing techniques*. France: Elsevier Inc. (2019). pp. 431–53. https://doi.org/10.1016/B978-0-12-815353-6.00016-1.

Niemira BA. Cold plasma decontamination of foods. *Annu Rev Food Sci Technol* (2012) 3:125–42. https://doi.org/10.1146/annurev-food-022811-101132.

Ninčević Grassino A, Ostojić J, Miletić V, Djaković S, Bosiljkov T, Zorić Z, et al. Application of high hydrostatic pressure and ultrasound-assisted extractions as a novel approach for pectin and polyphenols recovery from tomato peel waste. *Innov Food Sci Emerg Technol* (2020) 64:102424. https://doi.org/10.1016/j.ifset.2020.102424.

Niu D, Zeng XA, Ren EF, Xu FY, Li J, Wang MS, et al. Review of the application of pulsed electric fields (PEF) technology for food processing in China. *Food Res Int* (2020) 137:109715. https://doi.org/10.1016/j.foodres.2020.109715.

Oner ME, Demirci A. Ozone for food decontamination: Theory and applications. In: Lelieveld H, Gabric D, Holah J, editors. *Handbook of hygiene control in the food industry: Second edition*. Cambridge: Elsevier Ltd (2016). pp. 491–501. https://doi.org/10.1016/B978-0-08-100155-4.00033-9.

Pataro G, Carullo D, Falcone M, Ferrari G. Recovery of lycopene from industrially derived tomato processing by-products by pulsed electric fields-assisted extraction. *Innov Food Sci Emerg Technol* (2020) 63:102369. https://doi.org/10.1016/j.ifset.2020.102369.

Patil S, Valdramidis VP, Cullen PJ, Frias J, Bourke P. Inactivation of Escherichia coli by ozone treatment of apple juice at different pH levels. *Food Microbiol* (2010) 27(6):835–40. https://doi.org/10.1016/j.fm.2010.05.002.

Pavlić B, Pezo L, Marić B, Tukuljac LP, Zeković Z, Solarov MB, Teslić N. Supercritical fluid extraction of raspberry seed oil: Experiments and modelling. *J Supercrit Fluids* (2020) 157:104687. https://doi.org/10.1016/j.supflu.2019.104687.

Phan KTK, Phan HT, Brennan CS, Phimolsiripol Y. Nonthermal plasma for pesticide and microbial elimination on fruits and vegetables: An overview. *Int J Food Sci Technol* (2017) 52(10):2127–37. https://doi.org/10.1111/ijfs.13509.

Pinto D, De La Luz Cádiz-Gurrea M, Sut S, Ferreira AS, Leyva-Jimenez FJ, Dall'acqua S, et al. Valorisation of underexploited castanea sativa shells bioactive compounds recovered by supercritical fluid extraction with CO_2: A response surface methodology approach. *J CO2 Util* (2020) 40:101194. https://doi.org/10.1016/j.jcou.2020.101194.

Pinto L, Palma A, Cefola M, Pace B, D'Aquino S, Carboni C, et al. Effect of modified atmosphere packaging (MAP) and gaseous ozone pre-packaging treatment on the physico-chemical, microbiological and sensory quality of small berry fruit. *Food Package Shelf Life* (2020) 26:100573. https://doi.org/10.1016/j.fpsl.2020.100573.

Popovic V, Koutchma T, Pagan J. Emerging applications of ultraviolet light-emitting diodes for foods and beverages. In: Knoerzer K, Muthukumarappan K, editors. *Innovative food processing technologies*. Voctoria. (2021). pp. 335–44. https://doi.org/10.1016/B978-0-08-100596-5.22667-X.

Porto E, Alves Filho EG, Silva LMA, Fonteles TV, do Nascimento RBR, Fernandes FAN, et al. Ozone and plasma processing effect on green coconut water. *Food Res Int* (2020) 131:109000. https://doi.org/10.1016/j.foodres.2020.109000.

Preetha P, Pandiselvam R, Varadharaju N, Kennedy ZJ, Balakrishnan M, Kothakota A. Effect of pulsed light treatment on inactivation kinetics of Escherichia coli (MTCC 433) in fruit juices. *Food Control* (2021) 121:107547. https://doi.org/10.1016/j.foodcont.2020.107547.

Priyanka, Khanam S. Selection of suitable model for the supercritical fluid extraction of carrot seed oil: A parametric study. *LWT* (2020) 119:108815. https://doi.org/10.1016/j.lwt.2019.108815.

Qin L, Yu J, Zhu J, Kong B, Chen Q. Ultrasonic-assisted extraction of polyphenol from the seeds of allium senescens L. and its antioxidative role in Harbin dry sausage. *Meat Sci* (2021) 172:108351. https://doi.org/10.1016/j.meatsci.2020.108351.

Ricce C, Rojas ML, Miano AC, Siche R, Augusto PED. Ultrasound pretreatment enhances the carrot drying and rehydration. *Food Res Int* (2016) 89(1):701–8. https://doi.org/10.1016/j.foodres.2016.09.030.

Rios-Corripio G, Welti-Chanes J, Rodríguez-Martínez V, GuerreroBeltrán JÁ. Influence of high hydrostatic pressure processing on physicochemical characteristics of a fermented pomegranate (Punica granatum L.) beverage. *Innov Food Sci Emerg Technol* (2020) 59:102249. https://doi.org/10.1016/j.ifset.2019.102249.

Salea R, Veriansyah B, Tjandrawinata RR. Optimization and scale-up process for supercritical fluids extraction of ginger oilfrom Zingiber officinale var. Amarum. *J Supercrit Fluids* (2017) 120:285–94. https://doi.org/10.1016/j.supflu.2016.05.035.

Santos PH, Kammers JC, Silva AP, Vladimir J, Hense H. Antioxidant and antibacterial compounds from feijoa leaf extracts obtained by pressurized liquid extraction and supercritical fluid extraction. *Food Chem* (2020a) 344:128620. https://doi.org/10.1016/j.foodchem.2020.128620.

Santos OV, Lorenzo ND, Souza ALG, Costa CEF, Conceição LRV, Lannes SCda S, et al. CO_2 supercritical fluid extraction of pulp and nut oils from Terminalia catappa fruits: Thermogravimetric behavior, spectroscopic and fatty acid profiles. *Food Res Int* (2020b) 139:109814. https://doi.org/10.1016/j.foodres.2020.109814.

Silva EK, Meireles MAA, Saldaña MDA. Supercritical carbon dioxide technology: A promising technique for the non-thermal processing of freshly fruit and vegetable juices. *Trends Food Sci Technol* (2020) 97:381–90. https://doi.org/10.1016/j.tifs.2020.01.025.

Spilimbergo S, Bertucco A. Non-thermal bacteria inactivation with dense CO_2. *Biotechnol Bioeng* (2003) 84(6):627–38. https://doi.org/10.1002/bit.10783.

Starek A, Pawłat J, Chudzik B, Kwiatkowski M, Terebun P, Sagan A, et al. Evaluation of selected microbial and physicochemical parameters of fresh tomato juice after cold atmospheric pressure plasma treatment during refrigerated storage. *Sci Rep* (2019) 9(1):8407. https://doi.org/10.1038/s41598-019-44946-1.

Szadzinska J, Łechtańska J, Kowalski SJ, Stasiak M. The effect of high power airborne ultrasound and microwaves on convective drying effectiveness and quality of green pepper. *Ultrason Sonochem* (2017) 34:531–9. https://doi.org/10.1016/j.ultsonch.2016.06.030.

Tao Y, Han M, Gao X, Han Y, Show PL, Liu C, et al. Applications of water blanching, surface contacting ultrasound-assisted air drying, and their combination for dehydration of white cabbage: Drying mechanism, bioactive profile, color and rehydration property. *Ultrason Sonochem* (2019) 53:192–201. https://doi.org/10.1016/j.ultsonch.2019.01.003.

Temelli F, Saldaña MDA, Comin L. Application of supercritical fluid extraction in food processing. In: Pawliszyn J, editor. *Comprehensive sampling and sample preparation*. Vol. 4. Washington: Elsevier. (2012). pp. 415–40. https://doi.org/10.1016/B978-0-12-381373-2.00142-3.

Tiwari BK, Brennan CS, Curran T, Gallagher E, Cullen PJ, O' Donnell CP. Application of ozone in grain processing. *J Cereal Sci* (2010) 51(3):248–55. https://doi.org/10.1016/j.jcs.2010.01.007.

Thirumdas R, Sarangapani C, Annapure US. Cold plasma: A novel nonthermal technology for food processing. *Food Biophys* (2014) 10(1):1–11. https://doi.org/10.1007/s11483-014-9382-z.

Torres-Ossandón MJ, Vega-Gálvez A, López J, Stucken K, Romero J, Di Scala K. Effects of high hydrostatic pressure processing and supercritical fluid extraction on bioactive compounds and antioxidant capacity of cape gooseberry pulp (Physalis peruviana L.). *J Supercrit Fluids* (2018) 138:215–20. https://doi.org/10.1016/j.supflu.2018.05.005.

Van Loey IA, Smout C, Hendrickx M. High hydrostatic pressure technology in food preservation. In: Zeuthen P, Bogh-Sorensen L, editors. *Food preservation techniques.* Cambridge. (2003). pp. 428–48. https://doi.org/10.1533/9781855737143.3.428.

Wang L, Boussetta N, Lebovka N, Vorobiev E. Cell disintegration of apple peels induced by pulsed electric field and efficiency of bio-compound extraction. *Food Bioprod Process* (2020) 122:13–21. https://doi.org/10.1016/j.fbp.2020.03.004.

Wu L, Li L, Chen S, Wang L, Lin X. Deep eutectic solvent-based ultrasonicassisted extraction of phenolic compounds from Moringa oleifera L. leaves: Optimization, comparison and antioxidant activity. *Sep Purif Technol* (2020) 247:117014. https://doi.org/10.1016/j.seppur.2020.117014.

Xiang Q, Fan L, Zhang R, Ma Y, Liu S, Bai Y. Effect of UVC lightemitting diodes on apple juice: Inactivation of zygosaccharomyces rouxii and determination of quality. *Food Control* (2020) 111:107082. https://doi.org/10.1016/j.foodcont.2019.107082.

Chapter 2

High-Pressure Processing for Fruit and Vegetable-Based Beverages

Luis Sabillón and Sergio I. Martínez-Monteagudo

CONTENTS

2.1 INTRODUCTION

The market segment of drinkable products derived from fruits and vegetables has grown considerably during the past decade. Such growth is mainly driven by the perceived health benefits associated with the consumption of these commodities. Fruit- and vegetable-based drinks include the full spectrum of beverages, such as clarified and non-clarified juices, nectars, concentrates, and smoothies. Nowadays, many drinks are made of flavor blends, are low-calorie, and new bottle sizes (e.g., sports caps, wide-mouth bottles) are readily available in the supermarkets of North America, Europe, and Asia-Pacific (Ashurst & Hargitt, 2009).

The commercial production of fruit and vegetable drinks typically employs thermal pasteurization, where a predetermined combination of time and temperature is used to ensure

DOI: 10.1201/9781003222170-2

the microbial safety of the product (Ryan, 2014). Industrially, there are two types of continuous thermal processing – (1) high-temperature-short-time pasteurization (HTST) that inactivates vegetative bacterial cells and thus requires refrigeration after processing; and (2) ultra-high-temperature (UHT) pasteurization, where more intense conditions are used to inactivate both bacterial endospores and vegetative pathogenic cells, and the product can be stored under ambient conditions without the need of refrigeration, resulting in shelf-stable products (Martínez-Monteagudo, Yan, & Balasubramaniam, 2017). Although the different thermal treatments are carefully optimized, the different time-temperature combinations often cause changes in the fresh-like characteristics of the product, such as nutrient retention and development of off-flavor compounds (van Boekel et al., 2010).

Over the past 40 years, a number of emerging technologies have been investigated for their ability to reduce the microbial load (e.g., pathogenic and spoilage bacteria). Examples of such technologies are pulsed electric fields (Koch, Witt, Lammerskitten, Siemer, & Toepfl, 2022), UV processing (Pierscianowski et al., 2021), ozone (Panigrahi, Mishra, & De, 2022), ultrasound (Nyuydze & Martínez-Monteagudo, 2021), cavitation (Sim, Beckman, Anand, & Martínez-Monteagudo, 2021), high-pressure homogenization (Martínez-Monteagudo et al., 2017) and high-pressure processing (Balasubramaniam, Martínez-Monteagudo, & Gupta, 2015). Collectively, these technologies are also known as non-thermal technologies – a term coined

by Professor Dietrich Knorr in the 1990s, referring to food preservation methods that do not use external heat (Martínez-Monteagudo & Rathnakumar, 2021). Overall, non-thermal technologies offer the advantage of reducing the microbial load while retaining the fresh-like characteristics of fruits and vegetables.

Among the non-thermal technologies, high-pressure processing (HPP) has met consumers' rising demands for safety and quality, delivering a variety of innovative food products (Martínez-Monteagudo & Balasubramaniam, 2016). The term high-pressure processing (HPP) refers to the use of pressure levels in the range of 100 to 600 MPa, with or without the application of heat, to inactivate a number of pathogenic and spoilage bacteria, yeasts, molds, and viruses (Balasubramaniam et al., 2015). As a result, HPP offers the ability to extend the shelf life of food products to a level comparable to those that are thermally processed. In the case of fruits and vegetables, however, HPP presents notable differences in a number of quality parameters (e.g., color and vitamin retention, and fresh-like texture) (Martínez-Monteagudo & Balasubramaniam, 2016). Such positive benefits of HPP have been extensively documented in the scientific literature. Reviews on the impact of HPP on color, texture, and flavor of fruit- and vegetable-based food products can be found elsewhere (Indrawati Oey, Lille, Van Loey, & Hendrickx, 2008; Sánchez-Moreno, de Ancos, Plaza, Elez-Martínez, & Cano, 2009). This chapter aims at providing an overview of the operating principles of high-pressure processing, and a summary of recent advances in the production of fruit- and vegetable-based beverages.

2.2 PRINCIPLES OF HIGH-PRESSURE PROCESSING

2.2.1 Pressure and Hydrostatic Principle

Pressure is a thermodynamic variable that has technological implications in a number of food operations, where the application of pressure can range from ten to a hundred MPa. A summary of relevance of pressure in food operations can be found elsewhere (Balasubramaniam et al., 2015; Martínez-Monteagudo & Rathnakumar, 2021). In the case of HPP, the pressurization of a

TABLE 2.1   ILLUSTRATION OF THE INFLUENCE OF HYDROSTATIC PRESSURE ON SELECTED CHEMICAL BONDS

Type of interaction	Working distance (nm)	Distance dependence	Possible pressure effect
Coulomb	20	Proportional inversely	Highly affected
van der Waals	1–20	Optimum distance	Highly affected
Hydrogen bonding	0.2	Quadratic	Affected
Solvation	>2	Exponential	Affected
Hydrophobic	>2	Exponential	Affected
Covalent	0.2	Independent	Unlikely affected

Source: Reprinted by permission from Springer Nature: Martinez-Monteagudo & Saldaña (2014a).

material is carried out by employing a fluid–hydrostatic principle. This principle states that the pressure of a particle or body immersed in a fluid is equally distributed in all directions. This is because the force or pressure acting in the liquid is transported among the liquid itself without friction and acts without friction on the subsequent contact surface. Thus, material pressurized by means of HPP is not subjected to shear stress, and the effect of pressure is independent of time, space, and magnitude (Martínez-Monteagudo & Balasubramaniam, 2016). This unique characteristic has enabled process development, thus contributing to successful commercial applications (Balasubramaniam et al., 2015).

Overall, hydrostatic pressure alters the interatomic distance, having a strong influence on interactions such that their bonding energy depends on the distance (Martinez-Monteagudo & Saldaña, 2014a). Table 2.1 illustrates the concept of the role of pressure on distant-dependent in chemical bonds.

Interactions such as hydrogen bonding and van der Waals forces are distance-dependent, and to a larger extent on pressure. On the contrary, covalent bonds are unlikely to be affected by pressure because their bonding distance is minimally compressed with pressure. This has been demonstrated by Mozhaev, Heremans, Frank, Masson, and Balny (1994), who pressurized proteins up to 1,500 MPa and observed that covalent bonds were unaffected. The ability of hydrostatic pressure to keep covalent bonds unaffected has been the central hypothesis for the preservation of biological activity of functional compounds, such as ascorbic acid (Indrawati et al., 2004), folates (Butz et al., 2004), antioxidants (Matser, Krebbers, van den Berg, & Bartels, 2004), anthocyanins (Verbeyst, Oey, Van der Plancken, Hendrickx, & Van Loey, 2010), lycopene (Yan et al., 2017), and conjugated linoleic acid (Martínez-Monteagudo, Saldaña, Torres, & Kennelly, 2012).

2.3 PRESSURE-VOLUME-TEMPERATURE RELATIONS

A hydrostatically pressurized material reduces its internal energy and increases the amount of energy flowing out in the form of heat (Martínez-Monteagudo & Rathnakumar, 2021). This phenomenon occurs to counteract the mechanical energy work produced by the external pressure, and it is known as adiabatic heat or heat of compression. Table 2.2 exemplifies the heat of compression of selected foods. Overall, most foods exhibit a heat of compression similar to that of water (3°C/100 MPa). The exception to this generalization is found in foods containing

TABLE 2.2   VALUES OF HEAT COMPRESSION FOR SELECTED FOOD PRODUCTS

Food	Heat of compression (°C per 100 MPa)
Water	3.0
Milk	3.0
Yogurt	3.1
Honey	3.2
Salmon	3.2
Olive oil	8.7
Soy oil	9.1

Source: Reprinted by permission from Elsevier: Martinez-Monteagudo & Rathnakumar (2021).

relatively high fat content, where the heat of compression ranges from 8.7 to 9.1°C/100 MPa (Balasubramaniam, Barbosa-Canovas, & Lelieveld, 2016; Balasubramaniam et al., 2015).

The heat of compression is of technological relevance for assisting thermal processing. In a preheated sample, the application of pressure results in an increment of the product temperature to reach commercial sterilization temperatures (120°C) (Martínez-Monteagudo & Saldaña, 2014a).

Samples compressed hydrostatically rise their temperature pseudo-instantaneously and uniformly throughout the sample. Such rapid heating reduces the severity of traditional thermal process and the lack of temperature uniformity that occurs in a traditional thermal process. This process is known as pressure-assisted thermal processing (PATP) and it was first developed for the sterilization of low-acid foods (Sizer, Balasubramaniam, & Ting, 2002).

2.3.1 Le Chatelier's Principle and Transition State Theory

Le Chatelier's principle explains the volume reduction exhibited by systems subjected to elevated pressure. The adjustment of the volume is a response of the system to maintain the equilibrium. Thus, any phenomenon (e.g., phase transition, change in molecular configuration, chemical reaction) accompanied by a decrease in volume is enhanced by pressure (Martínez-Monteagudo & Rathnakumar, 2021).

Another important principle of HPP is the transition state theory, stating that the velocity of a given reaction can increase or decrease by changing pressure, according to whether the intermediate state is less or more voluminous (Wentorf & DeVries, 2003). This principle has been used to explain the effect of pressure on chemical and biochemical reactions and as well as physical processes (Martinez-Monteagudo & Saldaña, 2014a).

2.4 PROCESS DESCRIPTION

A typical operation of HPP starts with the traditional pre-treatment operations (e.g., cutting, mixing, and dissolving), followed by vacuum packaging of the product to be treated. Then, the

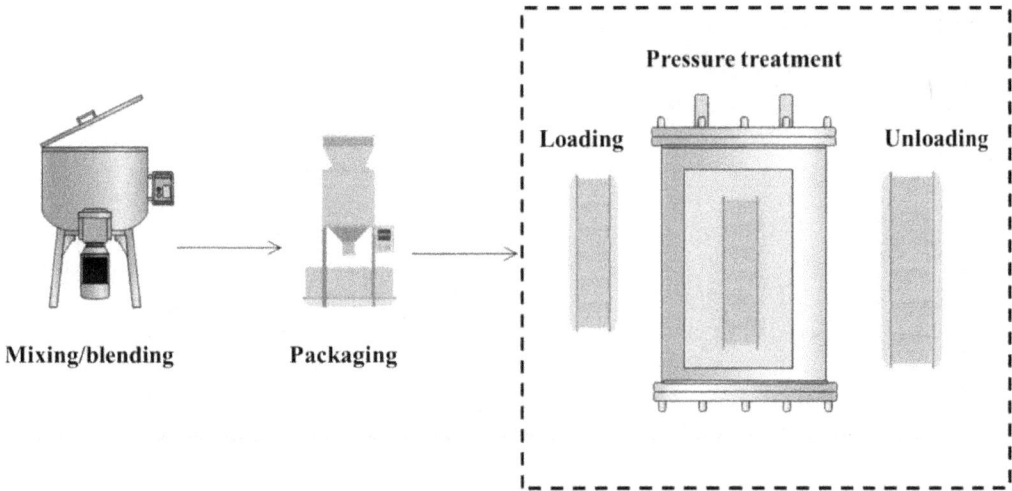

Figure 2.1   Schematic diagram of a typical high-pressure processing operation. (Reprinted by permission from Elsevier: Martinez-Monteagudo & Rathnakumar (2021).)

packaged product is loaded inside a carrier basket, and transferred into a stainless steel vessel (Martínez-Monteagudo & Rathnakumar, 2021). The pressure cylinder or vessel is closed with the end closures, and filled with the pressure transmitting fluid. Afterward, the vessel is pressurized to a target pressure through compression of the transmitting fluid due to a combined action of a high-pressure pump and an intensifier. Once the target pressure is reached, the vessel is held for a given time. After this time, the vessel is depressurized quickly, and the product is unloaded. A typical cycle time for the process is about ten minutes (Martínez-Monteagudo & Rathnakumar, 2021). Upon pressurization, foods containing high moisture experience a physical compression of about 15% of their volume. After decompression, the product returns close to its initial volume. Figure 2.1 illustrates the common steps during a typical high-pressure operation.

2.5 APPLICATIONS OF HIGH-PRESSURE PROCESSING

The chemical and physical effects induced by pressure have been the motivation factor for using HPP. Such effects can be summarized under three main categories: (1) changes in physical properties (e.g., melting point, solubility, and density); (2) effects on equilibrium processes (e.g., acid-base equilibria and ionization); and (3) effects on rates of processes (e.g., reaction rates and yields) (Balasubramaniam et al., 2015; Martinez-Monteagudo & Saldaña, 2014b). In general, changes in quality and safety attributes (e.g., microbiological inactivation, functionality, and nutritional content) are the result of how pressure impacts the physical properties, the equilibrium processes, and rates of processes. For instance, inactivation of microorganisms is a combination of changes in physical properties of membrane lipids, changes in the chemical equilibrium that modify the internal pH, and changes in the rate of specific physiological functions that cause irreversible or lethal damage on bacteria cells (Molina-Gutierrez, Stippl, Delgado, Gänzle, & Vogel, 2002).

TABLE 2.3   APPLICATIONS OF HIGH-PRESSURE PROCESSING FOR FRUIT- AND VEGETABLE-BASED PRODUCTS

Application	Typical conditions	Comments
Pasteurization	P = 400–600 MPa T = 25–50°C t = 3–6 min	Shelf life up to 4–6 weeks at 4°C and suitable for fruit juices, smoothies, and vegetable mixes
Sterilization	P = 600 MPa T = 100–120°C t = 5–15 min	Shelf life up to one year at ambient conditions and suitable for vegetable-based meals and products
Freezing and thawing	P = 200 MPa T = –20°C t = 5–15 min	Preservation of texture and suitable for individual fruits and vegetables
Cell permeabilization	P = 100–600 MPa T = 25–80°C t = 1–8 min	Suitable for extraction of components
Texture modification	P = 100–600 MPa T = 25–80°C t = 1–8 min	Preparation of fruit jam and texture modification of vegetables

Source: Adapted from Matser & Timmermans (2016).
P – pressure; T – temperature; t – time

Overall, a suitable application of HPP is determined by the magnitude of the pressure, the temperature, and the duration of the treatment. Table 2.3 exemplifies selected applications of HPP for fruit- and vegetable-based products.

HPP represents a $2.5 billion-dollar market, offering a wide range of fruit- and vegetable-based products, including juices, acidified avocado purée (guacamole), salsa dips and tomato sauces, beverage blends, jams, jellies, tropical fruits, fruit sauces, smoothies, and fruit desserts (Daryaei, Yousef, & Balasubramaniam, 2016). The industrial application of HPP has expanded over the past decade, providing opportunities to overcome the adverse effects of overprocessing.

2.5.1 Pasteurization

High-pressure processing has been effective in inactivating many pathogenic and spoilage vegetative cells, yeast, mold, and viruses (Martínez-Monteagudo & Balasubramaniam, 2016). Table 2.4 summarizes selected studies on the microbial inactivation of HPP on fruit- and vegetable-based products. Overall, the effectiveness of HPP during the inactivation of microorganisms strongly depends on the processing conditions, such as product temperature (20–60°C), pressure level (300–600 MPa), and holding time (3–10 minutes), as well as product composition, pH, and water activity (Balci & Wilbey, 1999). Mackey, Forestière, and Isaacs (1995) reported that high pressure caused physical changes to the cell structure, and that microbial cells in the stationary phase are more resistant than those in exponential phase. For instance, the authors observed that after HPP treatment at 400 MPa for ten minutes the viable numbers of *L. monocytogenes* cells in the stationary phase were reduced by only 1.3 log CFU/mL compared to more than 7 log CFU/mL for cells in the exponential growth phase.

TABLE 2.4   SUMMARY OF PUBLICATIONS DEALING WITH THE INACTIVATION OF MICROORGANISMS, YEASTS, AND MOLDS BY HIGH-PRESSURE PROCESSING

Bacteria	Matrix	Process conditions	Log$_{10}$ reduction	Reference
Salmonella	Navel and Valencia orange juices	P = 600 MPa T = 20°C t = 1 min	≤7.0	Bull et al. (2004)
Byssochlamys nivea (ascospores)	Grape juice (a$_w$ 0.97)	P = 700 MPa T = 70°C t = 30 min	4.0	Butz et al., (1996)
B. nivea (ascospores)	Bilberry jam (a$_w$ 0.84)	P = 700 MPa T = 70°C t = 30 min	<1.0	Butz et al., (1996)
Zygosaccharomyces bailii (vegetative cells)	Apple, orange, pineapple, cranberry, and grape juices	P = 700 MPa T = not reported t = 30 min	5.0	Raso et al., (1998)
Z. bailii (ascospores)	Apple, orange, pineapple, cranberry, and grape juices	P = 700 MPa T = not reported t = 30 min	0.5–1.0	Raso et al. (1998)

Source: Adapted from Matser & Timmermans (2016).
P – pressure; T – temperature; t – time

The pressure resistance of vegetative microbial cells varies considerably among different genera and species, with morphological characteristics playing a crucial role in the level of resistance. In general, rod-shaped bacterial cells, yeasts, and mold are more susceptible to pressure than spherical-shaped bacteria and bacterial endospores (Arroyo, Sanz, and Préstamo 1997). Yeasts are single-celled fungi and generally not associated with foodborne diseases, but are important in spoilage, especially in acidified foods. The sensitivity of yeasts to HPP makes this treatment effective for controlling yeast spoilage and extending the shelf life of acidic foods such as fruit-based products.

2.5.2 Sterilization

Table 2.5 summarizes selected studies on the inactivation of bacterial endospores due to Pressure-Assisted Thermal Processing (PATP) on fruit- and vegetable-based products. PATP, which refers to the combination of pressure and heating, involves preheating of food materials to about 75–90°C, and the subsequent application of high pressure for five to ten minutes. The rapid temperature increase during hydrostatic compression and subsequent cooling upon decompression is a unique feature of PATP.

Interestingly, PATP is somewhat analogous to in-container sterilization (retorts), where the food is packaged, loaded, and treated for a prescribed combination of time-temperature or time-pressure. Despite such similarities, their governing principles are quite different. Figure 2.2 attempts to provide a detailed description of a typical PATP treatment. The treatment time is the sum of the loading (t_1), compression (t_2), holding (t_3), decompression (t_4), and unloading (t_5) time. At the beginning of the treatment, a drop in the medium temperature (indicated by arrow

TABLE 2.5   SUMMARY OF PUBLICATIONS DEALING WITH THE INACTIVATION OF SPORES BY HIGH-PRESSURE PROCESSING

Bacterial endospores	Matrix	Process conditions	Log_{10} reduction	Reference
Clostridium botulinum nonproteolytic type B (strains ATCC 25765 and TMW 2.518)	Mashed carrots (pH 5.15)	P = 600 MPa T = 80°C t = not reported	>5.5	Margosch et al. (2004)
C. botulinum proteolytic type A (strain ATCC 19397)	Mashed carrots (pH 5.15)	P = 600 MPa T = 80°C t = 12 min	>5.0	Margosch et al. (2004)
C. botulinum proteolytic type B (strain TMW 2.357)	Mashed carrots (pH 5.15)	P = 600 MPa T = 80°C t = 60 min	<3.0	Margosch et al. (2004)
Bacillus coagulans ATCC 7050	Tomato juice (pH 4.5)	P = 392 MPa T = 45°C t = 10 min	5.0	Shearer et al. (2000)
G. stearothermophilus ATCC 7953	Mashed broccoli	P = 600 MPa T = 120°C t = 20 min	>6.0	Ananta et al. (2001)
Alicyclobacillus acidoterrestris DMS 2498	Orange juice	P = 700 MPa T = 80°C t = 20 min	6.0	Ananta et al. (2001)
Bacillus amyloliquefaciens TMW 2.479 Fad 82	Mashed carrots (pH 5.15)	P = 800 MPa T = 80°C t = 16 min	2.0	Margosch et al. (2004)
Bacillus licheniformis TMW 2.492	Mashed carrots (pH 5.15)	P = 600 or 800 T = 80°C t = 16 min	>7.0	Margosch et al. (2004)
Bacillus subtilis (laboratory strains)	Mashed carrots (pH 5.15)	P = 800 MPa T = 70°C t = 1 min	>6.0	Margosch et al. (2004)

Source: Reprinted by permission from Elsevier: Martinez-Monteagudo & Rathnakumar (2021).
P – pressure; T – temperature; t – time

(1)) occurs when the preheated sample is inserted in the high-pressure vessel. Arrow (1) also indicates the start of loading time, which is the time needed to insert the preheated sample, adjust the transmitted fluid volume, and close the high-pressure vessel. Then, the sample is pressurized until it reaches the target pressure. This period is known as compression or build-up time. Due to the adiabatic heating, the temperature of both sample and medium rises (arrow (2)). During t_2, the sample is subjected to non-isothermal and non-isobaric conditions over a relatively short time. The point at which the target temperature and pressure have been reached is considered to be the start of the holding time. The holding time is the only period at which the sample is in isothermal and isobaric conditions. At the end of the holding time, the decompression takes place which is characterized by a drop in the medium temperature indicated by arrow (3). This period is usually short and indicates the end of the treatment.

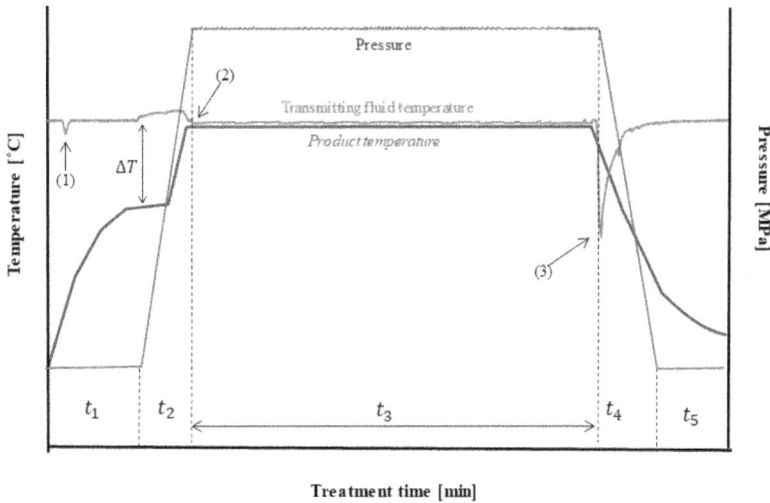

Figure 2.2 Illustration of a typical pressure-assisted thermal processing treatment. t_{1-5} is the loading, compression, holding, decompression, and unloading time. (Reprinted by permission from Elsevier: Martinez-Monteagudo & Rathnakumar (2021).)

In 2009, the FDA approved an industrial petition for the sterilization of mashed potato by PATP. Although low-acid products preserved by PATP are not yet marketed, this technology has the potential to deliver a variety of low-acid products, such as egg, and milk-based products, baby foods, vegetables, ready-to-eat foods, desserts, gravies, soups, and sauces. Further, units for PATP are primarily restricted to pilot scale (5–55 L). Since PATP utilize intensive pressure and heat, from the standpoint of material science and engineering, the process demands significant stress on the vessel and seals, potentially limiting the equipment's useful life. Another limitation of PATP technology is the preheating process, wherein food packages are typically heated (75–90°C) by conventional conduction and convection heat transfer at ambient pressures.

2.5.3 Pressure-Assisted Freezing

Pressure levels of 200 MPa cause a reversible depression of the freezing point of water from 0 to –21°C. Such a phenomenon, known as pressure-assisted freezing (PAF) and pressure-assisted thawing (PAT), has been exploited to rapidly freeze and thaw high moisture content foods, respectively (Balasubramaniam et al., 2015). In general, PAF consists of pressurizing a food matrix at about 200 MPa and reducing the temperature below 0°C, followed by decompression of the sample with the intention to induce rapid nucleation (LeBail, Chevalier, Mussa, & Ghoul, 2002). During PAF, the samples are cooled under pressure up to their phase change temperature at applied pressure. The product is frozen under pressure by supercooling at a faster ice-nucleation rate. This process helps in preserving the microstructure of food and biological materials. Conventional freezing of fruits and vegetables involves relatively slow freezing rates, resulting in the formation of extracellular ice and moisture movement from the original location to form large ice crystals. Samples frozen with PAF exhibited rapid and uniform nucleation leading to less tissue damage and higher quality retention (You, Habibi, Rattan, & Ramaswamy, 2016).

Mechanical damage of cells is less during fast freezing compared to that slow freezing. Fast freezing protects the structure of tissue due to the formation of a relatively large number of small ice crystals, which are uniformly distributed within the extracellular spacing. Thus, the movement of water is restricted, and the tissue is somewhat protected against breakage. As a result, cell walls retained their original structure and exhibited higher quality retention compared with conventional freezing. This has been exemplified by L. Otero, Martino, Zaritzky, Solas, and Sanz (2000), who treated peach and mango at 200 MPa and –20°C to reduce the structural damage when compared with air-blast freezing at –40°C. Otero et al. (2000) reported that PAF samples formed a relatively large number of small crystals that minimize the structural damage. Similarly, eggplant treated with PAF exhibited less structural damage compared with the conventional air-blast freezing (Otero, Solas, Sanz, de Elvira, & Carrasco, 1998). The efficiency of PAF in reducing structural damage depends on the type of product and pretreatment used. For instance, the texture and color of frozen potato were slightly different between the PAF and conventional freezing (Koch, Seyderhelm, Wille, Kalichevsky, & Knorr, 1996). Similarly, Van Buggenhout et al. (2006) reported no improvements in the texture and drip loss of strawberries treated with PAF. In contrast, Van Buggenhout et al. (2006) improved the texture of carrots by calcium soaking and blanching prior to PAF.

2.5.4 Pressure-Assisted Thawing

Pressure-assisted thawing (PAT) and pressure-induced thawing (PIT) have been used in the literature interchangeably (You et al., 2016). In the case of PAT, the phase change (solid to liquid) occurs under constant pressure, about 100 MPa. The heat of compression associated with the pressurization is responsible for the phase change. On the other hand, PIT consists of applying pressure levels of 200 MPa and the phase change starts during the pressurization and is continuous during the holding time under constant pressure. Both processes can help in reducing the thawing time and drip loss. Upon pressurization, the time needed for converting ice to water is significantly shorter than conventional thawing protocols due to the generated temperature gradient (Okamoto & Suzuki, 2002). Another potential application is the storage at freezing conditions (up to –20°C) without the transition from liquid to solid (You et al., 2016).

2.5.5 Pressure-Assisted Extraction

Extraction of valuable compounds from fruits and vegetables has been enhanced by high hydrostatic pressure (Martínez-Monteagudo & Rathnakumar, 2021). Table 2.6 summarizes selected studies on the extraction of bioactive compounds from fruits and vegetables. Extraction yields obtained by pressure mainly depend on the pressure, time, and type of solvent used. Mixtures of organic solvents have been tested under pressure to assist the extraction of antioxidants and other bioactive compounds.

Upon decompression, pressure can break weak chemical bonds of the extracting matrix, making some compounds available for extraction. HPP can enhance the extractability of bioactive compounds due to changes in cell permeabilization within the compounds and the solid matrix. Overall, pressure levels of 100 and 200 MPa induce crystallization of phospholipids in cell membranes, which results in higher permeability. This has been exemplified by Sánchez-Moreno, Plaza, de Ancos, and Cano (2004), who increased the extractability of Vitamin A by 24% on tomato puree treated at 400 Mpa and 25°C for 15 minutes. Castro, Saraiva, Domingues, and Delgadillo (2011) significantly improved the extraction of ascorbic acid from yellow bell peppers

TABLE 2.6   SUMMARY OF PUBLICATIONS DEALING WITH THE EXTRACTION OF BIOACTIVE COMPOUNDS BY PRESSURE-ASSISTED THERMAL PROCESSING

Matrix	Conditions	Remarks	References
Grape by-products	P = 600 MPa T = 70°C t = 60 min	Total phenolic content was increased 1.5-fold	Corrales et al. (2008)
Jerusalem artichoke	P = 100 MPa T = 50°C t = 24 hours	Pressure alone increased extractability	Kim et al. (2010)
Korean barberry	P = 500 MPa T = 30°C t = 30 min	Content increased 1.9-fold	Lee et al. (2010)
Litchi	P = 200 and 400 MPa T = 25°C t = 30 min	Extraction yield increased from 1.83% to 30%	Prasad et al. (2009)
Longan fruit pericarp	P = 200-500 MPa T = 30–70°C t = 2–30 min	Pressure alone increased extractability	Prasad et al. (2010)

Source: Reprinted by permission from Elsevier: Martinez-Monteagudo & Rathnakumar (2021).
P – pressure; T – temperature; t – time

by applying 200 Mpa at 20°C for 20 minutes. In the case of onions, pressure levels up to 300 Mpa induce significant loss of membrane integrity (Gonzalez, Anthon, & Barrett, 2010). However, there is no straight correlation between structural changes and extractability. Trejo et al. (2007) observed an increase in thickness of the cell wall in carrots treated at 400 MPa for two minutes.

A number of studies have been conducted to evaluate the effect of pressure on the extractability of tomato-based products. For instance, Garcia, Butz, and Tauscher (2001) applied pressure levels between 300 and 600 MPa and 20°C for up to 60 minutes to improve the extractability of total carotenoids in tomato puree. Pressure alters the structure from which the carotenoids are attached, exposing the hydrophilic structure and cell integrity. Tomato juice treated at 300–500 MPa and 25°C for ten minutes enhanced the extractability of lycopene and carotenoids increased by 62% (Hsu, 2008). Similarly, pressurized (700 MPa and 45°C for ten minutes) tomato juice improved the lycopene extractability up to 12% (Gupta, Balasubramaniam, Schwartz, & Francis, 2010). Keenan, Brunton, Gormley, and Butler (2011) reported that the extractability of polyphenolic compounds of fruit smoothies made with a blend of fruits depends on the processing conditions. For instance, HPP did not improve the extraction of total polyphenol content of Granny Smith apple puree (Landl, Abadias, Sárraga, Viñas, & Picouet, 2010).

2.6 IMPACT OF HIGH-PRESSURE PROCESSING ON QUALITY PARAMETERS

2.6.1 Texture

The application of HPP in combination with a moderate thermal treatment has been proposed to replace traditional blanching of fruits and vegetables (Sila et al., 2008). Overall, the impact of HPP on quality parameters (e.g., color, texture, and composition) of fruits and vegetables depends not only on the processing conditions but also on the characteristics of the product being treated. For instance, the amount of air in the product combined with its mechanical properties are important considerations for maintaining the texture after HPP. Strawberries are quite susceptible to structural damage due to the expansion of air during compression, resulting in puree (Matser et al., 2004). On the other hand, grapes and berries consisting mainly of water remain intact during pressurization at moderate temperatures. Matser, Knott, Teunissen, and Bartels (2000) reported the appearance of enzymatic browning after HPP due to the presence of air pockets that collapse the structure, facilitating the contact between the enzyme and substrate. Overall, plant cells partially lose their turgor during pressurization, resulting in a softer texture. This has been exemplified by Luscher, Schlüter, and Knorr (2005), who reported turgor loss of potato tissue as a result of membrane negatively influencing the texture. Similarly, Park, Balasubramaniam, and Sastry (2013) observed an increase in the electrical conductivity after HPP treatment due to permeabilization of membranes.

2.6.2 Color

In general, pigments that are responsible for the color of fruits and vegetables are relatively stable under high pressure at moderate temperatures. The stability of anthocyanins during HPP has been exemplified by Van der Plancken et al. (2012), who pressurized strawberries at ambient temperature. Similarly, chlorophyll has been shown to be stable during HPP treatments (Krebbers, Matser, Koets, & Van den Berg, 2002; Indrawati Oey et al., 2008). When compared with untreated samples, green beans exhibited a more intense green color after HPP due to leakage of chlorophyll into the intercellular space (Krebbers et al., 2002). In contrast, green beans treated by PATP exhibited an olive-green color due to partial degradation of the chlorophyll (Indrawati Oey et al., 2008).

Changes in color due to enzymatic reactions represent a critical quality parameter of an end-product. The inactivation of enzymes is a very complex phenomenon involving a reversible partial unfolding, followed by an irreversible modification that leads to the inactivation (Chourio, Salais-Fierro, Mehmood, Martinez-Monteagudo, & Saldaña, 2018). Oftentimes, HPP increases the enzymatic browning due to an incomplete inactivation of enzymes and changes in cell permeabilization resulting in increased contact between the enzyme and substrate. Chourio et al. (2018) evaluated the inactivation of polyphenol oxidase (PPO) and peroxidase (POD) in coconut water under PATP conditions. PPO and POD are responsible for the development of pink color in coconut water. The pink color after PATP treatment was not observed.

2.6.3 Composition

Changes in the composition of fruits and vegetables under pressure have been extensively evaluated. Table 2.7 summarizes studies on the effect of pressure on the stability of vitamins. An investigation of stability of tetrahydrofolate, 5-methyltetrahydrofolate, and 5-formyltetrahydrofolate in three different systems (individual standard solution, model orange juice, and freshly squeezed orange juice) showed that the combination of pressure and temperature synergistically influences the folate stability (Butz et al., 2004). However, the comparison between the different systems showed that other compounds such as ascorbic acid might enhance folate stability. It was found that the degraded amount of 5-formylfolate was comparable to the amount of 5, 10-methyltetrahydrofolate that was quantified after the use of 600 MPa and 80°C. Similarly,

TABLE 2.7   SUMMARY OF STUDIES DEALING WITH THE RETENTION OF BIOACTIVE COMPOUNDS DURING HIGH-PRESSURE PROCESSING

Matrix	Conditions	Remarks	Reference
Apple pectin	P = 500–700 MPa T = 90–115°C t = 0–100 min	Demethoxylation was stimulated by pressure	De Roeck et al. (2009)
Tomatoes and carrot juice	P = 150–250 MPa T = 25 and 35°C t = 5–15 min	Retention of ascorbic acid was higher than thermal treatment	Dede et al. (2007)
Tomato juice	P = 600 and 700 MPa T = 45 and 100°C t = 10 min	All-trans lycopene was found to be stable to isomerization	Gupta et al. (2010)
Orange, kiwi and carrot and asparagus juice	P = 100–700 Mpa T = 25–65°C t = 5–30 min	Degradation kinetics was described by first-order model	Indrawati et al. (2004)
Carrot juice	P = 300–500 Mpa T = 50–70°C t = 10 min	Pressure induced minimal changes	Kim et al. (2001)
Orange juice	P = 600 Mpa T = 40°C t = 4 min	Retention of ascorbic acid was higher for pressurized samples	Polydera et al. (2005)
Blood orange juice	P = 400–600 Mpa T = NR t = 15 min	99% and 94% for ascorbic acid and anthocyanin were retained	Torres et al. (2011)
Broccoli tissue	P = 0.1–600 MPa T = 25–45°C t = 30 min	Pressure yielded between 48 to78% of folate losses	Verlinde et al. (2008)
Strawberries and raspberries	P = 700 Mpa T = 60–110°C t = 0–120 min	Synergistic effect of temperature and pressure	Verbeyst et al. (2013)

Source: Reprinted by permission from Springer Nature: Martinez-Monteagudo & Saldaña (2014a).
P – pressure; T – temperature; t – time

Indrawati et al. (2004) evaluated the stability of 5-methyltetrahydrofolic acid towards pressure using a model system, orange juice, kiwi puree, carrot juice, and asparagus. The degradation rate constants in both model and food systems increased synergistically with pressure and temperature. The addition of ascorbic acid enhanced the stability of 5-methyltetrahydrofolic acid, independently of the processing conditions. A synergistic effect was reported for the degradation of (6R,S)-5-formyltethahydrofolic acid in model systems (Nguyen, Oey, Hendrickx, & van Loey, 2006).

A study on the effect of oxygen and ascorbic acid on the pressure-temperature stability of [6S]-5-methyltetrahydrofolate showed that ascorbic acid enhanced the stability of folate and, more importantly, they found the minimum concentration of ascorbic acid needed to prevent oxidation of folate (at least twice higher than the oxygen concentration) (Oey, Verlinde, Hendrickx, & Van Loey, 2006). The stability of anthocyanins upon PATP has been studied in raspberries and strawberries (Verbeyst, Crombruggen, Van der Plancken, Hendrickx, & Van Loey, 2011; Verbeyst et al., 2010). Another synergistic effect has been suggested by Corrales, Toepfl, Butz, Knorr, and Tauscher (2008). These authors reported that condensation reactions of anthocyanin are promoted by combinations of pressure and temperature. Their control treatment, model solutions of cyaniding-3-O-glucoside (Cy3gl), and pyruvate treated at 0.1 MPa and 70°C, showed around 5% of condensation, while significant degradation of Cy3gl (around 25%) was found at 600 MPa and 70°C. Peter Butz, Bognar, Dieterich, and Tauscher (2007) studied the effect of pressure and temperature on the stability of riboflavin, thiamine, and thiamine monophosphate in a model system and in minced pork. These authors found that vitamins were more stable in pork than in model systems, where the reduction rate was up to 30 times faster. These results suggest that the outcomes from model systems should be used with caution when translating to food systems. Butz et al. (2007) found that 600 MPa and 100°C for 15 minutes yielded a product with 10% of vitamin loss, while the traditional thermal treatment (121°C for 20 minutes) delivered a product with vitamin losses of up to 45%.

2.7 CONCLUSIONS

The benefits of HPP were demonstrated about 30 years ago, and the application of pressure levels in the range of 200 to 600 MPa is now an industrial reality. Overall, HPP is a versatile operation for assisting a number of processing steps, including pasteurization, sterilization, freezing, thawing, and extraction. Such versatility has created many commercial opportunities for fruit- and vegetable-based products, including clarified and non-clarified juices, nectars, concentrates, and smoothies.

REFERENCES

Ananta, E., Heinz, V., Schlüter, O., & Knorr, D. (2001). Kinetic studies on high-pressure inactivation of Bacillus stearothermophilus spores suspended in food matrices. *Innovative Food Science and Emerging Technologies, 2*(4), 261–272. https://doi.org/10.1016/S1466-8564(01)00046-7.

Arroyo, G., Sanz, P., & Préstamo, G. (1997). Effect of high pressure on the reduction of microbial populations in vegetables. *Journal of Applied Microbiology, 82*(6), 735–742.

Ashurst, P. R., & Hargitt, R. (2009). 1 - Product development of new soft drinks and fruit juices. In P. R. Ashurst & R. Hargitt (Eds.), *Soft drink and fruit juice problems solved* (pp. 1–19). Cambridge, UK: Woodhead Publishing.

Balasubramaniam, V. M., Barbosa-Canovas, G. V., & Lelieveld, H. L. M. (2016). High-pressure processing equipment for the food industry. In V. M. Balasubramaniam, G. V. Barbosa-Canovas, & H. L. M. Lelieveld (Eds.), *High pressure processing of food principles, technology and applications* (pp. 39–66). New York: Springer Science+Business.

Balasubramaniam, V. M., Martínez-Monteagudo, S. I., & Gupta, R. (2015). Principles and application of high pressure–based technologies in the food industry. *Annual Review of Food Science and Technology, 6*(1), 435–462. https://doi.org/10.1146/annurev-food-022814-015539.

Balci, A. T., & Wilbey, R. A. (1999). High pressure processing of milk-the first 100 years in the development of a new technology. *International Journal of Dairy Technology, 52*(4), 149–155. https://doi.org/10.1111/j.1471-0307.1999.tb02858.x.

Bull, M. K., Zerdin, K., Howe, E., Goicoechea, D., Paramanandhan, P., Stockman, R., … Stewart, C. M. (2004). The effect of high pressure processing on the microbial, physical and chemical properties of valencia and navel orange juice. *Innovative Food Science and Emerging Technologies, 5*(2), 135–149. https://doi.org/10.1016/j.ifset.2003.11.005.

Butz, P., Bognar, A., Dieterich, S., & Tauscher, B. (2007). Effect of high-pressure processing at elevated temperatures on thiamin and riboflavin in pork and model systems. *Journal of Agricultural and Food Chemistry*, *55*(4), 1289–1294. https://doi.org/10.1021/jf0626821.

Butz, P., Funtenberger, S., Haberditzl, T., & Tauscher, B. (1996). High pressure inactivation of Byssochlamys nivea ascospores and other heat resistant moulds. *LWT - Food Science and Technology*, *29*(5), 404–410. https://doi.org/10.1006/fstl.1996.0062.

Butz, P. B., Serfert, Y., Garcia, A. F., Dieterich, S., Lindauer, R., Bognar, A., & Tauscher, B. (2004). Influence of high-pressure treatment at 25°C and 80°C on folates in orange juice and model media. *Journal of Food Science*, *69*(3), SNQ117–SNQ121. https://doi.org/10.1111/j.1365-2621.2004.tb13380.x.

Castro, S. M., Saraiva, J. A., Domingues, F. M. J., & Delgadillo, I. (2011). Effect of mild pressure treatments and thermal blanching on yellow bell peppers (Capsicum annuum L.). *LWT - Food Science and Technology*, *44*(2), 363–369. https://doi.org/10.1016/j.lwt.2010.09.020.

Chourio, A. M., Salais-Fierro, F., Mehmood, Z., Martinez-Monteagudo, S. I., & Saldaña, M. D. A. (2018). Inactivation of peroxidase and polyphenoloxidase in coconut water using pressure-assisted thermal processing. *Innovative Food Science and Emerging Technologies*, *49*, 41–50. https://doi.org/10.1016/j.ifset.2018.07.014.

Corrales, M., Toepfl, S., Butz, P., Knorr, D., & Tauscher, B. (2008). Extraction of anthocyanins from grape by-products assisted by ultrasonics, high hydrostatic pressure or pulsed electric fields: A comparison. *Innovative Food Science and Emerging Technologies*, *9*(1), 85–91. https://doi.org/10.1016/j.ifset.2007.06.002.

Daryaei, H., Yousef, A. E., & Balasubramaniam, V. M. (2016). Microbiological aspects of high-pressure processing of food: Inactivation of microbial vegetative cells and spores. In V. M. Balasubramaniam, G. V. Barbosa-Cánovas, & H. L. M. Lelieveld (Eds.), *High pressure processing of food: Principles, technology and applications* (pp. 271–294). New York: Springer New York.

De Roeck, A., Duvetter, T., Fraeye, I., Plancken, I. V. D., Sila, D. N., Loey, A. V., & Hendrickx, M. (2009). Effect of high-pressure/high-temperature processing on chemical pectin conversions in relation to fruit and vegetable texture. *Food Chemistry*, *115*(1), 207–213. https://doi.org/10.1016/j.foodchem.2008.12.016.

Dede, S., Alpas, H., & Bayındırlı, A. (2007). High hydrostatic pressure treatment and storage of carrot and tomato juices: Antioxidant activity and microbial safety. *Journal of the Science of Food and Agriculture*, *87*(5), 773–782. https://doi.org/10.1002/jsfa.2758.

Garcia, A. F., Butz, P., & Tauscher, B. (2001). Effects of high-pressure processing on carotenoid extractability, antioxidant activity, glucose diffusion, and water binding of tomato puree (Lycopersicon esculentum Mill.). *Journal of Food Science*, *66*(7), 1033–1038. https://doi.org/10.1111/j.1365-2621.2001.tb08231.x.

Gonzalez, M. E., Anthon, G. E., & Barrett, D. M. (2010). Onion cells after high pressure and thermal processing: Comparison of membrane integrity changes using different analytical methods and impact on tissue texture. *Journal of Food Science*, *75*(7), E426–E432. https://doi.org/10.1111/j.1750-3841.2010.01767.x.

Gupta, R., Balasubramaniam, V. M., Schwartz, S. J., & Francis, D. M. (2010). Storage stability of lycopene in tomato juice subjected to combined pressure–heat treatments. *Journal of Agricultural and Food Chemistry*, *58*(14), 8305–8313. https://doi.org/10.1021/jf101180c.

Hsu, K.-C. (2008). Evaluation of processing qualities of tomato juice induced by thermal and pressure processing. *LWT - Food Science and Technology*, *41*(3), 450–459. https://doi.org/10.1016/j.lwt.2007.03.022.

Indrawati, Arroqui, C., Messagie, I., Nguyen, M. T., Van Loey, A., & Hendrickx, M. (2004). Comparative study on pressure and temperature stability of 5-methyltetrahydrofolic acid in model systems and in food products. *Journal of Agricultural and Food Chemistry*, *52*(3), 485–492. https://doi.org/10.1021/jf0349432.

Keenan, D. F., Brunton, N., Gormley, R., & Butler, F. (2011). Effects of thermal and high hydrostatic pressure processing and storage on the content of polyphenols and some quality attributes of fruit smoothies. *Journal of Agricultural and Food Chemistry*, *59*(2), 601–607. https://doi.org/10.1021/jf1035096.

Kim, D., Fan, J. P., Chung, H. C., & Han, G. D. (2010). Changes in extractability and antioxidant activity of Jerusalem artichoke (Helianthus tuberosus L.) tubers by various high hydrostatic pressure treatments. *Food Science and Biotechnology*, *19*(5), 1365–1371. https://doi.org/10.1007/s10068-010-0194-8.

Kim, Y.-S., Park, S.-J., Cho, Y.-H., & Park, J. (2001). Effects of combined treatment of high hydrostatic pressure and mild heat on the quality of carrot juice. *Journal of Food Science, 66*(9), 1355–1360. https://doi.org/10.1111/j.1365-2621.2001.tb15214.x.

Koch, H., Seyderhelm, I., Wille, P., Kalichevsky, M. T., & Knorr, D. (1996). Pressure-shift freezing and its influence on texture, colour, microstructure and rehydration behaviour of potato cubes. *Food/Nahrung, 40*(3), 125–131. https://doi.org/10.1002/food.19960400306.

Koch, Y., Witt, J., Lammerskitten, A., Siemer, C., & Toepfl, S. (2022). The influence of pulsed electric fields (PEF) on the peeling ability of different fruits and vegetables. *Journal of Food Engineering, 322*, 110938. https://doi.org/10.1016/j.jfoodeng.2021.110938.

Krebbers, B., Matser, A. M., Koets, M., & Van den Berg, R. W. (2002). Quality and storage-stability of high-pressure preserved green beans. *Journal of Food Engineering, 54*(1), 27–33. https://doi.org/10.1016/S0260-8774(01)00182-0.

Landl, A., Abadias, M., Sárraga, C., Viñas, I., & Picouet, P. A. (2010). Effect of high pressure processing on the quality of acidified Granny Smith apple purée product. *Innovative Food Science and Emerging Technologies, 11*(4), 557–564. https://doi.org/10.1016/j.ifset.2010.09.001.

LeBail, A., Chevalier, D., Mussa, D. M., & Ghoul, M. (2002). High pressure freezing and thawing of foods: A review. *International Journal of Refrigeration, 25*(5), 504–513. https://doi.org/10.1016/S0140-7007(01)00030-5.

Lee, H.-Y., He, X., & Ahn, J. (2010). Enhancement of antimicrobial and antimutagenic activities of Korean barberry (Berberis koreana Palib.) by the combined process of high-pressure extraction with probiotic fermentation. *Journal of the Science of Food and Agriculture, 90*(14), 2399–2404. https://doi.org/10.1002/jsfa.4098.

Luscher, C., Schlüter, O., & Knorr, D. (2005). High pressure–low temperature processing of foods: Impact on cell membranes, texture, color and visual appearance of potato tissue. *Innovative Food Science and Emerging Technologies, 6*(1), 59–71. https://doi.org/10.1016/j.ifset.2002.05.001.

Mackey, B. M., Forestière, K., & Isaacs, N. (1995). Factors affecting the resistance of listeria monocytogenes to high hydrostatic pressure. *Food Biotechnology, 9*(1–2), 1–11. https://doi.org/10.1080/08905439509549881.

Margosch, D., Ehrmann, M. A., Gänzle, M. G., & Vogel, R. F. (2004). Comparison of pressure and heat resistance of Clostridium botulinum and other endospores in mashed carrots. *Journal of Food Protection, 67*(11), 2530–2537. https://doi.org/10.4315/0362-028x-67.11.2530.

Martínez-Monteagudo, S. I., & Balasubramaniam, V. M. (2016). Fundamentals and applications of high-pressure processing technology. In V. M. Balasubramaniam, G. V. Barbosa-Canovas, & H. L. M. Lelieveld (Eds.), *High pressure processing of food* (pp. 3–17). New York: Springer, Springer Science+Business.

Martínez-Monteagudo, S. I., & Rathnakumar, K. (2021). 1.02 - High-pressure processing: Fundamentals, misconceptions, and advances. In K. Knoerzer & K. Muthukumarappan (Eds.), *Innovative food processing technologies* (pp. 19–38). Oxford: Elsevier.

Martinez-Monteagudo, S. I., & Saldaña, M. D. A. (2014a). Chemical reactions in food systems at high hydrostatic pressure. *Food Engineering Reviews, 6*(4), 105–127. https://doi.org/10.1007/s12393-014-9087-6.

Martínez-Monteagudo, S. I., & Saldaña, M. D. A. (2014b). Modeling the retention kinetics of conjugated linoleic acid during high-pressure sterilization of milk. *Food Research International, 62*, 169–176. https://doi.org/10.1016/j.foodres.2014.02.014.

Martínez-Monteagudo, S. I., Saldaña, M. D. A., Torres, J. A., & Kennelly, J. J. (2012). Effect of pressure-assisted thermal sterilization on conjugated linoleic acid (CLA) content in CLA-enriched milk. *Innovative Food Science and Emerging Technologies, 16*, 291–297. https://doi.org/10.1016/j.ifset.2012.07.004.

Martínez-Monteagudo, S. I., Yan, B., & Balasubramaniam, V. M. (2017). Engineering process characterization of high-pressure homogenization—From laboratory to industrial scale. *Food Engineering Reviews, 9*(3), 143–169. https://doi.org/10.1007/s12393-016-9151-5.

Matser, A. M., Knott, E. R., Teunissen, P. G. M., & Bartels, P. V. (2000). Effects of high isostatic pressure on mushrooms. *Journal of Food Engineering, 45*(1), 11–16. https://doi.org/10.1016/S0260-8774(00)00035-2.

Matser, A. M., Krebbers, B., van den Berg, R. W., & Bartels, P. V. (2004). Advantages of high pressure sterilisation on quality of food products. *Trends in Food Science and Technology*, 15(2), 79–85. https://doi.org/10.1016/j.tifs.2003.08.005.

Matser, A. M., & Timmermans, R. (2016). High-pressure effects on fruits and vegetables. In V. M. Balasubramaniam, G. V. Barbosa-Cánovas, & H. L. M Lelieveld (Eds.), *High pressure processing of food* (pp. 541–551). New York: Springer.

Molina-Gutierrez, A., Stippl, V., Delgado, A., Gänzle, M. G., & Vogel, R. F. (2002). In situ determination of the intracellular pH of *Lactococcus lactis* and *Lactobacillus plantarum* during pressure treatment. *Applied and Environmental Microbiology*, 68(9), 4399–4406. https://doi.org/10.1128/aem.68.9.4399-4406.2002.

Mozhaev, V. V., Heremans, K., Frank, J., Masson, P., & Balny, C. (1994). Exploiting the effects of high hydrostatic pressure in biotechnological applications. *Trends in Biotechnology*, 12(12), 493–501. https://doi.org/10.1016/0167-7799(94)90057-4.

Nguyen, M. T., Oey, I., Hendrickx, M., & van Loey, A. (2006). Kinetics of (6R,S) 5-formyltetrahydrofolic acid isobaric–isothermal degradation in a model system. *European Food Research and Technology*, 223(3), 325–331. https://doi.org/10.1007/s00217-005-0207-7.

Nyuydze, C., & Martínez-Monteagudo, S. I. (2021). Role of soy lecithin on emulsion stability of dairy beverages treated by ultrasound. *International Journal of Dairy Technology*, 74(1), 84–94. https://doi.org/10.1111/1471-0307.12731.

Oey, I., Lille, M., Van Loey, A., & Hendrickx, M. (2008). Effect of high-pressure processing on colour, texture and flavour of fruit- and vegetable-based food products: A review. *Trends in Food Science and Technology*, 19(6), 320–328. https://doi.org/10.1016/j.tifs.2008.04.001.

Oey, I., Verlinde, P., Hendrickx, M., & Van Loey, A. (2006). Temperature and pressure stability of l-ascorbic acid and/or [6s] 5-methyltetrahydrofolic acid: A kinetic study. *European Food Research and Technology*, 223(1), 71–77. https://doi.org/10.1007/s00217-005-0123-x.

Okamoto, A., & Suzuki, A. (2002). Effects of high hydrostatic pressure-thawing on pork meat. In R. Hayashi (Ed.), *Progress in biotechnology* (Vol. 19, pp. 571–576). Amsterdam: Elsevier.

Otero, L., Martino, M., Zaritzky, N., Solas, M., & Sanz, P. D. (2000). Preservation of microstructure in peach and mango during high-pressure-shift freezing. *Journal of Food Science*, 65(3), 466–470. https://doi.org/10.1111/j.1365-2621.2000.tb16029.x.

Otero, L., Solas, M. T., Sanz, P. D., de Elvira, C., & Carrasco, J. A. (1998). Contrasting effects of high-pressure-assisted freezing and conventional air-freezing on eggplant tissue microstructure. *Zeitschrift für Lebensmittel-untersuchung und –Forschung A*, 206(5), 338–342. https://doi.org/10.1007/s002170050269.

Panigrahi, C., Mishra, H. N., & De, S. (2022). Ozone treatment of ultrafiltered sugarcane juice: Process optimization using multi-objective genetic algorithm and correlation analysis by multivariate technique. *LWT*, 154, 112861. https://doi.org/10.1016/j.lwt.2021.112861.

Park, S. H., Balasubramaniam, V. M., & Sastry, S. K. (2013). Estimating pressure induced changes in vegetable tissue using in situ electrical conductivity measurement and instrumental analysis. *Journal of Food Engineering*, 114(1), 47–56. https://doi.org/10.1016/j.jfoodeng.2012.07.020.

Pierscianowski, J., Popović, V., Biancaniello, M., Bissonnette, S., Zhu, Y., & Koutchma, T. (2021). Continuous-flow UV-C processing of kale juice for the inactivation of E. coli and assessment of quality parameters. *Food Research International*, 140, 110085. https://doi.org/10.1016/j.foodres.2020.110085.

Polydera, A. C., Stoforos, N. G., & Taoukis, P. S. (2005). Quality degradation kinetics of pasteurised and high pressure processed fresh navel orange juice: Nutritional parameters and shelf life. *Innovative Food Science and Emerging Technologies*, 6(1), 1–9. https://doi.org/10.1016/j.ifset.2004.10.004.

Prasad, K. N., Yang, B., Shi, J., Yu, C., Zhao, M., Xue, S., & Jiang, Y. (2010). Enhanced antioxidant and anti-tyrosinase activities of longan fruit pericarp by ultra-high-pressure-assisted extraction. *Journal of Pharmaceutical and Biomedical Analysis*, 51(2), 471–477. https://doi.org/10.1016/j.jpba.2009.02.033.

Prasad, K. N., Yang, B., Zhao, M., Ruenroengklin, N., & Jiang, Y. (2009). Application of ultrasonication or high-pressure extraction of flavonoids from litchi fruit pericarp. *Journal of Food Process Engineering*, 32(6), 828–843. https://doi.org/10.1111/j.1745-4530.2008.00247.x.

Raso, J., Calderón, M. L., Góngora, M., Barbosa-Cánovas, G. V., & Swanson, B. G. (1998). Inactivation of zygosaccharomyces bailii in fruit juices by heat, high hydrostatic pressure and pulsed electric fields. *Journal of Food Science, 63*(6), 1042–1044. https://doi.org/10.1111/j.1365-2621.1998.tb15850.x.

Ryan, R. (2014). Safety of food and beverages: Soft drinks and fruit juices. In Y. Motarjemi (Ed.), *Encyclopedia of food safety* (pp. 360–363). Waltham, MA: Academic Press.

Sánchez-Moreno, C., de Ancos, B., Plaza, L., Elez-Martínez, P., & Cano, M. P. (2009). Nutritional approaches and health-related properties of plant foods processed by high pressure and pulsed electric fields. *Critical Reviews in Food Science and Nutrition, 49*(6), 552–576. https://doi.org/10.1080/10408390802145526.

Sánchez-Moreno, C., Plaza, L., de Ancos, B., & Cano, M. P. (2004). Effect of combined treatments of high-pressure and natural additives on carotenoid extractability and antioxidant activity of tomato puree (Lycopersicum esculentum Mill.). *European Food Research and Technology, 219*(2), 151–160. https://doi.org/10.1007/s00217-004-0926-1.

Shearer, A. E. H., Dunne, C. P., Sikes, A., & Hoover, D. G. (2000). Bacterial spore inhibition and inactivation in foods by pressure, chemical preservatives, and mild heat. *Journal of Food Protection, 63*(11), 1503–1510. https://doi.org/10.4315/0362-028x-63.11.1503.

Sila, D. N., Duvetter, T., De Roeck, A., Verlent, I., Smout, C., Moates, G. K., . . . Van Loey, A. (2008). Texture changes of processed fruits and vegetables: Potential use of high-pressure processing. *Trends in Food Science and Technology, 19*(6), 309–319. https://doi.org/10.1016/j.tifs.2007.12.007.

Sim, J. Y., Beckman, S. L., Anand, S., & Martínez-Monteagudo, S. I. (2021). Hydrodynamic cavitation coupled with thermal treatment for reducing counts of B. coagulans in skim milk concentrate. *Journal of Food Engineering, 293*, 110382. https://doi.org/10.1016/j.jfoodeng.2020.110382.

Sizer, C. E., Balasubramaniam, V. M., & Ting, E. (2002). Validating high-pressure processes for low-acid foods. *Food Technology, 2*, 36–42.

Torres, B., Tiwari, B. K., Patras, A., Cullen, P. J., Brunton, N., & O'Donnell, C. P. (2011). Stability of anthocyanins and ascorbic acid of high pressure processed blood orange juice during storage. *Innovative Food Science and Emerging Technologies, 12*(2), 93–97. https://doi.org/10.1016/j.ifset.2011.01.005.

Trejo Araya, X. I., Hendrickx, M., Verlinden, B. E., Van Buggenhout, S., Smale, N. J., Stewart, C., & John Mawson, A. (2007). Understanding texture changes of high pressure processed fresh carrots: A microstructural and biochemical approach. *Journal of Food Engineering, 80*(3), 873–884. https://doi.org/10.1016/j.jfoodeng.2006.08.014.

van Boekel, M., Fogliano, V., Pellegrini, N., Stanton, C., Scholz, G., Lalljie, S., . . . Eisenbrand, G. (2010). A review on the beneficial aspects of food processing. *Molecular Nutrition and Food Research, 54*(9), 1215–1247. https://doi.org/10.1002/mnfr.200900608.

Van Buggenhout, S., Lille, M., Messagie, I., Loey, A. V., Autio, K., & Hendrickx, M. (2006). Impact of pretreatment and freezing conditions on the microstructure of frozen carrots: Quantification and relation to texture loss. *European Food Research and Technology, 222*(5), 543–553. https://doi.org/10.1007/s00217-005-0135-6.

Van der Plancken, I., Verbeyst, L., De Vleeschouwer, K., Grauwet, T., Heiniö, R.-L., Husband, F. A., . . . Hendrickx, M. (2012). (Bio)chemical reactions during high pressure/high temperature processing affect safety and quality of plant-based foods. *Trends in Food Science and Technology, 23*(1), 28–38. https://doi.org/10.1016/j.tifs.2011.08.004.

Verbeyst, L., Bogaerts, R., Van der Plancken, I., Hendrickx, M., & Van Loey, A. (2013). Modelling of vitamin C degradation during thermal and high-pressure treatments of red fruit. *Food and Bioprocess Technology, 6*(4), 1015–1023. https://doi.org/10.1007/s11947-012-0784-y.

Verbeyst, L., Crombruggen, K. V., Van der Plancken, I., Hendrickx, M., & Van Loey, A. (2011). Anthocyanin degradation kinetics during thermal and high pressure treatments of raspberries. *Journal of Food Engineering, 105*(3), 513–521. https://doi.org/10.1016/j.jfoodeng.2011.03.015.

Verbeyst, L., Oey, I., Van der Plancken, I., Hendrickx, M., & Van Loey, A. (2010). Kinetic study on the thermal and pressure degradation of anthocyanins in strawberries. *Food Chemistry, 123*(2), 269–274. https://doi.org/10.1016/j.foodchem.2010.04.027.

Verlinde, P., Oey, I., Hendrickx, M., & Van Loey, A. (2008). High-pressure treatments induce folate polyglu-tamate profile changes in intact broccoli (Brassica oleraceae L. cv. Italica) tissue. *Food Chemistry*, *111*(1), 220–229. https://doi.org/10.1016/j.foodchem.2008.03.065.

Wentorf, R. H., & DeVries, R. C. (2003). High-pressure synthesis (chemistry). In R. A. Meyers (Ed.), *Encyclopedia of physical science and technology* (3rd ed., pp. 365–379). New York: Academic Press.

Yan, B., Martínez-Monteagudo, S. I., Cooperstone, J. L., Riedl, K. M., Schwartz, S. J., & Balasubramaniam, V. M. (2017). Impact of thermal and pressure-based technologies on carotenoid retention and quality attributes in tomato juice. *Food and Bioprocess Technology*, *10*(5), 808–818. https://doi.org/10.1007/s11947-016-1859-y.

You, J., Habibi, M., Rattan, N., & Ramaswamy, H. S. (2016). Pressure shift freezing and thawing. In V. M. Balasubramaniam, G. V. Barbosa-Cánovas, & H. L. M. Lelieveld (Eds.), *High pressure processing of food: Principles, technology and applications* (pp. 143–166). New York: Springer New York.

Chapter 3

Application of Cold Plasma Techniques for the Fruit and Vegetable Processing Industry

Kirty Pant, Mamta Thakur, and Vikas Nanda

CONTENTS

3.1 INTRODUCTION

In the past few years, consumers have become more aware and conscious of food selection by checking the quality and safety aspects of food produce – there is a high demand for those which are fresh or minimally processed such as fruits and vegetables (Dasan & Boyaci, 2018). Nevertheless, fresh agri-produce can act as a vehicle for the largescale spreading of pathogenic microbes due to their high moisture; hence nutritious content may lead to microbial proliferation and therefore damage their natural defense mechanism, which is in turn directly related to the high numbers of recorded outbreaks of public health scares and foodborne illnesses (Ma et al., 2017). For example, strawberries are extremely perishable and susceptible to physical damage during postharvest processing because of their soft and complex surface structure (Misra et al., 2016).

The FDA and WHO (Ramos et al., 2013), reported that approximately 1,100 different types of agri-produce are related to foodborne outbreaks due to different pathogenic microorganisms.

DOI: 10.1201/9781003222170-3

So, in order to overcome this issue, there is an increased demand to exploit technologies that can control the decaying mechanisms, and biological hazards, and also sustain the quality during post-harvesting, processing, and storage to extend the shelf life of fresh and minimally processed agri-produce.

Various conventional methods such as freezing (Xie et al., 2016; Li et al., 2018), cooling (Zhou et al., 2017; Zhu et al., 2018), and drying (Ma et al., 2017, Qu et al., 2017) are frequently utilized in the food industry to preserve food quality as well as safety. In addition, the intervention of various nonthermal physical processing techniques such as pulsed light application, high-pressure processing (HPP), irradiation, ultraviolet (UV) light, pulsed electric field (PEF), and ozone washing etc., are gaining attention from food scientists and manufactures across the globe to prolong the shelf life of fresh fruits and vegetables (Pan et al., 2017; Ramos et al., 2013). However, there are numerous difficulties and disadvantages related to these physical techniques.

On the other hand, cold plasma (CP) technology is an emerging and influential green nonthermal processing aid used in the food processing industry with the purpose of enhancing the quality and safety of food products hence improving shelf life, and reserving bioactive compounds in the food (Ekezie et al., 2017). The implementation of cold plasma as a powerful nonthermal processing tool can be seen in a variety of application systems where it is highly advantageous in the microbial decontamination of food products including spoilage/pathogenic microbes and sporulating; this is due to the adequate amount of reactive oxygen species (ROS) restricted in the quasi-neutral plasma gas (Min et al., 2017).

Due to this situation, cold plasma is booming as an essential part of the food processing sector by contributing to the quality and safety aspects of food products especially in the shelf life of fresh agri-produce by maintaining comparatively low temperatures plasma doses as well as pre-eminence in in-package food processing, thereby attracting more attention from food researchers and processors (Li & Farid, 2016). Furthermore, application of cold plasma technology can also proficiently degrade agrochemicals (Sarangapani et al., 2017b), toxins (Shi et al., 2017), allergens (Ekezie et al., 2019), and anti-nutrient compounds (Li et al., 2017).

In this chapter, we will discuss the application of cold plasma technology in fresh fruits and vegetable processing such as the effect of cold plasma treatment on the physico-chemical properties of fresh foods, the inactivation of microflora and mycotoxins, elimination of pesticide residue, safety aspects, current status of the industry, and future scope.

3.2 COLD PLASMA TECHNOLOGY – AN OVERVIEW

Plasma, referred to as the "fourth state of matter," is produced with a gas energy application which causes the ionization of the light-transmitting gas consisting of charged particles, free radicals, and UV rays. "Plasma" is often distinguished by the existence of positive, and sometimes negative ions and negative electrons. Various gases like air or argon, nitrogen, and helium depending on their enthalpy, efficiency, and economic feasibility are utilized for the production of cold plasma (Hoffmann et al., 2013). During cold plasma treatment an electron component of a matrix is basically governed by the non-uniform energy of the particles in which heat exchange takes place with the collision of electrons with heavy particles. Furthermore, consistent/uniform media temperature remains constant (35°C) which proves to be a distinctive feature desired for heat-liable food products. The inactivation of microbial cells in cold plasma is dependent upon the reactiveness of species rather than on the heat-induced inactivation of pathogens (Niemira & Gutsol, 2011).

Numerous cold plasma discharge procedures based on the unique mechanisms and desired target reactions have been used. An example is dielectric barrier discharge (DBD) plasma production is achieved by putting an insulating/dielectric substance among two electrodes which results in a self-pulsing action. Initially, the production of ozone gases was achieved using this procedure but currently this has enormous utility in the food industry (Romani et al., 2019a). On the other hand, corona-induced plasma can be produced by using a dielectric barrier discharge or pulsed streamer corona powered by an AC power supply – the process accentuates an activity of ions or energetic electrons involved in the plasma production. However, the characteristics of produced atoms are based on the nature of the gas mixture used and the polarity of discharge, whereas electron energies depend on the procedure of producing the corona-induced plasma laterally with the gas properties (Bussiahn et al., 2010). This process is described as a weakly luminous or non-uniform discharge, which has both positive and negative currents and appears at atmospheric pressure. It is reported by Fridman and Kennedy (2011) that a weakly luminous or irregular discharge, containing both positive and negative currents, is noticeable at atmospheric pressure. Similarly, in the case of glow discharge plasma, the generation of plasma is achieved by the channel of alternating current (AC) where gas acts as a medium, hence the plasma generation takes place by applying a high voltage input of 4.4 kV (60 Hz) among two electrodes. The samples are placed under a vacuum chamber and the production of plasma begins at a pressure of 2 Pa. (Romani et al., 2019).

In the case of cold plasma technology, the safety of food products is achieved in two ways: first by protecting bioactive constituents, and second by defending food from microorganisms. Microbial cells are damaged by the activity of cold plasma (Corradini, 2020); however, ionized gas generated by plasma does not show lethal effects on newly produced cells or tissues. The effect of plasma treatment on biological material is desirable as it avoids the possibility of mutation in cell during treatment (Gavahian & Khaneghah, 2020).

Many studies reported that the synergistic effect of reactive oxygen species (ROS), when subjected to UV radiation along with a combination of electrons and ions generated through plasma treatment, is responsible for devastating the intracellular material (like DNA and protein) of the microorganisms (Guo et al., 2017; Gok et al., 2019). The generation of these reactive species has a lethal effect on microbial cells by releasing stored energy over microorganisms and leading to cell death. Additionally, temperature and chemical configuration are the deciding factors for the quantity of energy transferred by plasma (Hosseini et al., 2020). Studies describe that CAP follows a distinct procedure for attenuating both gram-positive and gram-negative bacteria. Likewise, the cell leakage in gram-negative bacteria ensues due to ROS produced subsequent to CAP treatment which interacts with cellular components. However, ROS is also responsible for acute destruction to inner components of the cell but no leakage takes place in gram-positive bacteria (Han et al., 2020).

The application of cold plasma processing on fresh produce such as fruit and vegetable products including berries, cherries, apples, melons, kiwis etc. was investigated. The results of these studies illustrated that cold plasma treatments on the surface of fresh produce often change the pH and acidity of the food product. This variation happens when a reactive species of generated plasma counters with moisture on the superficial layer of a food product.

Studies reported by Misra et al. (2014) and Lacombe et al. (2015) found that application of CP on fresh and cut produce and blueberries proved effective against aerobic bacteria [39, 40]. Similarly, works on apple surfaces, melons, and mangoes show that there is a significant reduction in the counts of salmonella and *E. coli* after treatment with plasma. Other studies conducted by Tappi et al. (2014, 2016) demonstrated notable depletion in several forms of salmonella and *E.coli* on the surface of mangoes, apples, and melons with the application of plasma. Thus, the application of plasma is considered to be a constructive procedure in reducing the microbial load from the surface of fruits and vegetables with minor drawbacks during storage.

In a nutshell, to meet the requirements for food safety, standard regulations and documentation are currently demanded by the consumer, resulting in food industry professionals emphasizing the production of safe and quality food with a good shelf life via different innovative and novel techniques. Among various thermal and nonthermal procedures, the distinctness of the cold plasma technique has acquired the attention of scientists and researchers worldwide. Cold plasma could be an adaptable technique with varied applications favorable for the food industry.

3.3 PLASMA APPLICATIONS IN FRUIT AND VEGETABLE PROCESSING

3.3.1 Destruction of Microorganisms

Numerous investigations have proven that cold plasma technology could become a potential alternative treatment for the destruction of microbes. The application of cold plasma notably distresses the cell structure and cell material of the spoilage microorganism where plasma-generated particles cause physical impairment to the cell membrane of microbes (Lackmann et al., 2013). Ziuzina et al. (2014) studied the effect of atmospheric cold plasma treatment on fresh produce; they observed that cold plasma can efficiently reduce the population of bacterial cells in tomatoes up to an untraceable level within 10s, whereas in the case of strawberries, more than 300s of cold plasma treatment was required to achieve similar results for an *E. coli* population. In addition, the exposure to cold plasma was extensively effective in the case of impeding the growth of *Botrytis cinerea* and hence prolonging the shelf life of blueberries (Zhou et al., 2019). Zhuang et al. (2019) stated that cold plasma has the potential to be used as an in-package sterilization system because it has ability to generate a reactive species directly inside a package material with no postprocess contamination. Likewise, the experiment conducted by Patil et al. (2014) demonstrated the capability of DBD treatment to develop plasma inside sealed and packed strawberries for surface decontamination to reduce microflora up to 3.3 log cycles, without having any significant effect on the color, texture, and respiration rate (Figure 3.1).

In addition, PAW (liquid solution containing (NO_{3-}, NO_{2-}, and so on) used in indirect air plasma or gas plasma) was initially introduced by Kamgang-Youbi et al. (2007). PAW has a significant ability to sustain the quality of fresh commodities after harvesting. The antimicrobial properties of PAW emerged as a promising substitute for the destruction of *S. aureus* on the surface of strawberries reported by Ma et al. (2015); the results showed that long-term storage of strawberries for up to six days without any such treatment resulted in a layer of gray mold, whereas strawberries treated with sterile distilled water followed by plasma treatment for 20 minutes exposed no fungal spoilage. Xu et al. (2016) also suggested that PAW treatment was an outstanding approach for preserving the shelf life of button mushrooms after harvesting. Nevertheless, in PAW treatment the gas plasma interaction with liquid solutions leads to a rapid decline in the pH value of PAW solution and it attains a steady state, where the inactivation of microbes is achieved by oxidative stress caused by reactive species.

The inactivation efficiency of cold plasma treatment may be influenced by various factors, such as processing factors (i.e., plasma agent, pressure, air, input voltage etc.), environmental factors (i.e., oxidation-reduction potential, pH, conductivity, and, relative humidity) and intrinsic factors (i.e., enzymes, microorganism, and food matrix itself) (Tseng et al., 2012; Patil et al., 2014; Wan et al., 2017). On the other hand, the surface characteristics (such as surface roughness) of fresh commodities are also significantly associated with the efficiency of microbial inactivation. Joshi

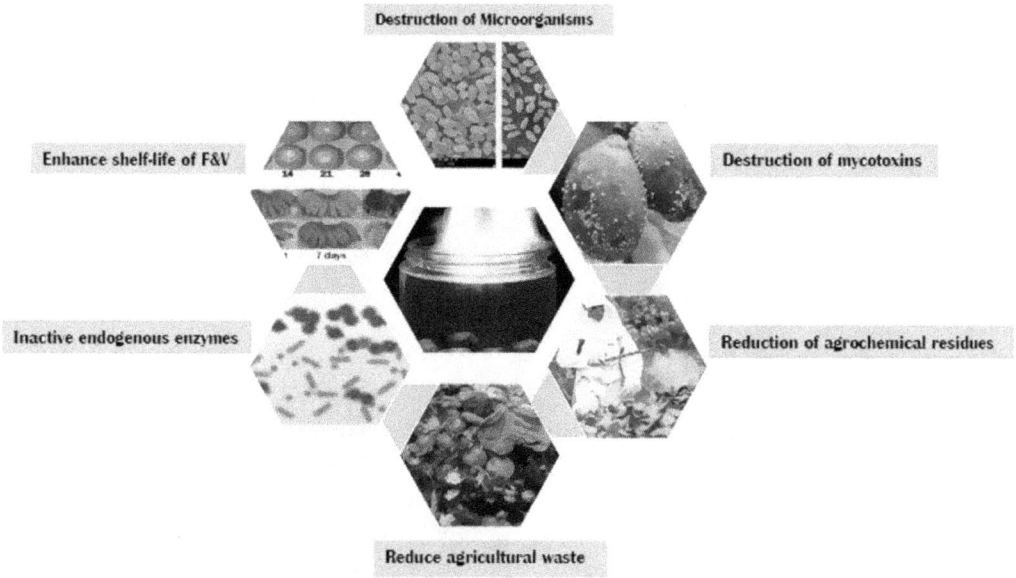

Figure 3.1 Advantages of cold plasma application in fruit and vegetable processing.

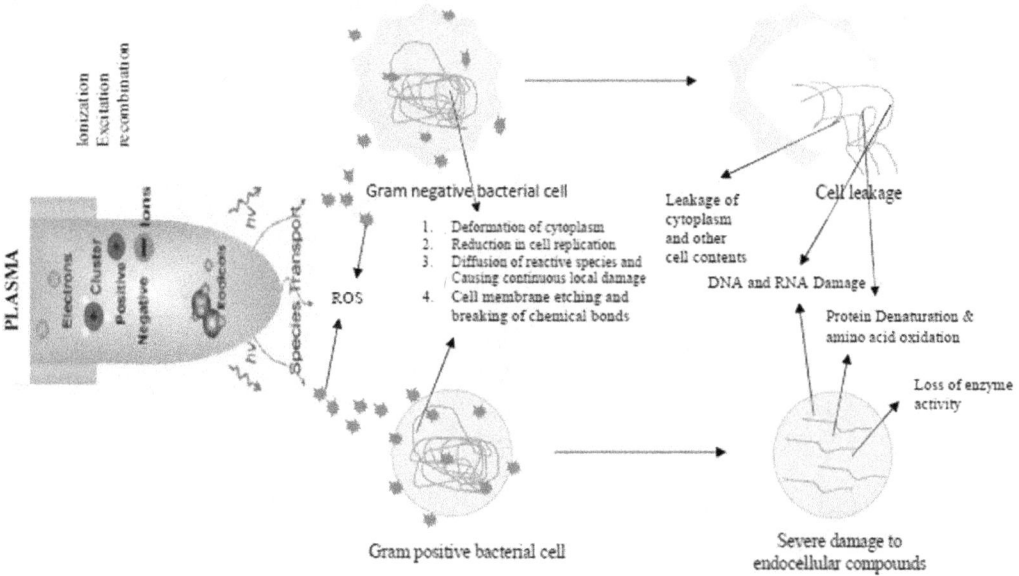

Figure 3.2 Inactivation of microbial cell by cold plasma treatment (Pant et al., 2021).

et al. (2018) conducted an experiment to examine the effect of surface roughness on the inactivation efficiency of *Enterobacter aerogenes* on the surface of fresh produce by applying a plasma-activated acidified buffer. The results demonstrated that the roughest surface (spiny gourd) saw an extensive reduction in the population of *E. aerogenes* by 2.52 ± 0.46 log CFU as compared to the smoothest surface (grape and tomatoes) by 5.31 ± 0.14 log CFU (Figure 3.2 and Table 3.1).

TABLE 3.1   EFFECT OF COLD PLASMA TREATMENT ON MICROBIAL DEACTIVATION IN FRESH PRODUCTS

Food products	Microorganism	Type of plasma	Treatment time	Processing conditions	Max log reduction (CFU/mL)	References
Orange juice	Salmonella enterica	Dielectric Barrier Discharge	120 s	Voltage: 90 kV, Gas: MA65, Frequency: 60 Hz	4.7 logs	Xu et al. (2017)
	Staphylococcus aureus Escherichia Coli Candida Albicans		12 s	Power: 1.14 W/cm2 Gas: air	>5 logs	Shi et al. (2011)
Orange juice tomato Juice	Escherichia Coli	Atmospheric Jet Plasma	120 s	Power: 650 W Gas: air Treatment amount: 11 ml	1.6 log / 1.4 log	Dasan and Boyaci (2018)
Sour Cherry Nectar					3.3 log	
Pasteurized tomato juice	Candida albicans Saccharomyces cerevisiae	Gliding Discharge Plasma	60 s / 5 min	Power: 40 W Voltage: 3.8 kV Gas: Nitrogen	3.7 log / 3.5 log	Starek et al. (2019)
Blueberry juice	Bacillus sp.	Atmospheric Jet Plasma	6 min	Frequency: 1000 Hz Voltage: 11 kV Gas: argon 1% oxygen Gas flow: 1 L/min	7.2 log	Hou et al. (2019)
Apple Juice	Escherichia coli O157:H7	Dielectric Barrier Discharge	40 s	Voltage: 9 kV Frequency: 100 Hz	>5 log	Montenegro et al. (2002)
Corn salad, cucumber, apples, and tomatoes	Escherichia Coli	Atmospheric pressure plasma jet	20 s	Frequency: 1.1 MHz, Voltage: 2 to 6 kV flow rate: 5 L/min	4.0 log	Baier et al. (2014)
Strawberries	S. aureus	PAW	-	work gas: Ar/O2 flow rate: 5 L/min	1.6 to 2.3 log	Ma et al. (2015)
Pomaine lettuce, baby carrots, and cocktail tomatoes	Escherichia Coli	APCP	30 s to 10 min	Voltage: 3.95 to 12.83 kV Frequency: 60 Hz	-	Bermudez-Aguirre et al. (2013)
Lettuce and cabbage	L. monocytogenes	Cold oxygen plasma (atmospheric air)	-	-	-	Srey et al. (2014)
Cabbage, lettuce, and dried figs		Microwave cold plasma	-	Voltage: 400 to 900 W Pressure: 667 Pa), work gas: He: O2 (99.8: 0.2) flow rate: <20 standard L/min	-	Lee et al. (2015)
Spinach	Salmonella sp.	Oxygen RF plasma	0 to 800s	Frequency: 13.56 MHz Voltage: 600 W	-	Zhang et al. (2013)

3.3.2 Destruction of Mycotoxins

Mycotoxins are considered to be the most potent chronic dietary hazard as compared to agrochemical residues, agricultural contaminants, environmental pollutants, and chemical additives (Goodman, 1998). For several food commodities, aflatoxins are comparatively resistant and stable during thermal treatments, such as cooking, ultra-high temperature processing, sterilization, and refrigerated temperature. It has been estimated that one-quarter of agricultural produce in the world is infested by mycotoxins or fungus to some extent (Fallah, 2010).

For example, blue (*Penicillium italicum*) and green (*Penicillium digitatum*) mold are often encountered during postharvest storage of fresh produce. Furthermore, once the infestation by fungus onsets, it is very hard to disinfect because fungal cells and spores have impervious structures.

Application of fungicide is commercial practice to avoid the fungal spoilage and inactivation of mold and yeasts after harvesting many fruits, especially those belonging to the citrus family to ensure the quality as well as safety. Several pieces of research have been conducted in order to investigate the role of plasma technology for mycotoxins and fungal inactivation in fresh produce and their products. Likewise, blueberries infested with both yeasts and mold were subjected to cold plasma by jet in air for a time duration ranging from 15 to 120s. The observation showed a reduction of 0.8 to 1.6 log CFU/g after one day while 1.5 to 2.0 log CFU/g reduction showed after seven days. Whilst there was a significant reduction in anthocyanins, color and firmness were found when subjected to plasma longer than 60s (Lacombe et al., 2015). Misra et al. (2014). Also reported was the sterilization efficiency of in-package plasma exposure exerted by atmospheric pressure. They found 60 kV plasma treatment for 5 minutes led to a reduction of 3.3 log cycle naturally existing yeast and mold on the surface of strawberries after 24 hours' storage in packaging.

In contrast Lacombe et al. (2015) observed no substantial changes in the firmness, color, and respiration rate of strawberries after being subjected to plasma treatment. Additionally, the investigation conducted by Won et al. (2017) indicated the significance of operating conditions where mandarins exposed to the nitrogen cold plasma treatment were powered by microwave for 10 minutes at 0.7 kPa pressure and 1 L/minute flow rate was found to inactivate the *Penicillium italicum* and decrease the occurrence of disease by 84%.

Ouf et al. (2015) presented the effectiveness of cold plasma to decrease the level of mycotoxins in date palm. They observed progressive reduction in the concentration of OTA and FuB2 with prolonged exposure of plasma as well as observing the complete elimination of FuB2 in date palm samples within 6 minutes of exposure to plasma treatment whereas 7.5 minutes was required to fully destroy the OTA from fruit samples via plasma treatment. Furthermore, Ouf et al. (2016) conducted a study in which they subjected wastewater collected after washing moldy cherries to plasma treatment for 9 minutes and observed successive degradation in the concentrating of fumonisins, ochratoxins, and aflatoxins by 66, 72, and 88% respectively.

Likewise, wastewater was collected after the cleaning of strawberries and grapes which were subjected to plasma treatment and resulted in the degradation of mycotoxin. The maximum rate of mycotoxin reduction was found for the wastewater generated after washing cherries. According to the report, the antioxidant capacity of wastewater from collected from strawberries (6.7 mmol Fe_{2+}/L) and grapes (2.1 mmol Fe_{2+}/L) was higher than the wastewater from cherry washing (0.7 mmol Fe_{2+}/L). Thus, this finding confirms the efficiency of the sample composition on the rate of mycotoxin degradation after plasma-induced treatment Ouf et al. (2016). A summary of other investigations is mentioned in Table 3.2.

TABLE 3.2   INHIBITION OF MYCOTOXIN EXISTING IN FRESH PRODUCE BY VARIOUS COLD PLASMA TREATMENT

Product	Processing conditions	Plasma source	Micro-organism	Results	References
Mandarin (fruit and peel)	Gas: Helium; N2+O2 mixture (4:1) Power: 400–900 W Pressure: 0.7 kPa Time: 2–10 min Flow rate: 1 L/min	Microwave plasma	*Penicillium italicum*	84% reduction organism with N2 cold plasma (900W, 10 min) Antioxidant and Phenolic activity enhanced in peels Quality retained during storage study	Won et al. (2017)
Onion powder	Gas: Helium Power: 400–900 W Pressure: 0.7 kPa Time: 10–40 min Flow rate: 1 L/min	Microwave plasma	*Aspergillus brasiliensis*	Spores count was reduced by 1.5 ± 0.2 log spores/cm2 at 400 W for 40 min, Color and antioxidant activity not significantly affected, Loss of volatile compound was observed	Kim et al. (2017)
Tomato seeds,	Gas: Air, SF6 Pressure: 500 mTorr Voltage: 20 kV Frequency: 1 kHz Power: 300 W Time: 5,10, 20 min	Inductively Coupled Plasma (ICP)	*Aspergillus spp., Penicillium spp*	15 min treatment allowed decreasing both species by 3 log10 Seed germination activity is retained after plasma processing	Selcuk et al. (2008)
Date palm	Gas: Argon Pressure: 1 atm Voltage: 20 kV (p-p) Time: 0.5–9.0 min Flow rate: 25 kHz Flow rate: 1.5–4.5 L/min	DBD plasma jet	*Aspergillus niger*	Complete inactivation of *A. niger* spores were observed after 9 min at 3.5 L/min argon flow rate	Ouf et al. (2015)
Saffron	Gas: oxygen Pressure: 13.5 mTorr Power: 10–90 W Time: 10 and 15 min	Radio-Frequency plasma	*Aspergillus sp.*	Complete inhibition of *Aspergillus sp* after 15 min treatment at 60 W. No changes in color, odor, flavor, and antioxidant activity of plasma-treated saffron.	Hosseini et al. (2018)
Apple juice	Pressure: 1 atm Gas: Air Input power: 90 W Time: 20–140 s	DBD	*Z. rouxii*	5 log deduction of microbial cells population in 140 s Observed changes in color, acidity, and pH No impact found on TSS, TPC, and reducing sugar after 140 s	Xiang et al. (2018)

3.3.3 Degradation of Agrochemical Residues

Agrochemicals such as pesticides are considered potent toxic compounds to living organisms including humans and it is necessary to take the utmost preventive measures throughout the application. However, in the past two decades, there has been a focus on food safety issues as a consistent global concern due to the unintended ingesting of pesticide residues related to fresh produce or minimally processed food products. Toxicity induced by pesticides (even existing at a very low level), may still create an issue of health concern, because of bioaccumulation over a period and the chances of adverse side effects at later stages. For example, in humans, a persistent exposure to dichlorvos may lead to hyperglycemia and can also cause disruption in glucose metabolism (Romero-Navarro et al., 2006).

These health alarms emphasize the development and improvement of various procedures for the deprivation of agrochemical residual in the food system, particularly in fresh agri-produce as well as drinking water. Nonthermal cold plasma technology emerged as a potential preventive measure for the degradation of pesticide residue in the food system. However, it has been predominantly utilized for the destruction of microorganisms, whereas recently few studies have reported its dissipation efficiency. A study reported by Sarangapani et al. (2017a) revealed that DBD cold plasma treatment given to blueberries at 80 kV for five minutes could efficiently reduce the concentration of boscalid and imidacloprid by 80.18% and 75.62%, respectively. No alteration in color or signs of any physical damage has been observed but firmness declined because of a breakdown of the internal structures and the surface softening during the storage period. An increase was seen in process parameters such as voltage and treatment time which enhances degradation efficiency. Likewise, dissipation efficacies were found for dichlorvos, malathion, and endosulfan, after being treated for 80 kV for eight minutes showed a reduction up to 78.98, 69.62, and 57.71% respectively, (Sarangapani et al., 2016). Also, 45.71% removal efficiency was attained for pyriproxyfen, azoxystrobin, fludioxonil, and cyprodinil in strawberries (Misra et al., 2014a).

Furthermore, preliminary research conducted by Zhu et al. (2010) revealed the efficiency of HeO_2-induced plasma on malathion. The mechanism induces oxidation of $P\frac{1}{4}S$ bond present in malathion oxygen species present in plasma into $P\frac{1}{4}O$, to form diethyl-2-(dimethoxyphosphoryl-thio) maleate (malaoxon). However, a study conducted by Bai et al. (2010) showed increased power doses had an inconsequential effect on the dissipation efficiency. Contrarily, no effect was detected on the concentration of boscalid, iprodione, and pyrimethanil when plasma treatment was given to strawberries (Misra et al., 2013). Other persuasive investigations authenticate that cold plasma is a promising technology for the degradation of agrochemical residues comprising organophosphate (Hu et al., 2013), 17b-Estradiol (Gao et al., 2013), dichlorvos/omethoate (Bai et al., 2009), and endosulfan (Reddy et al., 2014). At this stage, research intensively focuses on the clarification of the whole process mechanism and pathways and the identification of end products is compulsory to confirm that the process is harmless and does not interfere with the food quality. Some of the investigations have described the non-existence of toxic end products after cold plasma treatment (Boehm et al., 2016; Gao et al., 2013)

3.3.4 Inactivation Endogenous Enzymes

Besides microbial deterioration, endogenous enzymes also play a very significant role in causing self-deterioration and decaying of food products and which may eventually lead to undesirable or unacceptable sensorial alterations to the food system. In particular, fresh commodities such as freshly chopped fruits and vegetables and their juices (for example potato tissue, banana, sugarcane, strawberry, apple, mushroom, pear, Chinese yam, peach, lettuce etc.) are more likely to be prone to such enzymatic damages. Presently, the application of cold plasma as a trend to inactivate the endogenous enzymes mainly focuses on polyphenol oxidase (PPO), pectin methylesterase (PME), superoxide dismutase (SOD), peroxidase (POD), alkaline phosphatase, catalase (CAT), lysozyme, a-chymotrypsin, and lipase in various foods systems (Han et al., 2020). Indeed, PPO, SOD, PME, and POD are the common four enzymes which have been widely studied associated with fresh produce (Surowsky et al., 2013). Takai et al., (2012) examined the effect of low temperature atmospheric pressure plasma-induced for the deactivation of lysozyme in aqueous solution, and the observations revealed that plasma treatment hampers the activity of enzymes initially due to the alteration in secondary structure and the molecular weight. Furthermore, Bußler et al. (2017) studied the effect of cold plasma on fresh-cut apples and potatoes where they observed complete inhibition of PPO and POD enzymes as well as prolonging the storage period up to 19 days after posttreatment (mentioned in Table 3.3). Earlier, Surowsky et al. (2013) also provided a vision related to the deactivation mechanisms of POD and PPO in a food model. The application of plasma exerted a significant role in order to chemically alter the secondary structure of enzymes, with a transformation from the alpha-helix to beta-sheet regions hence eventually hindering the activity of PPO and POD enzymes. Moreover, (Surowsky et al., 2013) reported a higher activity of the POD enzyme as compared to PPO because the longer duration of plasma treatment (240s) could only inactivate 85% of the POD activity whereas a shorter duration of plasma treatment (180%) reduced the activity of PPO up to 90%.

Apart from PPO and POD, PME is also a significant enzyme, which plays a crucial part in the repining process as well as the shelf life of fruits and vegetables (Tappi et al., 2016). But the evidence of the effect of cold plasma treatment on deactivation of PME is very limited. In a recent study conducted by Tappi et al. (2016), they treated fresh-cut melon with DBD plasma treatment for 60 min and found no considerable deduction in the activity of PME enzyme. Xu et al. (2016) investigated the PAW treatment on button mushroom in order to reduce the SOD activity and sustain the shelf life, but posttreatment revealed varied activities of SOD enzyme.

They compared treated samples with the control, and showed that PAW treatment for different time laps followed by storage for a week resulted in the elevation of SOD content. Additionally, Zhou et al. (2019) observed that the aqueous ozone, UV-C, and cold plasma treatment raised the bioactivity of CAT and SOD in blueberries, but a gradual reduction was detected during storage for a few days after postprocessing, and claimed that plasma treatment showed better preservation impact on blueberries as compared with aqueous ozone treatments and UV-C. Henceforth, it was

implicit that different enzymes exert different resistances towards plasma treatments, because of variation in their structures and the existence of isoenzymes. Thus, cold plasma treatment is not suitable to inactivate all types of endogenous enzymes present in food systems. Therefore, further investigation is required to comprehend the interaction as well as the mechanisms among the endogenous enzymes in fruits and vegetables and cold plasma.

3.3.5 Effect on Antioxidant and Pigments

There are plentiful antioxidant compounds found in different varieties of fresh commodities. Antioxidants are considered very crucial parameters under the nutrient index including many pigment compounds in fresh produce. The antioxidant potential in fresh food products plays a significant role due to its sensorial qualities as well as its shelf-life extension. The application of cold plasma to authentic food commodities in place of pure antioxidant compounds leads to a

TABLE 3.3   EFFECT OF DIFFERENT COLD PLASMA TREATMENT ON ANTIOXIDANT AND ENZYME ACTIVITY

Product	Plasma conditions	Salient feature	Reference
Myrciaria dubia juice	CAP with aluminum electrodes at 24 kV and 200, to 960 Hz	Lower anthocyanin contents at 960 Hz Improved antioxidant compounds at 200 Hz Lower action of PPO and POD at 698 Hz.	de Castro et al. (2020)
	GPD for 10–30 min at 80 kHz	Enhanced color and anthocyanin content 18.5% improvement in ascorbic acid at a flow rate of 30 mL/min for 20 min with high (11–27%) bioavailability *in vitro*	
Fresh-cut apples	DBD, 15 kV, 10–30 min, 12.7 kHz	Linear decrease in activity Polyphenol oxidase with time.	Tappi et al. (2014)
Coconut water	CAP with 2 aluminum discs at 200 to 730 Hz and 15 and 20 kV for 15 min	Declined in enzymatic activity of POD with 28% of residual activity at 730 Hz The content of trans-zeatin riboside and procyanidin dimer is improved.	Porto et al. (2020)
Fresh-cut melon	DBD, 15 kV, 12.5 kHz, air 15 + 15, 30 + 30 min	Residual activity of peroxidase enzyme was 91% and 82% after 15 + 15 and 30 + 30 min whereas 94% decline in PME activity after 30 + 30 min treatment, respectively	Tappi et al. (2016)
Fresh-cut apples	DBD, 150 W, 15 + 15, 30 + 30 min, Air, 1.5 m/s	Variable effects on PPO activity Noticeable reduction in surface browning but not proportional to treatment time	Tappi et al. (2019)
Peppermint	RF of 13.56 MHz, low-pressure CAP at 20 to 60 W for 20 min	Improved antioxidant activity TPC from 262 to 292.53 mg GAE/g and FRAP- from 172.4 to 661 mmol Fe^{2+}/g	Kashfi et al. (2020)
Mushrooms	Plasma jet, 18 kV, 10 kHz, 98% Ar + 2% O_2, 5 L/min	SOD activity was higher after treatment during storage	Xu et al. (2016)
Fresh-cut kiwifruit	Atmospheric double barrier discharge plasma	No significant changes in antioxidants content and antioxidant activity were observed	Ramazzina et al. (2015)
Fresh-cut Pink Lady apples	DBD (Atmospheric air) at 150 W; 10, 20, and 30 min	PPO residual activity reduced gradually up to about 42% Browning reduced significant decrease up to about 65%	Tappi et al. (2014)
Fresh-cut potatoes and apples	Atmospheric air Microwave plasma combined with freeze drying or warm-air drying at 2.45 GHz; 1.2 kW; 2.5, 5, 7, 10 min	Sharp inactivation of PPO and POD, especially in potatoes that combined with freeze drying No change of browning index for potatoes regardless of treatment time. Increased browning index for apples	Bußler et al. (2017)

completely different scenario. Likewise, in various studies degradation or oxidation of antioxidant compounds have been observed due to the presence of active species in the plasma phase. The investigation conducted by Grzegorzewski et al. (2010, 2011), showed a reduction in both phenolic acids (chlorogenic and protocatechuic acid) and flavonoids (luteolin and diosmetin) after being treated by oxygen plasma; however, higher retention of phenolic acids was observed as compared to flavonoid compounds. Whilst a substantial rise in flavonoids (such as luteolin, diosmetin, and protocatechuic acid) in lamb's lettuce was reported by Grzegorzewski et al. (2010) within plasma treatment given for 120 s, additionally the diosmetin content was also improved (Grzegorzewski et al., 2011). Some of the investigations showed inconsiderable variations in the antioxidant compounds present in fresh fruits and vegetables after being treated by plasma. Guo et al. (2017) reported that the total amount of anthocyanins in grape were reserved after being treated by PAW for 30 minutes. In-packaged food exposed to plasma at intermittent mode revealed no significant alteration in the content of polyphenol, lycopene, and ascorbic acid as compared with untreated in-packaged cherry tomatoes and kumquats (Lee et al., 2018; Puligundla et al., 2018). Likewise, Misra et al. (2015) also stated anthocyanin and ascorbic acid are retained approximately the same after being introduced to plasma processing at the voltage of 60kV. Won et al. (2017) observed the accumulation of phenolic compounds in mandarin peel due to the UV radiation produced by plasma enhanced the mechanism of biosynthesis of phenolics. Actually, the consequences of the antioxidant compounds are influenced by several processing conditions such as treatment time, plasma intensity, and bioactive response of the fresh agri-produce. The study conducted by Wang et al. (2012) demonstrated a negligible reduction in the ascorbic acid content of fresh-cut carrots, cucumbers, and pears when exposed to shorter duration of time (4 s). Meanwhile, other findings revealed increased concentration of antioxidant compounds in fresh agri-produce followed by plasma treatment. During several studies it was found that plasma treatment at a moderate level can efficiently preserve the activity of antioxidant compound. However, it has also been reported that fresh produce treated with plasma leads to retention of maximum antioxidant compounds during storage. Ramazzina et al. (2015) found that the vitamin C content was significantly decreased by 7% after four days of storage, whereas total polyphenolic content reduced by 30% in plasma-treated kiwifruit slices (mentioned in Table 3.3). The results reported by Won et al. (2017) demonstrated rapid reduction (7 days storage) in newly developed phenolic compounds which were accumulated instantly after plasma treatment in peel. Blueberries treated with DBD plasma treatment for four minutes led to a significant increment in vitamin C content – up to 1.5 times higher as compared to untreated ones after 16 days of prolonged storage. Additionally, the content of anthocyanin in blueberries improved not only after introduced to plasma, but also reserved a higher amount when compared with the untreated samples with up to 20 days of storage (Dong & Yang, 2019). Thus, this showed that plasma treatment has the potential to extend the preservation time for Vitamin C and anthocyanins. Therefore, cold plasma treatment is a suitable technology for reserving the antioxidant compounds in fresh produce for longer periods.

3.3.6 Effects on Quality Attributes and Shelf-Life Extension

Cold plasma is blooming as a potential tool known for preserving the quality of food products by extending their shelf life. The plasma generated by various plasma systems are based on different mechanisms, so their impacts on food products are also different. Several investigations were conducted to explore the influence of cold plasma on the quality parameters of fresh produce (presented in Table 3.4). Pankaj et al. (2018) described that the presence of reactive species in

TABLE 3.4   SUMMARY OF EFFECT OF COLD PLASMA PROCESSING ON PHYSICO-CHEMICAL AND QUALITY ATTRIBUTES OF FRESH COMMODITIES

Product	Plasma type	Processing conditions	Quality attributes	References
Blueberries	Atmospheric pressure plasma jet	Frequency: 47 kHz Voltage: 549 W Air	Decline in tissue firmness after 60 s Effect on the surface color.	Lacombe et al. (2015)
	Dielectric barrier discharge	Frequency: 80 kV Electrode distance: 35 mm	Reduce firmness and variation in color Reduction in TPC, TFC, ascorbic acid, and anthocyanin at higher voltage and long exposure time Significantly eliminate pesticide residues with Removal efficiencies of 80% for Imidacloprid and 75% for Boscalid were observed	Sarangapani et al. (2017b)
Peas	Dielectric barrier discharge	Voltage: 6–12 kV Frequency:20 MHz/5 Min Electrode distance: 20 mm	Decline in the photosynthetic efficacy of seedlings Modification in the concentration of flavonoid glycosides observed	Bubler et al. (2015)
Cherry tomatoes	Dielectric barrier discharge	Voltage: 100 kV Time: 150 sec Electrode distance: 4.5 cm	No change observed in firmness, color, pH, and total soluble solids	Ziuzina et al. (2016)
Romaine lettuce	Dielectric barrier discharge	Voltage: 47.6 kV Time: 5min Electrode distance: 30 mm	No observation or sign of burns, wilting, and color variation	Min et al. (2016)
	Dielectric barrier discharge	Voltage: 42.6 kV Time: 10 min Electrode distance: 5.0 cm	No significant alteration in the respiration rate, surface morphology, color, and weight loss	Min et al. (2017)
Lamb lettuce	Atmospheric pressure plasma jet	Frequency: 27.12 MHz Time: 2 min Gas: argon	Reduction in phenolic acids and Flavonoids were observed	Grzegorzewski et al. (2011)
Radish Sprouts	Microwave plasma	Frequency: 2.45 GHz Voltage: 900 W Pressure: 669 Pa Time: 10 min Gas: N2	No significant changes observed in odor, color ascorbic acid content, and antioxidant activity Wilting occurred after 20 min of treatment	Oh et al. (2017)
Red chicory	Dielectric barrier discharge	Voltage: 19.15 V Time: 15min Electrode distance: 2.0 cm	No negative impact found on freshness texture and color Odor and overall acceptability slightly reduced during storage	Trevisani et al. (2017)

plasma can possibly affect the food components (carbohydrates, pigments, vitamins etc.) and physico-chemical properties (respiration rate, pH etc.) organoleptic attributes (color, taste, and texture) of various fresh agri-products. Numerous studies have been conducted to explore the effect plasma treatment which mostly affects the color properties of foodstuffs, because color is the most significant visual parameter which can directly affect the food acceptability by consumers (Pan et al., 2019). Apart from affecting color properties, plasma application also influences the chemical composition of fresh produce by causing oxidation and degradation of plant tissue directly. Moreover, plasma processing can also facilitate change in enzymatic activity which further causes the alteration in the respiration rate and photosynthesis of fruits and vegetables. Hence, these changes can lead to variations in the color, flavor, texture, and nutritional quality of fresh food products (Chen et al., 2020). In the study reported by Oehmigen et al. (2010) there was found to be no considerable loss in texture, color, and ascorbic acid content in the treated samples such as lettuce, tomato, cabbage, cherry, blueberries, strawberries, and kiwifruit but a decline in pH and acidity was observed which is mainly caused by the formation of some new chemical species (such as H_2O_2, HNO_3, and ONOOH) in the food product. In another study, there was a plasma application-induced partial degradation and slight loss of pigments, with improved color retention as well as a reduction in the development of the darkened area during storage in fresh-cut kiwifruit when compared with control samples (Ramazzina et al., 2015). Furthermore, the effect of cold plasma processing on strawberries under MAP (modified atmosphere packaging), showed no substantial incline in the rate of respiration (Misra et al., 2014). More remarkably, Chizoba et al. (2018) stated that the application of cold plasma can certainly influence the microstructure of natural enzymes and hence improve the shelf life.

3.4 CURRENT STATUS OF COLD PLASMA APPLICATION IN FRUIT AND VEGETABLE PROCESSING

As a result of the cold plasma treatment, fruit- and vegetable-based products are protected from microbes while retaining their functional components. Cold plasma has a deadly effect on bacterial cells, resulting in the death of microorganisms (Corradini, 2020). The cold plasma method is used to preserve a variety of fruits and vegetables, as well as their processed products, which protects the physico-chemical and organoleptic properties of products. According to recent research by Rana et al. (2020), the total bacterial count in fresh strawberries is reduced with the application of cold plasma. The usage of cold plasma also improved the quality and freshness of cut fruits (Mahnot et al., 2020). In carrots, mandarins, and strawberries, the cold plasma method lowers the degree of respiration rate in harvested produce (Bang et al., 2020; Mahnot et al., 2020; Ziuzina et al., 2020). As an added benefit to consumers, the bacteria cells and vegetative spores are killed without altering the organoleptic attributes of produce (Gavahian & Khaneghah, 2020).

The peroxidase inhibition in litchi fruit was examined by Wang et al. (2021) who found that the residual activity of peroxidase isolated from the pericarp was reduced to 47.16% after 10 minutes of treatment. As a result, it can be said that cold plasma can be utilized to extend the shelf life of fruits that are prone to browning. Further, ascorbic acid content was not significantly reduced after cold plasma treatment in most experiments, exposing the fresh-cut products to cold plasma. According to Ramazzina et al. (2015), Song et al. (2015), and Oh et al. (2017), the level of ascorbic acid in kiwifruit, lettuce, and radish sprouts was not significantly affected by cold plasma. Conversely, after plasma treatment of cut fruits and vegetables, the ascorbic acid level was reduced by up to 4% (Trivedi et al., 2019). The possible reason for degradation of ascorbic acid is its reactivity with ozone and other oxidizing plasma species during processing. The changes in ascorbic acid content was also affected by types of samples, variety of fruits and vegetables, processing duration, and plasma gas. According to a study by Leite et al. (2021), the ascorbic acid level of cashew apple juices increased after plasma treatment without altering the total phenolic content of juices.

It was shown by Bao et al. (2020) that the application of He and N_2 plasmas boosted the extraction yields of polyphenols by about 10% thus increasing the antioxidant potential of tomato pomace extracts with a mild alteration in the concentration profile of their phenolic compounds in the tomato pomace. The fresh-cut pitaya fruits were treated with cold plasma that stimulated the phenolic buildup and enhanced the antioxidant potential according to a study by Li et al. (2019). This study thus presents the cold plasma as a new method of preserving freshly cut fruits and vegetables. Giannoglou et al. (2021) investigated the effects of cold plasma in comparison with pulsed electric fields on the microbial, enzyme, and quality characteristics of fresh strawberries at 1 and 6°C. As a result, the total viable count of microorganisms was reduced by 0.9 log CFU/g when using cold plasma and there was an increase of 20.90% in phenolic compounds and 16.50% in antioxidant potential after application of cold plasma. Anthocyanin content, weight loss, color, and pH value were not significantly affected by either treatment. On the other hand, the phenolic content of juices treated with cold plasma has been found to increase, according to Garofuli et al. (2015) and Kovačević et al. (2016). The disturbances in plant cell wall during cold plasma may be responsible for the rise in phenolic components. When the plant cell is ruptured, the phenolic compounds are released into the media, increasing their concentration. Because the bioactive components are retained, the cold plasma treatment approach can be a more efficient way to produce functional food ingredients and nutraceuticals. It also helps to minimize the environmental issues associated with solvent-based treatments or waste management.

Thus, it can be concluded that cold plasma plays an important role in extending shelf life and improving antioxidant capacity as well as the biological qualities of fruits and vegetables. Nowadays, cold plasma processing is undergoing a lot of research in order to be widely used for healthy, fresh, and mildly processed fruit- and vegetable-based products. However, it is essential that more research is undertaken on the industrial uses of cold plasma technology, including product safety and scale-up studies.

3.5 SAFETY REGULATIONS

Numerous investigations included in this chapter explain the effectiveness of cold plasma processing on microbial inactivation and shelf-life extension. For its acceptance as a non-destructive main processing technique in the fruit and vegetable industry, safety regulations are crucial for adequate processing regulations. According to the literature review, cold plasma processing possesses negligible or little impact on the physico-chemical, nutritional, and sensory attributes of various products due to their nonthermal approach. Plasma processing generates highly reactive species such as molecular oxygen and ozone which are known as universal oxidizing agents (Joye et al., 2009). After the completion of plasma processing the active species practically vanish automatically, thus plasma processing is considered environmentally safe and also follows all the environmental safety requirements (Misra et al., 2011). But still there are very limited studies conducted on interactions between CP species and food components on the molecular level in terms of fruits and vegetables for precise knowledge of mechanism and quality control to establish its full potential on commercial scale (Pankaj et al., 2018). The regulatory approval and safety guidelines are another significant implementation needed to create a platform for plasma technologies in the food processing sector.

According to a regulatory viewpoint, it takes a large volume of data collection, data analysis, framing of safety regulations and standards, and time for approval of the plasma processing – whether for indirect or direct fresh food treatment. The overall understanding of the complexities of plasma chemistry, bio-chemical mechanisms, and other miscellaneous aspects needs to be fully investigated, prioritized, and evaluated.

Moreover, each country has its own plasma technologies and safety guidelines and approval process. For instance, in the US the Food and Drug Administration (FDA) has designated ozone as GRAS and no FDA supervisory review or approval is required for any device claiming a need for ozone generation. However, besides ozone, if other active gas species are generated, it will be based on the possible requirements and willingness of the industry to admit regulatory risk, followed by regulatory checks and final approval. Furthermore, the plasma community is also playing an active role in the review process of regulatory documentation and supports the commercialization of plasma technology (Keener & Misra, 2016).

In addition, there are several main technologies foreseen globally for the coming 5 years: Microwave Heating (MWH), High-Pressure Processing (HPP), and ultraviolet light (UV). Meanwhile, the European scenario is predicted to be vital for cold plasma and pulsed electrical field (PEF) for the next 10 years, while the aforementioned Microwave, HPP, and UV remain more relevant in North America (Jermann et al., 2015).

3.6 FUTURE PERSPECTIVES, CHALLENGES, AND LIMITATIONS

As broadly discussed in the literature, the efficiency of plasma technology is known to not only maintain but also enhance the postharvest quality attributes of fresh produce. There are numerous types of cold plasma treatment which have been established for microbial decontamination, elimination of mycotoxins and agrochemical residues, deactivation of endogenous enzymes and to well preserve the nutrient values and organoleptic properties of fruits and vegetables. Thus, cold plasma treatment should be considered a novel promising technology to decrease fresh agricultural waste by extending the shelf life of fruits and vegetables.

However, the application of cold plasma is still not explored much regarding fresh produce and limited only to the lab scale because each plasma system has unique plasma mechanisms, which would exert different impacts on the food products. Therefore, not every plasma technique is suitable for every food product. Furthermore, the cold plasma system has a very limited area of treatment which is a great challenge for extending its application widely on industrial platforms.

According to the industrial viewpoint, the technology transfer of cold plasma, installation, operation, and maintenance of processing unit costs are very high. Moreover, it is also highly necessary to produce plasma "cold" enough, or else the installation of an additional cooling system might be required, which will inevitably cause an upsurge in the cost of the system. That is why cold plasma processing is not so viable for commercial perception in the food sector, particularly in the fruit and vegetable industry.

The current chapter is mainly focused on the application of cold plasma technology on fresh produce, and its effects on food quality and safety. Various pieces of research conducted on quality variation driven by plasma-treated fresh products are still in their nascent stages and further investigations are required. Many experiments have been conducted with regard to the mutagenic and cytotoxicity potential of fresh food when subjected to cold plasma. Several trial studies have claimed that there is no development of any hazardous, toxic, and mutagenic compounds in food products after being treated by plasma whereas detailed investigations are a must in this sector in order to confirm the GRAS limit for the consumption of plasma-processed fruit and vegetable products.

The commercial trait and the various mechanisms of microbial and enzymes inactivation during cold plasma processing are still under exploitation. Likewise, the survival curves of microbes after being exposed to cold plasma, and the fundamental mechanisms for the repressive action of cold plasma against microorganisms are still not well understood yet; on the other hand, the exact mechanisms of plasma-induced inactivation and interactions with food components are still not completely defined due to several factors such as complexities of the food matrix, microorganisms, and plasma systems. As well as this, there is also a need to formulate standards and regulations with regard to plasma-treated food. In addition, the regulatory aspects of cold plasma treatment should be covered and fully ensure the health safety and quality of fresh food supplies.

Therefore, the forthcoming investigations should be specifically based on the effect of cold plasma on nutritional composition, quality parameters, functional characteristics, and toxicological behavior of fresh produce. As well as this, future studies should be intensified towards reproducibility of the outcomes according to the commercial aspects.

3.7 CONCLUSION

Cold plasma is considered an extremely reliable technology for microbial inactivation, pesticide elimination, quality preservation, and shelf-life extension of food materials. This technology is gaining recognition due to its unique features like low or ambient temperature processing for a tiny period of time which facilitates preservation of the freshness and the quality of fruits and vegetables. Several investigations also proved its positive effect on fresh food such as improvement of the bioactive profile, and the physico-chemical, rheology, and organoleptical properties.

This nonthermal technology generates various reactive species with various gas mixtures shows high efficiency and unique dominance over inhibition of microbial cells and inactivation of native enzymes as well as alters the barrier features or impartment of specific functionalities of packaging materials. In addition, there are no specific or minimal impacts are observed of plasma processing on the physico-chemical (like pH and acidity, respiration rate, photosynthesis rate, vitamins, antioxidants, anthocyanidin) and organoleptic (such as taste, odor, color, and texture) quality aspects of the treated food produce. However, despite these current developments, there are still research gaps existing regarding inactivating of antinutritional factors and allergens related to fresh Agri-produce, structural-characteristic modification, complete information on risk assessment of possible toxic by-products, toxicity, lethality, and biosecurity of plasma processing are still also scarce.

Furthermore, the ultimate objective behind the development of any novel technique is for the transformation of pilot results or prototypes into large-scale industrial and commercial applications and there is no exemption for cold plasma technology. Nevertheless, up to now, all the investigation and reviews presented are associated with fruits and vegetable processing have principally been at pilot or laboratory scale.

Safety regulatory approval, effective operational control, and authentication and validation lead toward consumer acceptance which is the main issue or challenge for large-scale industry uses of cold plasma as a processing technique; that is why more researchers aim for it to be widespread in its application on a global scale.

REFERENCES

Bai, Yanhong, Jierong Chen, Hui Mu, Chunhong Zhang, and Baoping Li. "Reduction of dichlorvos and omethoate residues by O_2 plasma treatment." *Journal of Agricultural and Food Chemistry* 57, no. 14 (2009): 6238–6245.

Bai, Yanhong, Jierong Chen, Yun Yang, Limei Guo, and Chunhong Zhang. "Degradation of organophosphorus pesticide induced by oxygen plasma: Effects of operating parameters and reaction mechanisms." *Chemosphere* 81, no. 3 (2010): 408–414.

Baier, Matthias, Mandy Görgen, Jörg Ehlbeck, Dietrich Knorr, Werner B. Herppich, and Oliver Schlüter. "Non-thermal atmospheric pressure plasma: Screening for gentle process conditions and antibacterial efficiency on perishable fresh produce." *Innovative Food Science & Emerging Technologies* 22 (2014): 147–157.

Bang, In Hee, Eun Song Lee, Ho Seon Lee, and Sea C. Min. "Microbial decontamination system combining antimicrobial solution washing and atmospheric dielectric barrier discharge cold plasma treatment for preservation of mandarins." *Postharvest Biology and Technology* 162 (2020): 111102.

Bao, Yiwen, Lavanya Reddivari, and Jen-Yi Huang. "Development of cold plasma pretreatment for improving phenolics extractability from tomato pomace." *Innovative Food Science & Emerging Technologies* 65 (2020): 102445.

Bermúdez-Aguirre, Daniela, Erik Wemlinger, Patrick Pedrow, Gustavo Barbosa-Cánovas, and Manuel Garcia-Perez. "Effect of atmospheric pressure cold plasma (APCP) on the inactivation of Escherichia coli in fresh produce." *Food Control* 34, no. 1 (2013): 149–157.

Boehm, Daniela, Caitlin Heslin, Patrick J. Cullen, and Paula Bourke. "Cytotoxic and mutagenic potential of solutions exposed to cold atmospheric plasma." *Scientific Reports* 6, no. 1 (2016): 1–14.

Bussiahn, R., R. Brandenburg, T. Gerling, E. Kindel, H. Lange, N. Lembke, K.-D. Weltmann, Th Von Woedtke, and T. Kocher. "The hairline plasma: An intermittent negative dc-corona discharge at atmospheric pressure for plasma medical applications." *Applied Physics Letters* 96, no. 14 (2010): 143701.

Bußler, Sara, Jörg Ehlbeck, and Oliver K. Schlüter. "Pre-drying treatment of plant related tissues using plasma processed air: Impact on enzyme activity and quality attributes of cut apple and potato." *Innovative Food Science & Emerging Technologies* 40 (2017): 78–86.

Bußler, Sara, Werner B. Herppich, Susanne Neugart, Monika Schreiner, Jörg Ehlbeck, Sascha Rohn, and Oliver Schlüter. "Impact of cold atmospheric pressure plasma on physiology and flavonol glycoside profile of peas (Pisum sativum 'Salamanca')." *Food Research International* 76 (2015): 132–141.

Chen, Ya-Qi, Jun-Hu Cheng, and Da-Wen Sun. "Chemical, physical and physiological quality attributes of fruit and vegetables induced by cold plasma treatment: Mechanisms and application advances." *Critical Reviews in Food Science and Nutrition* 60, no. 16 (2020): 2676–2690.

Chizoba Ekezie, Flora-Glad, Jun-Hu Cheng, and Da-Wen Sun. "Effects of mild oxidative and structural modifications induced by argon plasma on physicochemical properties of actomyosin from king prawn (Litopenaeus vannamei)." *Journal of agricultural and food chemistry* 66, no. 50 (2018): 13285–13294.

Corradini, Maria G. "Modeling microbial inactivation during cold atmospheric-pressure plasma (CAPP) processing." In D. Bermudez-Aguirre (Ed.), *Advances in cold plasma applications for food safety and preservation*, pp. 93–108. USA: Academic Press, 2020.

Dasan, Beyhan Gunaydin, and Ismail Hakki Boyaci. "Effect of cold atmospheric plasma on inactivation of Escherichia coli and physicochemical properties of apple, orange, tomato juices, and sour cherry nectar." *Food and Bioprocess Technology* 11, no. 2 (2018): 334–343.

de Castro, Debora Raquel Gomes, Josiana Moreira Mar, Laiane Santos da Silva, Kalil Araújo da Silva, Edgar Aparecido Sanches, Jaqueline de Araújo Bezerra, Sueli Rodrigues, Fabiano AN Fernandes, and Pedro Henrique Campelo. "Dielectric barrier atmospheric cold plasma applied on camu-camu juice processing: Effect of the excitation frequency." *Food Research International* 131 (2020): 109044.

Dong, Xiao Yu, and Yu Liang Yang. "A novel approach to enhance blueberry quality during storage using cold plasma at atmospheric air pressure." *Food and Bioprocess Technology* 12, no. 8 (2019): 1409–1421.

Ekezie, Flora-Glad Chizoba, Da-Wen Sun, and Jun-Hu Cheng. "A review on recent advances in cold plasma technology for the food industry: Current applications and future trends." *Trends in Food Science & Technology* 69 (2017): 46–58.

Ekezie, Flora-Glad Chizoba, Da-Wen Sun, and Jun-Hu Cheng. "Altering the IgE binding capacity of king prawn (Litopenaeus Vannamei) tropomyosin through conformational changes induced by cold argon-plasma jet." *Food Chemistry* 300 (2019): 125143.

Fallah, Aziz A. "Assessment of aflatoxin M1 contamination in pasteurized and UHT milk marketed in central part of Iran." *Food and Chemical Toxicology* 48, no. 3 (2010): 988–991.

Fridman, A., and L. A. Kennedy. "Nonequilibrium cold atmospheric pressure discharges." In *Plasma physics and engineering*, pp. 561–611. New York: CRC Press, 2011.

Gao, Lihong, Lei Sun, Shungang Wan, Zebin Yu, and Mingjie Li. "Degradation kinetics and mechanism of emerging contaminants in water by dielectric barrier discharge non-thermal plasma: The case of 17β-Estradiol." *Chemical Engineering Journal* 228 (2013): 790–798.

Garofulić, Ivona Elez, Anet Režek Jambrak, Slobodan Milošević, Verica Dragović-Uzelac, Zoran Zorić, and Zoran Herceg. "The effect of gas phase plasma treatment on the anthocyanin and phenolic acid content of sour cherry Marasca (Prunus cerasus var. Marasca) juice." *LWT-Food Science and Technology* 62, no. 1 (2015): 894–900.

Gavahian, Mohsen, and Amin Mousavi Khaneghah. "Cold plasma as a tool for the elimination of food contaminants: Recent advances and future trends." *Critical Reviews in Food Science and Nutrition* 60, no. 9 (2020): 1581–1592.

Giannoglou, Marianna, Zacharoula-Maria Xanthou, Sofia Chanioti, Panagiota Stergiou, Miltiadis Christopoulos, Panagiotis Dimitrakellis, Aspasia Efthimiadou, Evangelos Gogolides, and George Katsaros. "Effect of cold atmospheric plasma and pulsed electromagnetic fields on strawberry quality and shelf-life." *Innovative Food Science & Emerging Technologies* 68 (2021): 102631.

Gök, Veli, Simge Aktop, Mehmet Özkan, and Oktay Tomar. "The effects of atmospheric cold plasma on inactivation of Listeria monocytogenes and Staphylococcus aureus and some quality characteristics of pastırma—A dry-cured beef product." *Innovative Food Science & Emerging Technologies* 56 (2019): 102188.

Grzegorzewski, Franziska, Jörg Ehlbeck, Oliver Schlüter, Lothar W. Kroh, and Sascha Rohn. "Treating lamb's lettuce with a cold plasma–Influence of atmospheric pressure Ar plasma immanent species on the phenolic profile of Valerianella locusta." *LWT-Food Science and Technology* 44, no. 10 (2011): 2285–2289.

Grzegorzewski, Franziska, Sascha Rohn, Lothar W. Kroh, Martin Geyer, and Oliver Schlüter. "Surface morphology and chemical composition of lamb's lettuce (Valerianella locusta) after exposure to a low-pressure oxygen plasma." *Food Chemistry* 122, no. 4 (2010): 1145–1152.

Guo, Jian, Kang Huang, Xiao Wang, Chenang Lyu, Nannan Yang, Yanbin Li, and Jianping Wang. "Inactivation of yeast on grapes by plasma-activated water and its effects on quality attributes." *Journal of Food Protection* 80, no. 2 (2017): 225–230.

Han, Jin-Young, Won-Jae Song, Joo Hyun Kang, Sea C. Min, Sangheum Eom, Eun Jeong Hong, Seungmin Ryu, Seong bong Kim, Sangwoo Cho, and Dong-Hyun Kang. "Effect of cold atmospheric pressure plasma-activated water on the microbial safety of Korean rice cake." *LWT* 120 (2020): 108918.

Hoffmann, Clotilde, Carlos Berganza, and John Zhang. "Cold atmospheric plasma: Methods of production and application in dentistry and oncology." *Medical Gas Research* 3, no. 1 (2013): 1–15.

Hosseini, Seyed Iman, Naser Farrokhi, Khadijeh Shokri, Mohammad Reza Khani, and Babak Shokri. "Cold low pressure O_2 plasma treatment of crocus sativus: An efficient way to eliminate toxicogenic fungi with minor effect on molecular and cellular properties of saffron." *Food Chemistry* 257 (2018): 310–315.

Hosseini, Seyed Mehdi, Sajad Rostami, Bahram Hosseinzadeh Samani, and Zahra Lorigooini. "The effect of atmospheric pressure cold plasma on the inactivation of Escherichia coli in sour cherry juice and its qualitative properties." *Food Science & Nutrition* 8, no. 2 (2020): 870–883.

Hou, Yanan, Ruixue Wang, Zhilin Gan, Tao Shao, Xinxue Zhang, Mohe He, and Aidong Sun. "Effect of cold plasma on blueberry juice quality." *Food Chemistry* 290 (2019): 79–86.

Hu, Yingmei, Yanhong Bai, Hu Yu, Chunhong Zhang, and Jierong Chen. "Degradation of selected organophosphate pesticides in wastewater by dielectric barrier discharge plasma." *Bulletin of Environmental Contamination and Toxicology* 91, no. 3 (2013): 314–319.

Jermann, Colette, Tatiana Koutchma, Edyta Margas, Craig Leadley, and Valquiria Ros-Polski. "Mapping trends in novel and emerging food processing technologies around the world." *Innovative Food Science & Emerging Technologies* 31 (2015): 14–27.

Joshi, Isha, Deepti Salvi, Donald W. Schaffner, and Mukund V. Karwe. "Characterization of microbial inactivation using plasma-activated water and plasma-activated acidified buffer." *Journal of food protection* 81, no. 9 (2018): 1472–1480.

Joye, Iris J., Bert Lagrain, and Jan A. Delcour. "Use of chemical redox agents and exogenous enzymes to modify the protein network during breadmaking–A review." *Journal of Cereal Science* 50, no. 1 (2009): 11–21.

Kamgang-Youbi, Georges, Jean-Marie Herry, Marie-Noëlle Bellon-Fontaine, Jean-Louis Brisset, Avaly Doubla, and Murielle Naïtali. "Evidence of temporal postdischarge decontamination of bacteria by gliding electric discharges: Application to Hafnia alvei." *Applied and Environmental Microbiology* 73, no. 15 (2007): 4791–4796.

Kashfi, Atefeh Sadat, Yousef Ramezan, and Mohammad Reza Khani. "Simultaneous study of the antioxidant activity, microbial decontamination and color of dried peppermint (Mentha piperita L.) using low pressure cold plasma." *LWT* 123 (2020): 109121.

Keener, K. M., and N. N. Misra. "Future of cold plasma in food processing." In N. N. Misra, O. Schluter, & P. J. Cullen (Eds), *Cold plasma in food and agriculture*, pp. 343–360. London, UK: Academic Press, 2016.

Kim, Jung Eun, Yeong Ji Oh, Mee Yeon Won, Kwang-Sik Lee, and Sea C. Min. "Microbial decontamination of onion powder using microwave-powered cold plasma treatments." *Food Microbiology* 62 (2017): 112–123.

Kovačević, Danijela Bursać, Predrag Putnik, Verica Dragović-Uzelac, Sandra Pedisić, Anet Režek Jambrak, and Zoran Herceg. "Effects of cold atmospheric gas phase plasma on anthocyanins and color in pomegranate juice." *Food Chemistry* 190 (2016): 317–323.

Kuiper-Goodman, T. "Food safety: Mycotoxins and phycotoxins in perspective." In M. Miraglia, H. van Edmond, C. Brera, & J. Gilbert (Eds), *Mycotoxins and phycotoxins—Developments in chemistry, toxicology and food safety*, pp. 25–48. Fort Collins, CO: Alaken Inc., 1998.

Lackmann, Jan-Wilm, Simon Schneider, Eugen Edengeiser, Fabian Jarzina, Steffen Brinckmann, Elena Steinborn, Martina Havenith, Jan Benedikt, and Julia E. Bandow. "Photons and particles emitted from cold atmospheric-pressure plasma inactivate bacteria and biomolecules independently and synergistically." *Journal of the Royal Society Interface* 10, no. 89 (2013): 20130591.

Lacombe, Alison, Brendan A. Niemira, Joshua B. Gurtler, Xuetong Fan, Joseph Sites, Glenn Boyd, and Haiqiang Chen. "Atmospheric cold plasma inactivation of aerobic microorganisms on blueberries and effects on quality attributes." *Food Microbiology* 46 (2015): 479–484.

Lee, Hanna, Jung Eun Kim, Myong-Soo Chung, and Sea C. Min. "Cold plasma treatment for the microbiological safety of cabbage, lettuce, and dried figs." *Food Microbiology* 51 (2015): 74–80.

Lee, Taehoon, Pradeep Puligundla, and Chulkyoon Mok. "Intermittent corona discharge plasma jet for improving tomato quality." *Journal of Food Engineering* 223 (2018): 168–174.

Leite, Ana K. F., Thatyane V. Fonteles, Thaiz B. A. R. Miguel, Giselle Silvestre da Silva, Edy Sousa de Brito, Elenilson G. Alves Filho, Fabiano A. N. Fernandes, and Sueli Rodrigues. "Atmospheric cold plasma frequency imparts changes on cashew apple juice composition and improves vitamin C bioaccessibility." *Food Research International* 147 (2021): 110479.

Li, Xiang, and Mohammed Farid. "A review on recent development in non-conventional food sterilization technologies." *Journal of Food Engineering* 182 (2016): 33–45.

Li, Xiaoan, Meilin Li, Nana Ji, Peng Jin, Jianhao Zhang, Yonghua Zheng, Xinhua Zhang, and Fujun Li. "Cold plasma treatment induces phenolic accumulation and enhances antioxidant activity in fresh-cut pitaya (Hylocereus undatus) fruit." *LWT* 115 (2019): 108447.

Li, Junguang, Qisen Xiang, Xiufang Liu, Tian Ding, Xiangsheng Zhang, Yafei Zhai, and Yanhong Bai. "Inactivation of soybean trypsin inhibitor by dielectric-barrier discharge (DBD) plasma." *Food Chemistry* 232 (2017): 515–522.

Li, Dongmei, Zhiwei Zhu, and Da-Wen Sun. "Effects of freezing on cell structure of fresh cellular food materials: A review." *Trends in Food Science & Technology* 75 (2018): 46–55.

Mahnot, Nikhil Kumar, Lih-Peng Siyu, Zifan Wan, Kevin M. Keener, and N. N. Misra. "In-package cold plasma decontamination of fresh-cut carrots: Microbial and quality aspects." *Journal of Physics D: Applied Physics* 53, no. 15 (2020): 154002.

Ma, Ji, Da-Wen Sun, Jia-Huan Qu, and Hongbin Pu. "Prediction of textural changes in grass carp fillets as affected by vacuum freeze drying using hyperspectral imaging based on integrated group wavelengths." *LWT-Food Science and Technology* 82 (2017): 377–385.

Ma, Ruonan, Guomin Wang, Ying Tian, Kaile Wang, Jue Zhang, and Jing Fang. "Non-thermal plasma-activated water inactivation of food-borne pathogen on fresh produce." *Journal of Hazardous Materials* 300 (2015): 643–651.

Min, Sea C., Si Hyeon Roh, Brendan A. Niemira, Glenn Boyd, Joseph E. Sites, Joseph Uknalis, and Xuetong Fan. "In-package inhibition of E. coli O157: H7 on bulk Romaine lettuce using cold plasma." *Food Microbiology* 65 (2017): 1–6.

Min, Sea C., Si Hyeon Roh, Brendan A. Niemira, Joseph E. Sites, Glenn Boyd, and Alison Lacombe. "Dielectric barrier discharge atmospheric cold plasma inhibits Escherichia coli O157: H7, Salmonella, Listeria monocytogenes, and Tulane virus in Romaine lettuce." *International Journal of Food Microbiology* 237 (2016): 114–120.

Misra, N. N., Tamara Moiseev, Sonal Patil, S. K. Pankaj, Paula Bourke, J. P. Mosnier, K. M. Keener, and P. J. Cullen. "Cold plasma in modified atmospheres for post-harvest treatment of strawberries." *Food and Bioprocess Technology* 7, no. 10 (2014): 3045–3054.

Misra, N. N., S. K. Pankaj, P. Bourke, and P. J. Cullen. "In-package nonthermal plasma treatment degrades contemporary pesticides on strawberries." In R. Singh Rawat & Paul Lee (Eds), *International conference on plasma science and applications (ICPSA2013)*, pp. 91–92. Singapore: Nanyang Technological University: National Institute of Education, 2013.

Misra, N. N., S. K. Pankaj, J. M. Frias, K. M. Keener, and P. J. Cullen. "The effects of nonthermal plasma on chemical quality of strawberries." *Postharvest Biology and Technology* 110 (2015): 197–202.

Misra, N. N., S. K. Pankaj, Annalisa Segat, and Kenji Ishikawa. "Cold plasma interactions with enzymes in foods and model systems." *Trends in Food Science & Technology* 55 (2016): 39–47.

Misra, N. N., B. K. Tiwari, K. S. M. S. Raghavarao, and P. J. Cullen. "Nonthermal plasma inactivation of food-borne pathogens." *Food Engineering Reviews* 3, no. 3–4 (2011): 159–170.

Montenegro, J., R. Ruan, H. Ma, and P. Chen. "Inactivation of E. coli O157: H7 using a pulsed nonthermal plasma system." *Journal of Food Science* 67, no. 2 (2002): 646–648.

Niemira, Brendan A., and Alexander Gutsol. "Nonthermal plasma as a novel food processing technology." *Nonthermal Processing Technologies for Food* (2011): 272–288.

Oehmigen, K., M. Hähnel, R. Brandenburg, Ch Wilke, K.-D. Weltmann, and Th Von Woedtke. "The role of acidification for antimicrobial activity of atmospheric pressure plasma in liquids." *Plasma Processes and Polymers* 7, no. 3–4 (2010): 250–257.

Oh, Yeong Ji, A. Young Song, and Sea C. Min. "Inhibition of Salmonella typhimurium on radish sprouts using nitrogen-cold plasma." *International Journal of Food Microbiology* 249 (2017): 66–71.

Ouf, Salama A., Abdulrahman H. Basher, and Abdel-Aleam H. Mohamed. "Inhibitory effect of double atmospheric pressure argon cold plasma on spores and mycotoxin production of Aspergillus niger contaminating date palm fruits." *Journal of the Science of Food and Agriculture* 95, no. 15 (2015): 3204–3210.

Ouf, Salama A., Abdel-Aleam H. Mohamed, and Wael S. El-Sayed. "Fungal decontamination of fleshy fruit water washes by double atmospheric pressure cold plasma." *CLEAN–Soil, Air, Water* 44, no. 2 (2016): 134–142.

Pan, Yuanyuan, Da-Wen Sun, and Zhong Han. "Applications of electromagnetic fields for nonthermal inactivation of microorganisms in foods: An overview." *Trends in Food Science & Technology* 64 (2017): 13–22.

Pan, Yuanyuan, Jun-Hu Cheng, and Da-Wen Sun. "Cold plasma-mediated treatments for shelf life extension of fresh produce: A review of recent research developments." *Comprehensive Reviews in Food Science and Food Safety* 18, no. 5 (2019): 1312–1326.

Pankaj, Shashi K., Zifan Wan, and Kevin M. Keener. "Effects of cold plasma on food quality: A review." *Foods* 7, no. 1 (2018): 4.

Pant, Kirty, Thakur Mamta, and Nanda Vikas. "Applications of cold plasma technology in grain processing." In M. Selvamuthukumaran (Ed.), *Non-thermal processing technologies for the grain industry*, pp. 95–120. Florida: CRC Press, 2021.

Patil, S., T. Moiseev, N. N. Misra, P. J. Cullen, J. P. Mosnier, K. M. Keener, and P. Bourke. "Influence of high voltage atmospheric cold plasma process parameters and role of relative humidity on inactivation of Bacillus atrophaeus spores inside a sealed package." *Journal of Hospital Infection* 88, no. 3 (2014): 162–169.

Porto, Elaine, Elenilson G. Alves Filho, Lorena Mara A. Silva, Thatyane Vidal Fonteles, Ronnyely Braz Reis do Nascimento, Fabiano AN Fernandes, Edy Sousa de Brito, and Sueli Rodrigues. "Ozone and plasma processing effect on green coconut water." *Food Research International* 131 (2020): 109000.

Puligundla, Pradeep, Taehoon Lee, and Chulkyoon Mok. "Effect of intermittent corona discharge plasma treatment for improving microbial quality and shelf life of kumquat (Citrus japonica) fruits." *LWT* 91 (2018): 8–13.

Qu, Jia-Huan, Da-Wen Sun, Jun-Hu Cheng, and Hongbin Pu. "Mapping moisture contents in grass carp (Ctenopharyngodon idella) slices under different freeze drying periods by Vis-NIR hyperspectral imaging." *LWT* 75 (2017): 529–536.

Ramazzina, Ileana, Annachiara Berardinelli, Federica Rizzi, Silvia Tappi, Luigi Ragni, Giampiero Sacchetti, and Pietro Rocculi. "Effect of cold plasma treatment on physico-chemical parameters and antioxidant activity of minimally processed kiwifruit." *Postharvest Biology and Technology* 107 (2015): 55–65.

Ramos, Bárbara, F. A. Miller, T. R. S. Brandão, P. Teixeira, and C. L. M. Silva. "Fresh fruits and vegetables— An overview on applied methodologies to improve its quality and safety." *Innovative Food Science & Emerging Technologies* 20 (2013): 1–15.

Rana, Sudha, Deepak Mehta, Vasudha Bansal, U. S. Shivhare, and Sudesh Kumar Yadav. "Atmospheric cold plasma (ACP) treatment improved in-package shelf-life of strawberry fruit." *Journal of Food Science and Technology* 57, no. 1 (2020): 102–112.

Reddy, P. Manoj Kumar, S. K. Mahammadunnisa, and Ch Subrahmanyam. "Catalytic non-thermal plasma reactor for mineralization of endosulfan in aqueous medium: A green approach for the treatment of pesticide contaminated water." *Chemical Engineering Journal* 238 (2014): 157–163.

Romani, Viviane Patrícia, Bradley Olsen, Magno Pinto Collares, Juan Rodrigo Meireles Oliveira, Carlos Prentice-Hernández, and Vilasia Guimaraes Martins. "Improvement of fish protein films properties for food packaging through glow discharge plasma application." *Food Hydrocolloids* 87 (2019): 970–976.

Romero-Navarro, Guillermo, Teresita Lopez-Aceves, Alberto Rojas-Ochoa, and Cristina Fernandez Mejia. "Effect of dichlorvos on hepatic and pancreatic glucokinase activity and gene expression, and on insulin mRNA levels." *Life Sciences* 78, no. 9 (2006): 1015–1020.

Sarangapani, Chaitanya, R. Yamuna Devi, Rohit Thirumdas, Ajinkya M. Trimukhe, Rajendra R. Deshmukh, and Uday S. Annapure. "Physico-chemical properties of low-pressure plasma treated black gram." *LWT-Food Science and Technology* 79 (2017): 102–110.

Sarangapani, Chaitanya, N. N. Misra, Vladimir Milosavljevic, Paula Bourke, Finbarr O'Regan, and P. J. Cullen. "Pesticide degradation in water using atmospheric air cold plasma." *Journal of Water Process Engineering* 9 (2016): 225–232.

Sarangapani, Chaitanya, Grainne O'Toole, P. J. Cullen, and Paula Bourke. "Atmospheric cold plasma dissipation efficiency of agrochemicals on blueberries." *Innovative Food Science & Emerging Technologies* 44 (2017): 235–241.

Selcuk, Meral, Lutfi Oksuz, and Pervin Basaran. "Decontamination of grains and legumes infected with Aspergillus spp. and Penicillum spp. by cold plasma treatment." *Bioresource Technology* 99, no. 11 (2008): 5104–5109.

Shi, Hu, Klein Ileleji, Richard L. Stroshine, Kevin Keener, and Jeanette L. Jensen. "Reduction of aflatoxin in corn by high voltage atmospheric cold plasma." *Food and Bioprocess Technology* 10, no. 6 (2017): 1042–1052.

Shi, Xing-Min, Guan-Jun Zhang, Xi-Li Wu, Ya-Xi Li, Yue Ma, and Xian-Jun Shao. "Effect of low-temperature plasma on microorganism inactivation and quality of freshly squeezed orange juice." *IEEE Transactions on Plasma Science* 39, no. 7 (2011): 1591–1597.

Song, Ah Young, Yeong Ji Oh, Jung Eun Kim, Kyung Bin Song, Deog Hwan Oh, and Sea C. Min. "Cold plasma treatment for microbial safety and preservation of fresh lettuce." *Food Science and Biotechnology* 24, no. 5 (2015): 1717–1724.

Srey, Sokunrotanak, Shin Young Park, Iqbal Kabir Jahid, and Sang-Do Ha. "Reduction effect of the selected chemical and physical treatments to reduce L. monocytogenes biofilms formed on lettuce and cabbage." *Food Research International* 62 (2014): 484–491.

Starek, Agnieszka, Joanna Pawłat, Barbara Chudzik, Michał Kwiatkowski, Piotr Terebun, Agnieszka Sagan, and Dariusz Andrejko. "Evaluation of selected microbial and physicochemical parameters of fresh tomato juice after cold atmospheric pressure plasma treatment during refrigerated storage." *Scientific Reports* 9, no. 1 (2019): 1–11.

Surowsky, Bjoern, Axel Fischer, Oliver Schlueter, and Dietrich Knorr. "Cold plasma effects on enzyme activity in a model food system." *Innovative Food Science & Emerging Technologies* 19 (2013): 146–152.

Takai, Eisuke, Katsuhisa Kitano, Junpei Kuwabara, and Kentaro Shiraki. "Protein inactivation by low-temperature atmospheric pressure plasma in aqueous solution." *Plasma Processes and Polymers* 9, no. 1 (2012): 77–82.

Tappi, Silvia, Annachiara Berardinelli, Luigi Ragni, Marco Dalla Rosa, Adriano Guarnieri, and Pietro Rocculi. "Atmospheric gas plasma treatment of fresh-cut apples." *Innovative Food Science & Emerging Technologies* 21 (2014): 114–122.

Tappi, Silvia, Giorgia Gozzi, Lucia Vannini, Annachiara Berardinelli, Santina Romani, Luigi Ragni, and Pietro Rocculi. "Cold plasma treatment for fresh-cut melon stabilization." *Innovative Food Science & Emerging Technologies* 33 (2016): 225–233.

Tappi, Silvia, Luigi Ragni, Urszula Tylewicz, Santina Romani, Ileana Ramazzina, and Pietro Rocculi. "Browning response of fresh-cut apples of different cultivars to cold gas plasma treatment." *Innovative Food Science & Emerging Technologies* 53 (2019): 56–62.

Trevisani, Marcello, Annachiara Berardinelli, Chiara Cevoli, Matilde Cecchini, Luigi Ragni, and Frederique Pasquali. "Effects of sanitizing treatments with atmospheric cold plasma, SDS and lactic acid on verotoxin-producing Escherichia coli and Listeria monocytogenes in red chicory (radicchio)." *Food Control* 78 (2017): 138–143.

Trivedi, Maharshi H., Kanishka Patel, Hanako Itokazu, Ngoc Anh Huynh, Mykola Kovalenko, Gary Nirenberg, Vandana Miller, Alexander A. Fridman, Gregory Fridman, Jacob Lahne, and Jasreen K. Sekhon. "Enhancing shelf life of bananas by using atmospheric pressure pulsed cold plasma treatment of the storage atmosphere." *Plasma Medicine* 9, no. 1 (2019): 23–38.

Tseng, Shawn, Nina Abramzon, James O. Jackson, and Wei-Jen Lin. "Gas discharge plasmas are effective in inactivating Bacillus and Clostridium spores." *Applied Microbiology and Biotechnology* 93, no. 6 (2012): 2563–2570.

Wan, Zifan, Yi Chen, S. K. Pankaj, and Kevin M. Keener. "High voltage atmospheric cold plasma treatment of refrigerated chicken eggs for control of Salmonella Enteritidis contamination on egg shell." *LWT-Food Science and Technology* 76 (2017): 124–130.

Wang, R. X., W. F. Nian, H. Y. Wu, H. Q. Feng, K. Zhang, J. Zhang, W. D. Zhu, K. H. Becker, and J. Fang. "Atmospheric-pressure cold plasma treatment of contaminated fresh fruit and vegetable slices: Inactivation and physiochemical properties evaluation." *The European Physical Journal D* 66, no. 10 (2012): 1–7.

Wang, Yijie, Zichong Ye, Jiahui Li, Yan Zhang, Yingxi Guo, and Jun-Hu Cheng. "Effects of dielectric barrier discharge cold plasma on the activity, structure and conformation of horseradish peroxidase (HRP) and on the activity of litchi peroxidase (POD)." *LWT* 141 (2021): 111078.

Won, Mee Yeon, Seung Jo Lee, and Sea C. Min. "Mandarin preservation by microwave-powered cold plasma treatment." *Innovative Food Science & Emerging Technologies* 39 (2017): 25–32.

Xiang, Qisen, Xiufang Liu, Junguang Li, Shengnan Liu, Hua Zhang, and Yanhong Bai. "Effects of dielectric barrier discharge plasma on the inactivation of Zygosaccharomyces rouxii and quality of apple juice." *Food Chemistry* 254 (2018): 201–207.

Xie, Anguo, Da-Wen Sun, Zhiwei Zhu, and Hongbin Pu. "Nondestructive measurements of freezing parameters of frozen porcine meat by NIR hyperspectral imaging." *Food and Bioprocess Technology* 9, no. 9 (2016): 1444–1454.

Xu, Lei, Allen L. Garner, Bernard Tao, and Kevin M. Keener. "Microbial inactivation and quality changes in orange juice treated by high voltage atmospheric cold plasma." *Food and Bioprocess Technology* 10, no. 10 (2017): 1778–1791.

Xu, Yingyin, Ying Tian, Ruonan Ma, Qinghong Liu, and Jue Zhang. "Effect of plasma activated water on the postharvest quality of button mushrooms, Agaricus bisporus." *Food Chemistry* 197 (2016): 436–444.

Zhang, Ming, Jun Kyun Oh, Luis Cisneros-Zevallos, and Mustafa Akbulut. "Bactericidal effects of nonthermal low-pressure oxygen plasma on S. typhimurium LT2 attached to fresh produce surfaces." *Journal of Food Engineering* 119, no. 3 (2013): 425–432.

Zhou, Dandan, Zhuo Wang, Sicong Tu, Shaoxia Chen, Jing Peng, and Kang Tu. "Effects of cold plasma, UV-C or aqueous ozone treatment on Botrytis cinerea and their potential application in preserving blueberry." *Journal of Applied Microbiology* 127, no. 1 (2019): 175–185.

Zhou, Shengnan, Zhiwei Zhu, Da-Wen Sun, Zhongyue Xu, Zhihang Zhang, and Qi-Jun Wang. "Effects of different cooling methods on the carbon footprint of cooked rice." *Journal of Food Engineering* 215 (2017): 44–50.

Zhuang, Hong, Michael J. Rothrock Jr, Kelli L. Hiett, Kurt C. Lawrence, Gary R. Gamble, Brian C. Bowker, and Kevin M. Keener. "In-package air cold plasma treatment of chicken breast meat: Treatment time effect." *Journal of Food Quality* 2019 (2019): 1–7.

Zhu, Wen-Chao, Bai-Rong Wang, Hai-Ling Xi, and Yi-Kang Pu. "Decontamination of VX surrogate malathion by atmospheric pressure radio-frequency plasma jet." *Plasma Chemistry and Plasma Processing* 30, no. 3 (2010): 381–389.

Zhu, Zhiwei, Xinwei Wu, Yi Geng, Da-Wen Sun, Haiyang Chen, Yongjun Zhao, Wenqing Zhou, Xianguang Li, and Hongzhun Pan. "Effects of modified atmosphere vacuum cooling (MAVC) on the quality of three different leafy cabbages." *LWT* 94 (2018): 190–197.

Ziuzina, Dana, N. N. Misra, P. J. Cullen, Kevin M. Keener, J. P. Mosnier, Ivan Vilaró, Edurne Gaston, and Paula Bourke. "Demonstrating the potential of industrial scale in-package atmospheric cold plasma for decontamination of cherry tomatoes." *Plasma Medicine* 6, no. 3–4 (2016): 397–412.

Ziuzina, Dana, N. N. Misra, Lu Han, P. J. Cullen, T. Moiseev, Jean-Paul Mosnier, K. Keener, E. Gaston, I. Vilaró, and P. Bourke. "Investigation of a large gap cold plasma reactor for continuous in-package decontamination of fresh strawberries and spinach." *Innovative Food Science & Emerging Technologies* 59 (2020): 102229.

Ziuzina, Dana, Sonal Patil, Patrick J. Cullen, K. M. Keener, and Paula Bourke. "Atmospheric cold plasma inactivation of Escherichia coli, Salmonella enterica serovar Typhimurium and Listeria monocytogenes inoculated on fresh produce." *Food Microbiology* 42 (2014): 109–116.

Chapter 4

Irradiation

A Non-Thermal Processing Approach for the Fruit and Vegetable Industry

Amanda Malik, Mamta Thakur, and Vikas Nanda

CONTENTS

DOI: 10.1201/9781003222170-4

4.1 INTRODUCTION

Out of all the methods of preserving food, irradiation was the most controversial prior to its implementation. In the UK and many other countries, people have been suspicious of irradiation generally due to a negative and ill-informed media. An absurd situation has arisen in the UK as authorities have granted permission for carrying out irradiation of foods but it is seldom done. A lot of items other than food materials namely pharmaceuticals, body-care essentials, biomedical equipment, and polymers are irradiated with the aid of robots. Still, the usage life and safety of many freshly harvested commodities are safeguarded in one way or the other, yet causing little to no damage to the end products. Food irradiation is not a novel idea as the chances of control of microbial load were taken into consideration after the exploration of X-rays in the latter part of the 19th century. The security and "wholesomeness" of the foods treated with radiation have been the main focus of many scientists for more than 50 years and this has been to their satisfaction.

One such option is offered by food irradiation where the procedure for exposing the pre-packaged fruits and vegetables to γ-rays, X-rays, or electron beams is carried out. Since no energy gets dissipated in the form of virtual heat, irradiation can also be called "cold-sterilization". The food processing industry utilizes some of the common irradiation sources like γ-rays given out

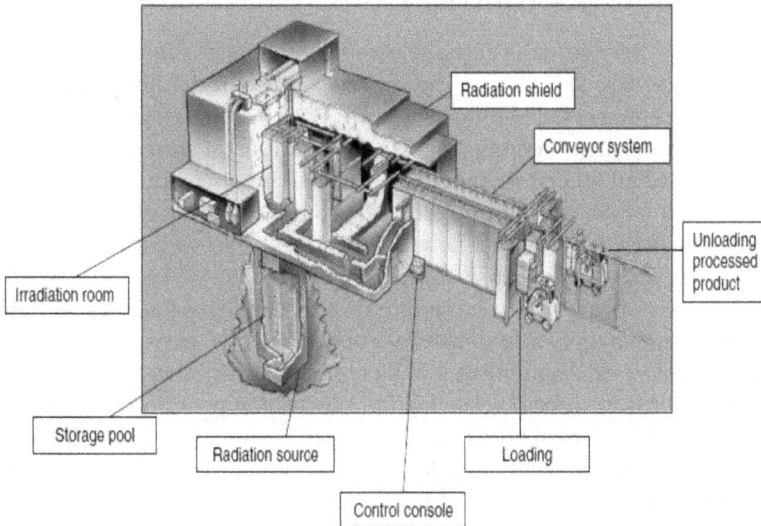

Figure 4.1 A food irradiation facility. (Image source: WHO, FAO, and the IAEA guidelines for food irradiation.)

by radioisotopes (Cesium-137 and Cobalt-60), X-ray producing devices, or accelerators of electrons (Arvanitoyannis et al., 2009). Figure 4.1 gives detailed information of a food irradiation facility. Irradiation has been termed a secure and non-chemical type methodology by international health organizations such as WHO, the FAO, and the International Atomic Energy Agency (IAEA) for application in the irradiation of fruits and vegetables. With a limit of 1 kGy (100krad), the United States gave the green light to the usage of irradiation in the processing of fruits and vegetables in April of 1986. The storage period by sterilization and upkeep of freshly harvested commodities was enhanced through ionizing radiations. A host of advantages – namely maturity and ripening delay, decreased breathing and ethylene generation, lowering of membrane breakage and oxidation of lipids – are offered by this technique of irradiation, and they finally prove effective in enhancing the storage life of these perishable products. A large number of findings suggested that consumption of irradiated food from different radiation sources is secure and documents supporting the claims have been verified.

As the demands and desires to restrict the usage of chemical entities has increased, the characteristic of irradiation to leave no traces after processing gives it an edge over other methods. The effective destruction of DNA through the application of ionizing radiations forms the principle of food processing by irradiation. Cells get deactivated prior to irradiation, and hence, microbes, gametes of insects, and meristems of plants are not permitted to reproduce and therefore different safe-keeping consequences of absorbed irradiation can be seen. Radiation instigates chemical conversions in various foods which also tend to get reduced to their lowest degree (Thayer, 1990). It also benefits in terms of reduction in losses after the product is harvested, thus checking microbes and insects during storage. This can be for the inspection of sprouting potatoes, destroying grain pests, altering certain components, and carrying out changes in the physicochemical and sensory characteristics of food (Wang and Chao, 2002). And for the purpose of eliminating contamination from food or ensuring the sterility of dry horticultural products, γ-irradiation has always been the prominent choice for food processors (Yang et al., 1993).

4.2 MECHANISM OF IRRADIATION

The literal meaning of irradiation is actually exposing the food to radiation. γ-rays, X-rays, or highly energized E-beams (particles) are the various kinds of radiation sources that have been utilized for the safe-keeping of food products and are known as ionizing rays. The three radiation types are very affluent in their ionizing tasks and stimulation of atoms in the end product. But their strength is limited and therefore they are not able to interact with the radio nuclei to cause radioactivity. As both γ-rays and X-rays form part of the electromagnetic spectrum, they are somewhat similar in their functional attributes, although their origins are different (Brennan et al., 2006). Electron volts (eV), or more importantly MeV (1 MeV = 1.602×10⁻¹³ J), is allotted for denotation of the power of photons or characteristic particles of the ionizing rays. The movement energy gained by the moving particles when they are accelerated through a potential difference of one voltage is termed 1-eV. An important and confusing distinction lies in the terms of power and dose. When power is absorbed by the food product upon penetration of the radiation into the food, it is called the "absorbed dose" and denoted by Grays (Gy), where 1Gy is the amount of absorbed power of magnitude one J per kilogram. The power/dose absorbed changes with the penetration depth of radiation, the time it falls on a surface, and the constituents of the food, even when the energy of radiation is the same for characteristic radiations. Since γ-rays are generated from radioisotopes, they possess fixed energies and Co-60 is the main source of this. As the isotope is the first choice for radiation, it also becomes problematic as radioactive waste material. Cs¹³⁷ is the isotope of choice as it gets produced as a waste of reprocessing radioactive fuel and is not as popular as Co-60. There are radiation sources that are powered by electricity, these are E-beam and X-rays. Therefore, a continuous energy band is showcased by them as per the requirements of the equipment. The big advantage of this method over others when compared with the isotopic method is that they can be turned ON/OFF and they are also not connected to nuclear companies (Brennan et al., 2006).

Ionizing radiations act in the following manner in three phases:

- primary function of physically irradiating atoms;
- chemical consequence of the above primary function;
- consequences of irradiation to living cells and microbial contaminants.

4.3 SOURCES OF IRRADIATION USED IN FOOD PRESERVATION

We have a large number of choices when it comes to irradiation but only a handful of them can be used for their role in food preservation. Three types of radiations are most widely used and these are namely accelerated electron beams (E-beams), γ-irradiation with Cobalt-60 or Cesium-137, and X-ray generating machines. Out of these, γ-irradiation is the most widely used and commercialized to a large extent. Table 4.1 describes various types of radiation and their sources along with their characteristics.

4.3.1 *Electron Beams as Radiation Source in Food Processing*

For the treatment of fruits or vegetables to remove microbial contaminants, electron-beam irradiation (EBI) utilizes low-dose ionizing radiations. Devices that are efficient enough to accelerate the electrons to nearly the speed of light are used for the production of electron beams. Figure 4.2 gives an overview of the working of electron beam irradiation for food processing.

TABLE 4.1   DIFFERENT TYPES OF IONIZING RADIATION, THEIR SOURCES AND NATURE

Type	Gamma rays		Electron beams	X-rays
Source	Cobalt-60	Cesium-137	Van de Graaff generators or linear accelerators	Bombarding metal plate with high energy electron beams
Uses	Food irradiation, sterilization, radiotherapy	Sterilization, disinfestation	Irradiating large volumes of small food items like grains	Insecticide for grains, produce, or spices, shelf-life extension, and delays ripening
Nature	Water insoluble	Water soluble	Less penetrating than gamma rays	High penetration depth and high dose uniformity
Comprised of	Photons		Electrons	Photons
Mass	None		Yes	None
Electric charge	None		Yes	None
Penetration	Good/very good		Limited	Good/very good
Consequence	Products of low and medium density can be treated in cartons, drums, or pallets		Products of low density can be treated in cartons	

Source: Radomyski et al. 1994, Ashraf et al. 2019, USFDA.

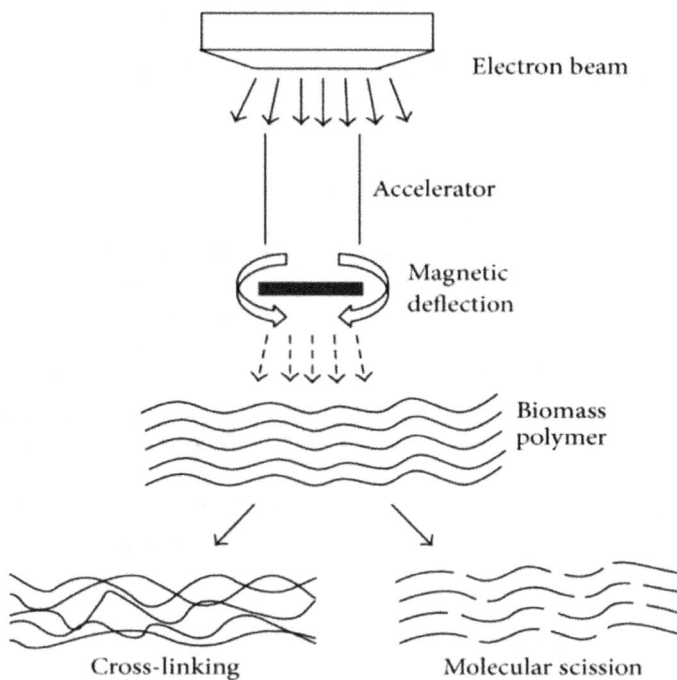

Figure 4.2   Working of an electron beam irradiation unit for food processing. (Image source: Anita Saini et al., 2015.)

These devices are machine sources such as a linear accelerator (Linac) or a Van de Graaff generator which are powerful enough to accelerate the electrons almost to the speed of light. For the quantum of the light utilized, 10 MeV should be the limit. The Linac of the electron beam employs energetic flares of about 10,000 watts and modulates the electrons to potentials in the range of 5 to 10 million electron volts. The penetration depth of electrons in food is only about 3 cm, leaving the only option for the food to be not thicker than 3cm in order to be treated properly. Commercial electricity supply acts as a power source for the generation of electron beams and can be operated easily. The electron gun releases streams of electrons which are fed into a speeding tube in which under vacuum conditions the electrons are produced and sped up by the machine. Owing to their small mass, the electrons do not have enough capacity to penetrate deep into the products. When compared with other methods of irradiation like γ-rays or X-rays, the range and power of the electron beam are limited due to its low penetration ability (<2 inches) and is suitable only for the treatment of thin products (Wilkinson and Gould, 1996). The US Food and Drug Administration (USFDA) has approved the EBI for its use in place of chemical treatments to carry out food irradiation. A wide variety of foodborne contaminants and pathogens are controlled and killed by the EBI in order to effectively preserve foods and enhance their storage life and keep them fresh for longer durations. For further enhancement of the effectiveness of EBI, it can be utilized as one of the steps in the hurdle technique along with other conventional or non-conventional methods to improve food protection (Meisberger et al., 1996). Due to various merits, electron accelerators prove to be better suited for irradiating food instead of others like γ-rays: (1) electron beams can be generated in any amount and at any time; (2) no need for power refilling like with other methods, such as Co-60; (3) no danger of harmful wastes containing radioactive substances; and (4) usability in applications/processes needing dense-flow and heavy doses of irradiation (Clemmons et al., 2015). However, there are certain shortcomings of the technique like superficial penetrative powers.

4.3.2 X-Ray Irradiation for Food Processing and Preservation

These are electromagnetic radiations which are produced when metal atoms undergo transition and these radiations are less powerful than γ-rays. X-rays of different energy levels are generated by mechanistic devices and these devices engineered for preserving food products are highly powerful when compared with the ones used for X-ray imaging. Generally speaking, the production of X-rays is done when electron beams go through a slender plate of metal (gold or tungsten). Upon hitting the metal plate by the electrons from the electron beam, a huge amount of energy is produced and the majority of this energy gets lost in the form of heat. Only a fraction of the energy is taken up by the electrons in the metal plate and the energy subsequently gets changed into X-rays. But this efficiency of X-ray production can be enhanced significantly through the help of heavier atoms or by improving the mass of the target only by increasing the power of the electron beam (Kume et al., 2009). When compared with electrons, a greater deal of penetrating power and penetration depth (about 15 inches) is offered by the X-rays, thereby allowing greater ease of food filling into big containers during transport by themselves. And also, the machine can be turned ON/OFF easily just like the electron beam assembly and no threat of harmful radiation or case of radioactivity is there.

4.3.3 γ-Rays for Food Processing and Preservation

Use of radioactive substances is also called "radioisotopes", and is one of the most widely used methods for the generation of γ-rays. Cobalt-60 (^{60}Co) and Cesium-137 (^{137}Cs) are the authorized sources for the generation of γ-rays for irradiating food substances. Energy levels of 1.17 and 1.33 MeV (^{60}Co) and 0.662MeV(^{137}Cs), respectively, are contained within them. Spontaneous fragmentation of radionuclides is carried out in order to generate distinct power γ-rays. As γ-rays are an ionizing type of radiation, they can remove electrons from atoms. The DNA of the viable cells are destroyed by the free electrons as they are able to take part in chemical reactions. This forms the principle for the destruction of microorganisms through irradiation. The source of irradiation, irradiation cabinet (biological shield), source panel, source for storing water, source for carrying product, and transition system for source, warehousing facility, and automated control systems are taken together for the formation of an irradiation facility based on γ-rays. The food irradiation is finished inside. A biological shield (a room made of two-meter thick concrete) is used to carry out the irradiation process of food materials. The radiation-producing systems are kept beneath a depth of about 600cm water inside a pool. The radiation sources are taken out by use of a pneumatic piston device during treatment and sandwiched using tote bins in which food material to be irradiated is placed. A conveyer belt is used to pass the prepacked materials stationed inside the tote bins. The entry and movement of the boxes around the radiation point is in a halt-and-go system such that the irradiation of the product is carried out in an efficient manner by irradiating the product in many positions and finally conveying them out of the system through automated belts. The whole operation of the irradiation facility is monitored through automation in which the supply of radiation and conveying system are both guided by computers (SVR Reddy et al., 2018).

4.4 IRRADIATION: EFFECTS ON VEGETABLES AND FRUITS

Horticultural products when faced with doses of ionizing radiations, undergo alterations in various attributes namely physiological, biochemical, cellular forms and structural and genetic arrangements, cellular and tissue microbiota, and attributes (Afify et al., 2013). Various organoleptic and dietary attributes of the horticultural commodities undergo modifications when ionized radiations strike them and the modification range depends on the quality of raw material quality and specifications of irradiation treatment quality (also known as dose range or specified dose range) (Figure 4.3), and irradiation source (Alothman et al., 2009). The ascertainment of irradiation consequences of the horticultural commodities could be verified by examining the degree of mitigation in the constituents and their extent.

4.4.1 Effects of Irradiation on Physical Properties of Fruits and Vegetables

4.4.1.1 Effect of Irradiation on Texture of Fruits and Vegetables

The textural characteristics of the horticultural commodities whether harvested anew or in storage, are dictated principally by the solidarity of the cell wall and its framework and also by cellular hydrostatic pressure (Mahto and Das, 2013). The alterations in the ion-mobility attributes of corpuscle membrane influence food conductance characteristics have been linked with treatment of the foods by ionizing radiations. It has been advised that variations in the ion transport characteristics of cell membrane, due to which electrical conductivity of food changes, occurs due to radiation treatment of food. Saccharide-rich commodities or fruits suffer greater surface fragmentation than their horticultural counterpart or vegetables as a result of the excessive presence of moisture in them. Enzymic catalysis of pectin in the company of stimulants like PME (pectin methyl esterase) and PG (poly-galactarunase) eventually proceeds to the degradation of

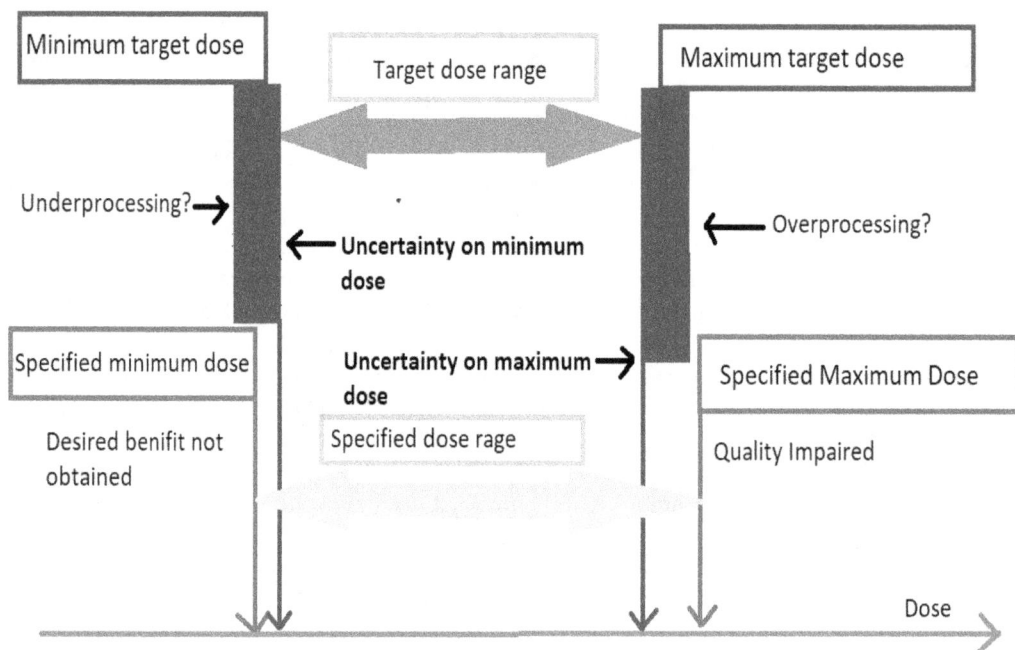

Figure 4.3 Specified dose range and target dose range of radiation to be used in food processing. (Image source: WHO, FAO, and the IAEA guidelines for food irradiation.)

compactness of stored fruits. Freshly slashed *Ananas comosus* L; "Dusehri" and "Fazli"; *Rubus idaeus* L; the apple variety known as "Red Delicious" which were processed through the application of ionizing rays – all these saw a surprising drop in the toughness of their structure (Mostafavi et al., 2012) – and as a consequence, ionizing rays carrying energy in excess of 0.4 kGy for application as phytosanitary treatment produced high losses of the compactness in cells/tissues structural organization (Jung et al., 2014). Puncture analysis of the cell wall would yield adjustments in its biochemical make-up after supplying diminishing irradiation strengths (<2 kGy) (Lattore et al., 2010). By utilizing the SEM-based direct gauzing technique, the microstructural deformities in the king-oyster-mushrooms, arising after irradiation, could be visualized (Akram et al., 2012). Electrolytic seepage was the alternate methodology for quantification of cytomembrane variability, thereby affirming safety after heavy irradiation of crucifers in contrast to freshly harvested onion, carrots, and celery (Fan and Sokorai, 2008). Textural attributes of the irradiated horticultural commodities are a salient attribute and utilized for gain in the storage period of the commodities so as to improve their market value.

4.4.1.2 Effect of Irradiation on Weight Loss of Fruits and Vegetables

Diverse inspections have theorized that poundage loss in horticultural commodities incurs as a consequence of treatment by ionizing radiations. The irradiated specimens upon their storage period squandered their weights as a consequence of moisture drop. Thomas and Moy (1986) disclosed about the mass squandering due to desiccation as about 15–16% and 27–28% during safekeeping of irradiated (one-sixteenth of 1 kGy) potato-tuber and onions, respectively. In a similar manner, bulbs of freshly harvested onions processed at irradiation energy of 0.06–0.076

kGy squandered around 5.7% of their mass as of shrinkage and sprouting during their safekeeping for 150 days in contrast to 23.2% of non-treated ones (Arvanitoyannis et al., 2009). *Valenciana sintetica* radiated with 0.03 kGy of radiation dose underwent 43.3% squandering in weight in contrast with non-irradiated ones that squandered 22.8% of weight during their 270 days of safekeeping. γ-irradiation (0–0.3 kGy) of grapefruit (*Citrus paradise*) did not yield any noticeable variation in their TSS contents when they were safeguarded in two different surroundings (212°F, 392°F) for safekeeping (Vanamala et al., 2007).

4.4.1.3 Effect of Irradiation on Overall Acceptability (Aroma, Flavor, and Color)

Food qualities (aroma, flavor, and color) normally get enhanced by the duration of irradiation as conclusion of the breakdown of glycosidic linkages and more favorable glycone formation (Tripathi et al., 2016). Irradiated vegetables showcased minimal differences in the flavoring elements after storage; however, radiation-processed onions exhibited increased clustering of pyruvates, flavor-zymes in both radiation-treated and untreated bulbs at a later period of their safekeeping (Ceci et al., 1991). Blooming and shrinkage of aroma and taste due to cleavage of the sulfhydryl moieties from Sulphur-AAs (amino-acids) at soaring dose (10 kGy) of radiation altered the sensory sites in Allium species leading to loss in attributes (Swallow, 1991). Retardation in the dosing of irradiation would cause an additional augmentation of organoleptic attributes as radiation-mediated degradation relies on dosing amounts (Lacroix et al., 2009). Table 4.2 describes the prescribed dose limit for various food products and horticultural produce.

Overall acceptability of fruits and vegetables may additionally increase as in carrot at 0.13 kGy, *Litchi chinensis* and *Dimocarpus longan* at 0.25–0.4 kGy, pre-cut *Citrullus lanatus* at 1 kGy; as additionally limit as in tomato at 0.08 kGy, *Pyrus malus* at 3 kGy; or besides altered common acceptability as in *Coriandrum sativum*, *Mentha spicata*, *Lactuca sativa*, and *Nasturium officinale* up to 1 kGy, mangoes at 0.6 kGy; pre-cut *Cucumis melo* treated with doses more than 1 kGy (Mohacsi-Farkas et al., 2006), and *Rubus idaeus* at 0–1.5 kGy (Verde et al., 2013) when treated with low doses of irradiation (1–10 kGy). Mami et al. (2013) cautioned that low doses of irradiation would possibly make bigger ordinary acceptability in terms of visibility as a reason for elevation of brightness levels of desiccated tubers (Wang and Du, 2005) and truffle. Therefore, it used to be observed that radiation dose limit of 0–1.5 kGy is appropriate to achieve the marketable acceptability of processed fruits and vegetables in tomato, carrot, mango, and melon. Trigo et al. (2009) stated that in most green vegetables, a dose of more than 2 kGy reduced the average acceptability of the products in terms of appearance (color and texture), aroma, and flavor.

4.4.2 Effect of Irradiation on Changes in Chemical Properties

4.4.2.1 Effects of Ionizing Radiation on Protein Characteristics

Irradiation treatment has immediate or secondary control over the dietary proteins. Treatment of foods with ionizing radiations chiefly impacts the polypeptides in horticultural commodities due to the adversity of S-linkages and H-bonds on peptidyl-linkages. As per Audette-Stuart et al. (2005) horticultural commodities sustain chemo-transfiguration due to physiology of proteins and parameters of irradiation of foods (environment conditions). Electrons, H-atom, and -OH species get synthesized when ionizing radiations communicate with proteins as a result of polymerization (dimerization) and fragmentation mechanisms. It was flagged that a tremendous dose of radiation power was once used for protein fragmentation. By employing gel filtration such as SDS-PAGE assay, the connections across molecules or segregation of proteins by irradiation was once evident by using the place separation of irradiated proteins as based on

TABLE 4.2   ADVISORY TECHNOLOGICAL DOSE LIMITS (KGY)

Food classes→ / Purpose↓	Class 1 (Bulbs, roots, tubers):	Class 2 (Fresh fruits, vegetables):	Class 3 (Cereals, dried fruits etc.):	Class 4 (Fish, seafood):	Class 5 (Raw poultry, meat):	Class 6 (Dry vegetables, spices, etc.):	Class 7 (Dried, of animal origin):	Miscellaneous:	Class 8: including honey, space food, hospital meals, military rations, liquid egg, spices, thickener
Sprout inhibition	0.2	-	-	-	-	-	-	-	
Ripening delay	-	1.0	-	-	-	-	-	-	
Insect disinfestation	-	1.0*	1.0*	-	-	1.0*	1.0*	-	
Quarantine measures*	-	1.0*	-	-	-	-	-	-	
Shelf-life extension	-	2.5	-	3.0	3.0	-	-	-	
Pathogen reduction**	-	-	-	5.0**	7.0**	>10**	-	-	
Parasite elimination	-	-	-	2.0**	3.0**	-	-	-	
Other	-	-	-	-	-	-	-	>10**	

Source: WHO, FAO, and the IAEA guidelines for food irradiation.

* Minimum doses for particular pests, fruit flies > 0.15 kGy

** Minimum doses to ensure hygienic quality

the concept of molecular size differentiation (Kuan et al., 2011). The presence of proteins in radiated products is dependent on the tolerance of AAs in the radiated doses. Abundance of AAs such as Ser, Lys, Leu, Asp, Thr, Arg, Ala, and Ile enhanced in the irradiated commodities but Glu, Pro, Met, and Phe slumped after the irradiation of the commodities (Ussuf and Nair, 1972). The carbonyl associations of peptide linkages are predicted as highly stable entities towards free radicals generated prior to irradiating food as compared with the carbonyl linkages of the ester connections of the glycosidic bonds between the different sugar units in polysaccharides. Thus, even after radiating food by severe irradiation doses of up to 10 kGy, many peptide bonds could be seen left incessant. It was, however, observed that alterations in the protein skeleton as an outcome of ionizing radiations are virtually insignificant for dosing up to 10 kGy (Kushwah and Daheriya, 2012). It shows the versatility of irradiation dosing up to 10 kGy for treatment and processing of products which block structural alteration of proteins.

4.4.2.2 Effects of Ionizing Radiation on Enzymatic Changes

Enzymatic proteins in the horticultural commodities would be susceptible to chemical responses in the absence of antioxidating agents which otherwise remain tolerant to radiation doses of five to ten times as that required for destruction of microbiota (Swallow, 1991; Jongen, 2005). Few scientists have calculated consequences of quantity of radiation doses on food enzymes involved in cell wall alteration, cell membrane modification, browning, and other enzymatic reactions and oxidative enzymes. Cell wall enzyme activity such as β-galactosidase, pectin-methyl-esterase, and poly-galacturonase was drastically affected in tomatoes when irradiated at 0.7–2.2 kGy radiation dose. Polyphenol oxidase enzyme activity was reduced in freshly cut celery irradiated at 1 kGy dose of radiation (Lu et al., 2005); freshly cut lettuce irradiated at 1 kGy dose of radiation; and irradiated (0–0.75 kGy) minimally processed eggplant (Husain et al., 2015). Polyphenol oxidase gets inactivated in 100 minutes, whilst peroxidase is fully damaged in only 15 minutes in different variants of apple juices, when treated by using elevated-pressure Hg-light of 400W (Falguera et al., 2011). Total anthocyanins in strawberries were also witnessed to improve after treatment with blue light. Also, activities of the given components also got enhanced e.g., G-6-P, shikimate-dehydrogenase, Tyr-NH_3-lyase, Phe-NH_3-lyase, Cinnamate-4-hydroxylase, 4-Coumarate/Co-A-ligase, Dihydroflavonol-4-reductase, Chalcone-synthase, Flavanone-3-β-hydroxylase, Anthocyanin-synthase, and UDP-glucose flavonoid-3-O-glucosyltransferase (Xu et al., 2014). The concentration of antioxidant enzymes such as peroxidase increased when irradiated at dose ranges of less than 10 kGy in apricots; α-resorcylic acid in RTC-ash gourd; enzymatic activity of following enzymes in blueberries-superoxide-dismutase, catalase, and peroxidase in blueberries.

4.4.2.3 Effects of Ionizing Radiation on Properties of Carbohydrates

When starches from cereal grains and tubers were irradiated with γ-irradiation of doses up to 6.2 kGy per hour, it caused precipitation of the following aldehydes: malonaldehyde, HCHO, and CH_3CHO, carboxylic acid HCOOH, and H_2O_2 as primary breakdown products of irradiation (Stefanova et al., 2010). Chemically, depolymerization of macromolecular carbohydrates takes place due to food irradiation (Jongen, 2005) and as a consequence, glycosidic linkage breakdown occurs due to glycone formation, and as a role it disintegrates vital cytological wall functionalities. As a result of heavy moisture irradiation dosing, carbohydrates showcased linear dissimilitude. Therefore, ionizing radiations elevated the companionship of carbonyl and H_2O_2 compounds with radiating energies. Certain specific functionalities could be positively grasped by amending the composition of starch typically via physical, chemical, and enzymatic strategies (Jongen, 2005). The conformation and functionalization of the amylum converted more rapidly by ionizing rays than conventional methodologies as irradiation does not demand the presence of any stimulant and undergoes minuscule temperature variations (Battaeib et al., 2014). Physicochemical and rheological attributes like flow characters, H_2O solubility, tartness of the solution, increment in amylose contents, transitions in gelatinization temperatures, and diminishing of the pasting characteristics of amylum constituents of cassava, rice, and tubers as per findings. Minute starch granules of the irradiated tubers also accumulated at certain stages of their safekeeping (Ashwar et al., 2014).

Sugary carbohydrates were also affected by irradiation of the horticultural produce during their storage period. The physicochemical alterations produced in the saccharides when they were treated at doses below 1 kGy proved to be more minute than modifications produced after conventional sterilization methods (Dauphin and Saint-Lebe, 1977). It is noted that sugars in tuber vegetation escalate (glucose, fructose, and sucrose) upon irradiation then hit their maximum charts and shrank after storage. The fructose/glucose proportions were slightly greater in the standards, which in contrast were perpetual in radiation-treated tubers. Among all the sugars, it was found that temperature and irradiation appreciably affected the fructo-oligosaccharides in the bulbs. In certain potato cultivars, it was established that reducing sugars accelerated at 10.2%, but non-reducing ones also diminished at around 12.75% (Rezaee et al., 2013). Reducing sugars flourished after radiation treatment which resulted in starch deterioration. It was once observed that UV-C radiation treatment reduced sugars in *Ananas comosus* L which had been previously minimally processed (Pan and Zu, 2012) whereas γ-radiation elevated saccharide content in certain popular mango varieties (Mahto and Das, 2013). It was evaluated that processing using ionizing radiations at elevated ranges did not significantly alter carbohydrate content in the horticultural commodities (Skala et al., 1987). The feedback of saccharides, both reducing as well as non-reducing ones, thus depends on the type of horticultural produce and their treatment and safekeeping exploits.

4.4.2.4 Effects of Ionizing Radiation on Properties of Lipids

Lipid contents in edible oils are affected by ionizing radiations in the same fashion as naturally occurring ones. Generally, lipids degrade and give rise to intermediary commodities like FFAs and diacylglycerols with immediate segregation of triglycerides, when the cells are exposed to ionizing radiations. The lipid modification mechanisms utilizing ionizing radiation include essential-ionization, accompanied by transfer of positive charges in the carboxylic family and the dual bond linkages in the proximity of carbonyl associations (Nawar, 1972) (Afify et al., 2013). Treatment with ionizing radiations causes breakdown of lipids primarily by carboxyl or dual bonds to synthesize oxides of carbon and dihydrogen along with carboniferous and =CO compounds. After irradiating the potatoes, crude fat and phospholipids were shown to diminish (Mondy and Gosselin, 1989). With minuscule (0.5–2 kGy) irradiation doses, unsaturation of the fatty acids escalated in crucifers owing to transport of omega-6 fatty acid (18:3) from mono galactosyl diacylglycerol; whereas with greater doses (0–7.5 kGy) it diminished in the seeds of soybean, peanut, and sesame. Fatty acids, particularly linolenic acid in the irradiated (60Gy) garlic bulbs, slumped during their safekeeping timeline of eight months with the assistance of irradiation (Perez et al., 2007). Elimination of oxygen might be needed to diminish off-odors generated by the wobbly compounds formed from the radiolysis of the commodities due to lipid amendment as a result of irradiation processing. Radiation in the exclusion of air may also oxidize lipids through the polymerization of decarboxylated and USFA (unsaturated fatty acids) (Afify et al., 2013). As neither air nor water can be stamped out entirely from the horticultural commodities through human use, thus, commercial doses of irradiation could not prevent the reduction of unsaturation from fatty acids produced. PUFA (polyunsaturated fatty acids) have been declared as instrumental for human well-being and are somewhat prone to the dangers of peroxidation. Hence, the durability of these factors could be assessed for the standardization of irradiation protocols (Erkan and Ozden, 2007).

4.4.2.5 Effects of Ionizing Radiations on Properties of Vitamins

Food vitamins are better safeguarded from ionizing radiations in their solution forms due to the protecting characteristics of dietary elements. As per Fellows (2001), various essential micronutrients (whether lipid or water solvable) have exceptional susceptibility to ionizing radiations. During proceedings for horticultural commodities, chiefly B1 or Vit-C in the category of water-dissolvable and Vit-E and K in the fat-dissolvable category have elevated radio-sensitivity. It was discovered earlier that the portions of γ-irradiation higher than 1 kGy generated free radicals, resulting in a reduction of Vit-C, and various antioxidants of varieties of spinach; whereas folate, α-tocopherols, phylloquinones, and neoxanthins were faintly modified by elevated ionizing energy doses (Lester et al., 2010). In the course of treatment with ionizing rays, forfeiture in the quantity of vital micronutrients was presumed as class-dependent (horticultural varieties), depending on their origins and the thermal characteristics, the extent of O_2 removal/existence, and the quantity of ionizing rays to be used. A considerable amount of research has been done for identifying the consequences of radiation (ionizing) on the Vit-C availability in horticultural produces. Vit-C has an elevated zone of reactivity towards irradiation and gets transformed into dehydro-Vit C when faced with severe doses of ionizing radiation. It was further implied that dietary vitamins also mortify in radiation-treated commodities at a certain point in their safe-keeping timeline (De Silva, 2012). Vitamin content could also extend as located in γ-irradiated onions and Fuji apples; UV-C radiated freshly cut *Ananas comosus*; RTC pumpkin. Decrease in important dietary constituents could also be traced in γ-radiated onions; potatoes and "Rio Red" fruit; and/or no fluctuation in vast amounts in case of irradiated tuber plants; carrot plus cucumber; *Shatang mandarin*; clean pomegranate at 1 kGy after irradiation (Shahbaz et al., 2014). Thus, the effects of radiating foods at low enough doses do not cause any significant forfeiture of vitamins but sporadically, free radicals see a significant loss in some of their effectiveness during the radiation of antioxidizing vital-amines (Stefanova et al., 2010).

4.4.2.6 Effects of Ionizing Radiations on Properties of Phytonutrients

As per Castagna et al. (2013), a diet containing enough quantity of horticultural commodities reduces the risk of cardiovascular illnesses caused by the deficiency of phytochemicals in a diet with carotenoids. Phytonutrients exist adequately in berries and leafy vegetables at distinct aggregation structured around constituents of the horticultural produce in the phytochemical section. Treatment with ionizing rays could achieve the potential of particular horticultural commodities to generate outstanding phytonutrients at good enough levels (Alothman et al., 2009); for example, Maillard response merchandise developed in γ-irradiation processing synthesized antioxidants (Chawla et al., 2007). Phytochemicals manufacturing ordinarily banks upon the strength of ionizing rays (whether minuscule, moderate, or towering), the phytochemical's feedback for reactive species generated by irradiation, and the effects of radiating fruits and vegetables on their constituents. The behavior of phenols, antioxidants, and flavonoids has been quantified with the aid of a number of researchers in irradiated fruits and vegetables earlier than and after the irradiation. Phenolic compounds have been accountable for the deceleration of microbiological contaminants of horticultural commodities and it was once established that enzymes responsible for polyphenol degradation could have their functional attributes modified through treatment with ionizing rays (Oufedjikh et al., 2000). Stimulation of phenolic compounds has been done using ionizing radiations for processing of fruits (Maraei and Elsawy, 2017). Table 4.3 describes the effects of ionizing radiations on the phytochemicals of various fruits and vegetables.

Higher antioxidant activity exhibited by means of irradiated clean produce can also be attributed to increase in the number of phenols (Fan et al., 2005) as located in irradiated kale juice (Song et al., 2007); *A. bisporus*; *Benincasa hispida*; cauliflower. Antioxidants have additionally increased after irradiation in peach; minimally processed red beetroot; RTC-pumpkin as a mechanism of oxidation forces the phenolic compounds to act in a similar manner as antioxidants by curtailing the oxidants. Phenols and antioxidants' presence may additionally reduce after irradiation in the case of cv. Red Delicious apple; *Rubus idaeus*; pre-cut tomatoes and sliced carrots at 2 kGy irradiation, and fresh pomegranate juice (Shahbaz et al., 2014). Flavonoids typically multiplied by way of a small amount one day after processing in fruits and vegetables. However, they might extend upon irradiation as in minimally processed (0.8 kGy) onions or clementines; or may not be affected as in irradiated fresh-cut watercress at 5 kGy (Pinella et al., 2016); RTC-cauliflower (Vaishnav et al., 2015). Therefore, with few exceptions, phytochemicals present in fruits and vegetables generally increase after irradiation processing.

4.5 APPLICATION OF IONIZING RADIATION FOR THE TREATMENT OF FRUITS AND VEGETABLES

The wide functionality of ionizing radiations for a range of sanitary operations and safekeeping renders them the most sought-after technique for processing a long array of horticultural products (Isanullah and Rashid, 2017). Thus, the approach helps many shoppers to accept authentic organoleptic, rheological, and nutritional attributes. In order to minimize the impact on product quality, a large series of searches were often conducted in order to determine the best place of irradiation application for clean produce (Ma et al., 2017). The health regulatory body of the USA confined the maximum irradiation dose for fresh horticultural commodities to 1 kGy for decontamination and to postpone senescence of all ripened products (Shahbaz et al., 2014). Also, FSSAI, 2018 labeled a low dose of irradiation (<1 kGy) in a number of dose stages for its application in fruits and vegetables in accordance with their respective uses such as delaying senescence and ripening, sprout inhibition, and insect disinfection. The application of ionizing radiations for various food applications is given in Table 4.4.

4.5.1 Effects of Ionizing Radiations for Controlling Sprouting of Vegetables

Tuber vegetables such as onion, garlic, and potato undergo various activities such as formation of sprouts, rotten exteriors and interiors, loss of water from leaves, generation of heat, and microbiological contaminants appear as a whole upon catabolism when stored at 4–20°C, plus an excessively humid (>85%) state. Aeration could decrease contamination capacity of microorganisms; additionally, sprouting privation might be exacerbated by sprouted bulbs transpiring more rapidly, which causes them to shatter. According to Rezaee et al. (2013) and Elias (1987), minuscule amounts of ionizing rays (0.15 kGy) prove beneficial for diminishing sprouting effects in potatoes (Figure 4.4) upon efficient application in the course of the passivity. The physicochemical properties of produce are no longer believed to be affected by the dosages of irradiation needed

TABLE 4.3   EFFECTS OF IONIZING RADIATIONS ON PHYTOCHEMICALS IN FRUITS AND VEGETABLES

Sample	Treatment	Effect	References
Strawberries	UV-C at 0.25 and 1.0 kJ/m2	Anthocyanin content increased	
Grape pomace	γ-rays at 0–9 kGy	High doses decreased anthocyanin content (below 2 kGy) while low doses prevented loss	
Grapes	UV-B, UV-C irradiation	Resveratrol content increased, but no significant change in other polyphenolic compounds was observed	
Clementine peel	γ-rays of 0.3 kGy	Irradiated fruit accumulated significantly less of the compounds analyzed after storage at 3°C for 49 days	Oufedjikh et al. (2000)
Strawberries	γ-rays (1–10 kGy)	Cinnamic, p-coumaric, gallic, and hydroxybenzoic acids degrade	
Artichoke	Γ-rays (0, 10, 20, 30 kGy)	No differences in flavonoids, tannins, or phenols, but b-carotene content significantly decreased	
Fresh-cut vegetables	γ-rays (0, 0.5, 1, and 2 kGy)	Increase in total phenol content and antioxidant capacity with prolonged treatment	
Pomegranate arils	UV-C at 0.56–13.62 kJ/m^2	A nominal change in anthocyanins as well as antioxidant capacity	
Tomato	γ-rays (2, 4, and 6 kGy)	Phenolic compounds concentrations dropped markedly	
Peppers	UV-C at 7 kJ/m^2	Light-exposed peppers exhibited a lower total phenolic content and a higher antioxidant capacity (DPPH scavenging)	
Broccoli florets	UV-C treatment (4-14 kJ/m^2)	In comparison with control samples, broccoli florets contained lower levels of total phenols and total flavonoids	
Brazilian mushroom	γ-rays at 2.5–20 kGy	Antioxidating capacity enhanced	
Carrot and kale juice	10 kGy of γ-rays	Phenolics and antioxidating capacity elevated	
Fresh-cut mangoes	UV-C for 0, 10, 20, and 30 min	Enhancement of phenolics and flavonoids with increase of process duration, whereas β-carotene, Vit-C degraded	
Apple fruit	UV-B treatment	Antioxidants of peel enhanced but no deviation in phenolics of pulp	
Citrus unshiu pomaces	37.9 kGy EBI	Elevated phenolics and reduction capacity	
Blueberries	2 or 4 kJ/m^2 UV-C	No deviation in phenolics but elevation in anthocyanins	
Strawberries	UV-C (1, 5, and 10 min)	Elevation of antioxidants, phenolics, and anthocyanins	
Mango	EBI at 1–3.1 kGy	No deviation in phenols but flavonoids enhanced, Vit-C reduced, no deviation for carotenoids	
Broccoli	UV-C (8 kJ/m^2)	Enhanced total phenols and Vit-C and antioxidating ability	

TABLE 4.4   APPLICATIONS OF IRRADIATION TO FOOD AND INDICATIVE DOSE RANGE

Indicative dose range (kGy)[b]	Effects	Dose (kGy)[a]	Examples
Low dose (Up to 1.0 kGy)	Sprouting inhibited	0.08–0.14	Potatoes and yams
	Sprouting inhibited	0.04–0.10	Ginger
	Sprouting inhibited	0.03–0.12	Shallots, onions, garlic
	Sprouting inhibited	0.2	Chestnuts
	Insects unable to reproduce (phytosanitary treatment)	0.15–0.50	Fresh produce
	Insect reinfestation	0.2–0.7	Dried fruits and vegetables and in nuts
	Insects killed	0.15–0.50	Dried fruits and legumes
	Parasites inactivated (helminths and protozoa)	0.15–0.50	Fresh fruit and vegetables
	Ripening delayed	0.50–1.0	Banana, mangoes, and papaya
	Shelf-life extension	Up to 1.0	Fresh vegetables (prepared cut, soup greens)
Medium dose (1–10 kGy) sterilized by gamma rays at doses between 15 and 25 kGy	Number of spoilage organisms reduced	1.0–5.0	Strawberries, mangoes, and papayas
	Shelf-life extension	1.0–3.0	Refrigerated and ready-to-eat meals
	Shelf-life extension	1.0–4.0	Bananas, litchis, pickled mango (achar), avocados frozen fruit juices, green beans
	Control of ripening	1.0–4.0	Tomatoes, brinjals, soya pickle products, ginger, vegetable paste
	Insect disinfestation	1.0–4.0	Bananas (dried), almonds, various spices, various dehydrated vegetables
	Decontamination from spoilage and pathogenic organisms	3.0–10.0	Spices, dry vegetable seasonings, herbs, starch, protein concentrates, and commercial enzyme preparations
	Improving technological properties of food	2.0–7.0	Grapes (increasing juice yield), dehydrated vegetables (reduced cooking time), etc.
High dose (10–50 kGy)[b]	Reduce microorganisms to the point of sterility	30–50	Hospital diets, emergency rations, and food for astronauts
	Sterilization of long-life packaging material	15–25	Fruit juices as well as milk and milk products
	Decontamination of certain food additives and ingredients	10–50	Spices, enzyme preparations, natural gum, etc.

Source: FAO/WHO/CODEX STAN-106-1983.

a: Gy: gray – unit used to measure absorbed dose.

b: Only used for special purposes. The Joint FAO/WHO Codex Alimentarius Commission has not yet endorsed high-dose applications.

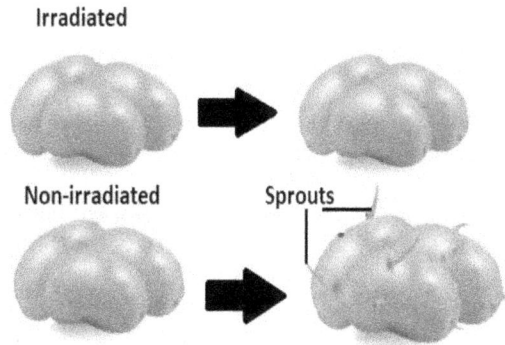

Figure 4.4 Use of irradiation for sprout inhibition in potatoes. (Image source: WHO, FAO, and the IAEA guidelines for food irradiation.)

for sprout inhibition; however, brownish discoloration in the inner buds or growth centers do occasionally occur. γ-irradiated (60 Gy) onions of various varieties like reddish and whitish and yellowish, saved for 180 days, showcased no symptoms of sprouting for bulbs processed through ionizing rays, whilst 34.3% sprouting was clearly visible in non-treated samples (Tripathi et al., 2011).

Similarly, sprouting was controlled by way of a γ-dose of 0.12 kGy for four months in onion tubers of var. PG-18 under ambient conditions with minimum storage losses and superior great maintenance (Sharma et al., 2020). Similar sprout and rot inhibition in potatoes had been obtained in case of Agria and Marfona types of potatoes upon irradiating at 0.05 kGy and 0.1 kGy doses (Rezaee et al., 2013) and mango at 0.3–10 kGy dose (Mahto and Das, 2013). After irradiation, bluish-grey discoloration of a few varieties of Indian potatoes was observed when irradiated tubers had been boiled or french-fried.

4.5.2 Effects of Ionizing Radiations for Controlling Ripening in Fruits

Irradiating the fruits using γ-irradiation and effects on their physiology such as ripening of fruits like bananas, pears, and tomatoes of the climacteric team and astringency of persimmons and satsuma oranges of the non-climacteric group have been investigated. It was once recognized that the irradiation dose for the length of ripening used to be distinct in the form of fruits and the levels of maturity and it was 0.5 kGy for mature green bananas, 2.0–2.5 kGy for mature green tomatoes and for slightly yellowish pears. When the fruits of banana were treated by using radiation doses of 0.5 kGy, the results showed a relationship between the prolongation of the common pattern of carbon dioxide production and the distribution of sugars in them. With pears, it was once observed that irradiation with the dose of 2.5 kGy delayed the ethylene production at some stage in ripening as well as carbon dioxide production, and further investigations have been carried out on the changes of respiratory metabolism, some natural acids, protein N and RNA at some point of the ripening of Bartlett pears in order to find out about the mechanism of management of ripening through γ-rays. Otherwise, when the astringent persimmon's ripening rate accelerated with the use of 1.5–2.5 kGy of ionizing radiation doses, it suppressed the ripening of fruits of the climacteric group: the fruits irradiated with the doses of 1.5–2.5 kGy, ripening within five days of irradiation, and becoming soft and much less astringent with the boosted ripening – with the exception of the 10.0 kGy fruits which the enzyme activities associated with ripening and would have been reduced by using irradiation. Tannin content lowered substantially in irradiated and ripened fruits.

4.5.3 Effects of Ionizing Radiations for Shelf-Life Extension of Fruits and Vegetables

By retarding the growth of freshly harvested vegetables and fruits by inhibition of senescence or ripening, or by inhibiting spoilage microorganisms, irradiation increases the shelf life of horticultural produces like apples, litchis, plums, strawberries, etc. (Figure 4.5). Conventionally, Co-60 radioisotopes provide the irradiation supply used for the enhancement of the shelf life of freshly harvested fruits and vegetables (Farkas and Mohacsi-Farkas, 2011).

Irradiation effects on the storage life and quality of fresh horticultural produce basically depend on cultivar, maturity, and pre- and post-harvest losses. In some types of produce, irradiation extends the shelf life for a number of days, sometimes weeks, whilst in some by months. The shelf life of immature leafy vegetables such as coriander, mint, and lettuce is elevated by two, three, and four days respectively after irradiation treatment at 0.5 kGy (Triago et al., 2009); by three days and four days in mango cultivar "Dusehri" and "Fazli", respectively (Mahto and Das, 2013); by four days in mint leaves dealt with at 2 kGy; by two to ten days in mushrooms (at 2 kGy) (Jiang et al., 2010); by seven days in light reduce watercress at 1, 2, and 5 kGy (Pinela et al., 2016); by a number of days in king oyster mushrooms at 1 kGy (Akram et al., 2012); by seven days in RTC ash gourd at 2 kGy; by seven days in RTC-cauliflower at 0.5 kGy (Vaishnav et al., 2015); by 21 days in RTC-pumpkin at 1 kGy; for 28 days in Litchi fruit (*Litchi chinensis*) at 0.5 kGy (Hajare et al., 2010), and for three months in dried Sangari dealt with at 5.0 kGy (Joshi et al., 2011). The shelf lives of fruits were prolonged at doses of 0.2–0.4 kGy; however, greater doses led to fruit decaying (Zhang et al., 2014). About 39% decay of untreated blueberries was checked after 14 days of storage at 4°C, whilst irradiating the fruits at 2 kGy and 3 kGy resulted in only 8% and 3% decay, respectively, showcasing the nice interrelationship between the dosage of radiation and the storage period of blueberry fruits (Kong et al., 2014). Due to their immense perishability, mushrooms' storage period is extremely short. The reason behind this is the case of PPO (poly-phenol oxidase) related browning, loss of weight of the commodity, and conversion in the textural properties. When untreated samples were kept along the ionization treated samples for comparison, EBI

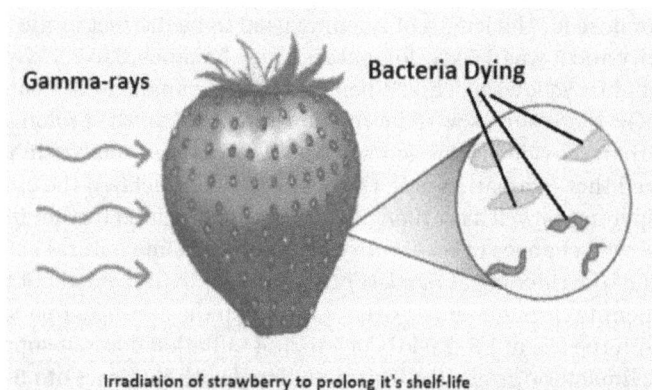

Irradiation of strawberry to prolong it's shelf-life

Figure 4.5   Irradiation treatment of strawberries to prolong their shelf life. (Image source: IAEA.)

treatment at 2 kGy brought about greater lightness, highest complete antioxidating capacity, and minimum discharge of electrolytes of the mushroom; although enormous reduction of Vit-C had been previously estimated (Mami et al., 2013).

4.5.4 Effects of Ionizing Radiations against Phytosanitary and Sanitary Control Measures

Demand for ready-to-cook food with the removal of additives such as preservatives is growing among consumers. The process of skin removal due to cell damage and wastage of integrity of tissue structure during minimal transformation of horticultural commodities speeds up infection by microbes (Fellows, 2001). Chemo-techniques have been utilized since ancient times to minimize the threat of microbes in horticultural produce; however, only superficial sterility is provided by them and not for deep fissures. Take the case of 200-ppm chlorine solution: it could not remove the microbial contamination caused by *E. coli* from surface of leafy and green vegetables (Niemira, 2008). Chemo-methods in association with cold processing slowed bodily functions and multiplication of microorganisms; however, they were unable to slow down the multiplication mechanism of psychotrophs that go on to start diet-mediated sickness (Vaishnav et al., 2015). The shelf life of fresh-cut vegetables and fruits is reduced by transforming methodologies through a chain of visible events like loosening of cell bundles, reduction in the browning of cut surfaces, reduction of nutritive values, detection of bad flavors, and damage by microbes during the keeping period (Ma et al., 2017). Therefore, to overcome such hurdles in preservation, γ-irradiation is recommended as a choice method that results in the profitable sterilization of minimally processed and fresh produce without affecting the nutritional, textural, and sensory traits (Hussain et al., 2014). Irradiation offers the advantage of inactivating foodborne pathogens in ambient temperature conditions, and therefore, it is referred to as a cold sterilization technique (Vaishnav et al., 2015). The dose of radiation cure for microbial inactivation should be decided on the groundwork of D10 value, preliminary wide variety, and diversification of microorganisms present. Use of irradiation for phytosanitary applications is commercially known due to its relevance to box up products with elevated resistance to radiating doses (Hallman, 2017). Table 4.5 describes the relative tolerance of various fruits and vegetables to ionizing radiations of energy doses <1 kGy (Bustos-Griffin et al., 2012).

Irradiation doses of up to 1 kGy could be resisted by the majority of the vegetables and fruits processed minimally besides pleasant modifications and deterioration (Fan and Sokorai, 2008). The treatment of fungal spores by irradiation and subsequent chemical and physiological changes blocks their metabolic functions (Jeong et al., 2016). In the past, contamination of fruits and vegetables by bacteria was easily reduced by low to medium doses, but viruses or toxins were not. Thus, in order to produce sterile food (mainly referred to as specific reason food), stronger radiation does not need to be used (above 10 kGy) as these are used by patients with AIDS or many forms of cancer, astronauts, and the armed forces. Whole ingredients, vegetables, and most animal-derived food products are suitable for irradiation processing to bring about complete destruction of the microbiota.

4.5.5 Effect of Irradiation on Food Safety and Stability Enhancement

Microbes are prone to radiation exposure primarily due to the presence of free radicals in their DNA and RNA. DNA is much larger compared to other factors in a cell wall which contributes to its susceptibility to ionizing radiation. (Scott and Suresh, 2004). The disruption of H-bonds

TABLE 4.5 RADIATION TOLERANCE OF VARIOUS FRESH FRUITS AND VEGETABLES AT DOSES <1 KGY		
High	**Moderate**	**Low**
Apple	Apricot	Avocado
Banana	Custard Apple	Broccoli
Carambola	Grapefruit	Cauliflower Cucumber
Cherry	Loquat	Green Beans
Blueberry	Mandarin	Leafy Greens
Dragon Fruit	Passion Fruit	
Fig	Pear	
Grape	Pineapple	
Guava	Plum	
Longan	Soursop	
Lychee	Tangelo	
Mango		
Mangosteen		
Manzano		
Pepper		
Melon		
Papaya		
Peach		
Persimmon		
Sapote		
Raspberry		
Rambutan		
Strawberry		
Sweet Lime		

among the double-stranded DNA molecules takes place by free radicals that forestall replication leading to loss of living tissues whilst causing minimum effect on non-living ones (IFST-2006).

For most remedies and irradiation, the spore-forming endospores are highly resistant. 10 kGy doses and beyond that are used for pasteurization food can also solely supply a 2–3 log reduction in spore count which is not satisfactory for the production of shelf-stable ingredients (Patterson, 2005, Koopmans and Duizer, 2004). Bidawid et al. (2000) reported the irradiation efficacy upon inactivation of hepatitis A virus on berries and leafy vegetables with less than a 10 kGy dose. A linear destruction of the virus count was witnessed upon increasing the doses of irradiation for every product. At 10 kGy, a >3 log reduction befell in both strawberries and lettuce. Rajkowski and Thayer (2000) observed salmonella was no longer produced from sprouts of alfalfa with 0.5 kGy irradiation even though the seeds yielding the sprouts contained variable stages of the pathogen and concluded that by irradiation pathogen populations in sprouts could be minimized.

The whole colony number of bell peppers was reduced by 1.0 kGy about four log CFU/g through four days of storage at abrupt temperatures (10 or 15°C), giving further growth, yet a refrigerated temperature (4°C) maintained the preliminary efficiency of the treatment (Farkas et al., 1997). A 1.0 kGy dose was successful in killing *L monocytogenes* (>5 logs) from chopped celery (Prakash et al., 2000a). Having been handled with 1 kGy, carrots demonstrated a reduction in plate numbers of 4 log CFU when packed, and 4.5 log mark downs when stored under alteration of the external environment (Lafortune et al., 2005).

Wang et al. (2006) investigated survival of microbes in Golden Empress cantaloupe juice prior to Cobalt-60 radiation and determined that *E. coli* microorganisms are especially sensitive to irradiation; they should be reduced with 7-log cycles of 1 kGy, although the microbial count and spore count of successful juice increased its resistance to irradiation as indicated by D10 numbers of 0.9908 and 1.1923 kGy. On the grounds of acceptable off-odor, the authors reported that γ-irradiation is not successful in inactivating the colony count or spore bacteria in cantaloupe juice. Anthracnose and *Colletotrichum gloeosporioides* lesion sizes were reduced by 0.75–1 kGy γ-radiation after application to papaya fruit (*Carica papaya*), but no longer protected the fruit 24–72 hours after application. The doses retarded *C. gloeosporioides* conidia germination and mycelia growth with enhanced fungal sporulation. The fruits were kept at 25°C, 80% RH for a duration of seven days (Cia et al., 2007).

According to Hussain et al. (2008), it is pleasant to consume Ambri, Golden Delicious, and Royal Delicious apples that have been γ-irradiated (0.1–0.5 kGy) at an ambient temperature (15°C) and refrigerated (3°C). Three kinds of apple under three types of storage were treated with doses of 0.2, 0.3, and 0.5 kGy of irradiation that resulted in their average exceptional keeping. γ-irradiating apples significantly reduced yeast and mold counts after storage. Shurong *et al.* (2006) defined the D10 values of *E. coli*, *Listeria innocua*, and *Salmonella enteritidis*. *E. coli* inoculated in cherry tomatoes and carrots demonstrated D10 values of 0.08 and 0.13 kGy, respectively. Similarly, the values of D10 of *Salmonella enteritidis* inoculated in cherry tomatoes, pre-cut carrots, and a mixture of blanched celery and peanuts spanned 0.24–0.33 kGy. Less than 2 kGy irradiation dose ensured a 5-log reduction of the most resistant pathogen *S. enteritidis*. However, ionizing radiations effectively controlled the development of pathogens at some point in the storage period. Song et al. (2007) reported total aerobic microbes and coliform counts found in the carrot juice as 106 CFU/ml, and those of the kale juice were 107 CFU/ml. Aerobic microbes and coliforms were efficiently destroyed with 3 kGy irradiation, and the D10 values of the microflora in carrot juice were about 0.5 kGy. Radiation doses of 5 kGy might not be enough to eliminate all the microbes in clean kale juice. The D10 values in the kale juice were once higher than 1.0 kGy.

4.6 ECONOMICS: RADIATION PROCESSING AND RELATED FACILITY

The plant layout and thus the economics of processing the food by using ionizing radiations depending upon the characteristics linked with the process. They include type of food material, its density, dosing range, effectiveness of processing equipment, yearly running time, output, requirement of the Co-60 load and its replenishment, and the workers engaged for the processing.

4.6.1 Commercial Viability of Radiation Processing Plants

The desire of an entrepreneur for erecting a food irradiation unit raises the demand when determining if a potential project or related businesses are feasible.

4.6.2 Plant Utilization for Irradiation of Foods

The goal to make the plant operate throughout the year to ensure year-round product accessibility becomes challenging if the facility is only suitable for processing with the use of small doses of radiation thus limiting the volume and range of products which can be processed. The case is true for products of seasonal type like potatoes and onions. Plus, there are places restricted by a variety of products that are being produced therein. The chances of better functionality of the plant are also diminished due to such things (AERB-SS-6). The work for the radiation processing of food is placed in three categories:

- Low dose consisting of doses below 1.0 kGy
- Medium dose consisting of doses between 1.0 and 10 kGy
- High dose consisting of doses above 25 kGy

4.6.3 Capital Cost or Project Cost for Running of Irradiation Unit

An irradiation plant with conveyer type processing would require a cost of about Rs.1000 lakhs for its completion (without the cost of the radiation devices). Presently, the cost of irradiating foods using Co-60 as a processing tool will be around Rs.70/Ci (Curie) (AERB-SC-IRRAD).

4.6.3.1 Operating Costs

Transportation and running costs for such kinds of plants will consist of the:

- Replenishing costs for Co-60
- Distributing wages to the staff
- Amount of interest on the capital
- Utility (electricity & water) costs
- Costs for repairs and maintenance

The rate of degradation of Cobalt is about 12.3% annually and this demands for their refueling at similar intervals, and annually after that for the second or third year. The decay costs for a processing facility operating at energy consumption of 500 kCi will be about Rs. 43 lakhs per year.

The big point of increment in running costs apart from Co-60 refueling will be providing salaries to the staff. It would cost around 1–2% of the plant's cost for carrying out the repairing and maintenance work of the entire facility. The cost of utilities for running or maintaining such a full-fledged facility could be totaled to around Rs. 10 to 12 lakhs annually (AE Rules, 1996).

4.7 ADVANTAGES AND DISADVANTAGES RELATED TO IRRADIATION OF FRUIT AND VEGETABLE PROCESSING

4.7.1 Advantages of Irradiation of Fruits and Vegetables

4.7.1.1 Using Irradiation against Insect Disinfestation of Dried Spices, Vegetables, and Fruits

Insect infestation of vegetables, dried spices, and fruits are frequent cases and thus they are regularly disinfected with methyl bromide. Irradiation can also be used for removal of insects by using doses of 1.0 kGy. Similar cases of efficiency and maintenance of wholesomeness are applicable and the fact that the weight of dried spices ingested by the insects is a minuscule fraction of the full diet. If larger quantities of ionizing radiations have to be used, the microbiological load can be eliminated from the given products and this method has generated a lot of enthusiasm among the industrialists.

4.7.1.2 Using Irradiation against Insect Disinfestation of Fresh Fruit

The irradiation can be applied for the instant and complete removal of insects from freshly harvested fruits. But phytotoxic damage to the freshly harvested fruits and nutritional component degradation will eventually add to overall losses when the processed radiation amounts are fully taken in by the products. The use of irradiation for the processing of freshly harvested fruits in order to eliminate insects has the hidden advantage in that there is no need to use insecticides at the time of harvesting the horticultural produce. The demand for the application of insect disinfecting methodologies of individual choices expanded significantly in the USA after the restriction on the application of ethylene-dibromide took place along with cases of attack by fruit flies in California. Insects and bugs that attack horticultural crops produced in mainland USA are not discovered in other countries and the opposite side of the statement also holds well. Take the case of one fruit fly *Anastrepha suspensa* L from a Caribbean island that eats and damages grapefruit along with other pests by the name codling moth *Cydiu pomonella* L attacking cherries, which have never been found on the Japanese mainland. Thus, these horticultural products can be transported to Japan after they have been irradiated using ionizing radiations after they are harvested (Burditt, 1982). Restrictions on the growth of insects like naval orange worms, fruit flies, spider mites, coding moths, scale insects, mango weevils, and potato tuber worms can be done by application of ionizing radiations below 1 kGy doses. The majority of the freshly harvested fruits can withstand ionizing radiations of 0.25 kGy doses without undergoing any or minuscule changes in their quality (CAST 1984). Minute phytotoxicity damages happen in horticultural produces like mangoes, guavas, apples, dates, cherries, nectarines, peaches, tomatoes, raspberries, strawberries, and papayas (CAST 1984). However, inconsistent or detrimental effects have been bought with merchandise such as avocados, grapefruit, grapes, kumquats, lemons, limes, oranges, pears, peppers, pineapples, and tangerines (CAST 1984).

Bruhn and Noel (1987) performed a watchfully steered research on irradiation of papaya fruit through recording feedback from consumers for its application in California (Bruhn and Noel, 1987). Commonly used double-dipped (in hot water) variety and well irradiated by ionizing radiation (0.45 kGy) papayas, both having equivalent sizes and characteristic qualities were given to the customers. The consumers were more interested in purchasing the fruits which were ripened on the tree (irradiated using ionizing rays) than the fruit products that have been only double-dipped in hot water. The taste of the commodities processed using ionizing radiations was also the demand of the buyers. Most of the buyers shockingly (60–75%) were not aware of the fact that papaya fruits have to be treated for controlling insects and pests before they could be brought into mainland USA from the island of Hawaii. A warm dip in benomyl solution along with the ionizing radiation of doses in the range of 0.75–6.0 kGy helped best safekeeping of the mango fruits and had negligible phytotoxicity effects (Jessup et al., 1988). Only minuscule degradation of dietary components like β-carotene and few of the B-complex vitamins have been reported due to treatment of freshly harvested fruits with ionizing radiations (CAST 1984). According to the variety of the product being processed, time under radiation treatment, nature of the product and its storage temperature, the degradation of the Vit-C varied from 0% to 95%. As per the dietary point of view, both ascorbic acid and dehydroascorbic acid are important and need to be considered as most of the ascorbate gets converted into dehydro-ascorbate through metabolic pathways inside human cells (Kraybill, 1982). When ionizing radiations have been used, a strange anaphase appeared in beans (Bradley et al., 1968), decaying of tissues of carrots also expanded (Holsten et al., 1965), and the amount of the microorganism *Salmonella typhimurium* revertant also increased (Beyers et al., 1983).

4.7.1.3 Using Irradiation against Inhibition of Sprouting in Tubers and Bulbs

In order to stop sprouting cases in white potato varieties, the ionizing radiations in the dose range of 0.05–0.15 kGy were applied to the products in order to get them accepted into the USA in 1964. In Japan, irradiation has been used mainly for commercial applications. For inhibiting cases of sprouting, radiation doses of at least 0.02–0.03 kGy would be required for onions and a dose of 0.03 kGy would be required for potatoes. In addition to this, little radiation doses could be helpful in causing stimulation of sprouting. Dangerous results may happen if the irradiation of the tubers and bulbs is done at doses greater than 0.15 kGy (Matsuyama and Umeda, 2018). In order to prevent discoloration of the desired vegetable, the fundamentals that hold good are cautious handling of the commodity, careful selection, prevention of mechanical damages, irradiating them to avoid dormancy, and ideal conditions of storage. Heavy losses of ascorbic-acid content are usually prevented when onions and potatoes are irradiated with sprout-inhibiting doses before they are stored, as compared to the non-irradiated ones. The ionizing radiations when applied into aqueous mixes, caused degradation of ascorbic acid into dehydroascorbic acid. After irradiating the tubers prior to their storage and instantaneously after storage is over, the content of ascorbic acid was calculated. Approximately 15% of the ascorbic acid was degraded at irradiation doses suitable for prevention of sprouting; however, after delayed storage, the changes in Vit-C content among radiation treated and untreated potato samples were not highly significant. Conclusive reports by Matsuyama and Umeda (2018) stated that even when free amino acids of the irradiated (0.15 kGy) potatoes were adapted for radiation doses suitable for stopping sprouting of the tuber, the adaptations diluted eventually and diminished during its storage. In the irradiation treated potato samples, no major adaptations were identified in the amino acids bound with proteins or among the biological values of the proteins.

4.7.1.4 Using Irradiation against Controlling Postharvest Ripening and Senescence in Fruits

Equivalent or just enough doses of ionizing radiations which help in restricting pests and insects, in certain cases will help in changing ripening and senescence of fruits that happen after they are harvested from fruits. For example, the duration of bananas ripening at 0.20 kGy, mangoes at 0.10–0.25 kGy, and papayas at 0.75 kGy. A 3 kGy dose of ionizing radiation will prolong the storage life of candy cherries and a 2 kGy dose of ionizing radiation for strawberries, by significantly restricting the majority of the pathogens of fungal origins (Akamine and Moy, 2018). The changes in the sensory as well as nutritional characteristics could be brought about by applying the same radiation doses as for the disinfestation of insects. Table 4.6 describes the advantages and disadvantages of ionizing radiations on food products.

4.7.2 Disadvantages of Irradiated Fruits and Vegetables

4.7.2.1 Disadvantages of Textural, Organoleptic, and Storage Characteristics of Fruits

In certain fruit products like sparkling fruit salads and non-pasteurized type juices, cases of food-mediated outbreaks were disclosed (Lynch et al., 2009; Vojdani et al., 2008). The effects of ionizing radiations on a fruit product and other varieties impact its applicability due to specificity of the dose (Arvanitoyannis et al., 2009). Most of the time, superficial disinfection of the microbiological contaminants of freshly harvested fruits, pre-water cleaned plus packed vegetables, and RTE-safely processed commodities with certain checkpoints of chemicals could prove least effective (at most 2 log-reduction) instead of radiating food (Arvanitoyannis et al., 2009). Treatment of

TABLE 4.6   MAJOR ADVANTAGES AND DISADVANTAGES OF FOOD IRRADIATION

S. No.	Advantages	Disadvantages
1.	By irradiating food, foodborne illnesses can be prevented	Food irradiation is a costly process
2.	Food exposure to insects is controlled	Changing the nutritional content of foods containing B vitamins
3.	Process that regulates the ripening and sprouting of fruits and vegetables	As a result of irradiation, new molecules are formed
4.	Using food irradiation as a means of sterilization	Bacteria can sometimes develop resistance to radiation
5.	Food irradiation levels above a certain level can destroy all contaminants	Currently, our food standards are inconsistent worldwide
6.	There's no radioactivity left on food after it is irradiated	Irradiated food does not replace good growing and processing practices
7.	Foods fresh and frozen can be processed by this technology	Foods will no longer show spoilage warnings

fruits with ionizing radiations caused a decrease in firmness values as pointed out by the majority of the findings. For treatment of fruits and vegetables, 1–2 kGy is the maximum limit for an irradiation dose that can be applied. However, the variety of products can accommodate the new and resistive varieties depending upon the extent of radiation dose applied (Zhu et al., 2009). For example, certain vegetables used for salad making could resist up to 4 kGy of ionizing radiations and prevent physico-chemical losses and high-quality degradations (Nunes et al., 2008). Some products on the other hand cannot tolerate radiation doses above 0.6 kGy for similar changes to take place, e.g., lettuce leaves and apple fruit (Gala and Fuji variety) (Dionisio et al., 2009; Niemira et al., 2002). Therefore, certain commodities have a very brief storage life. For example, if the tip cap of vegetable mushrooms opens within a period of 24 hours at 10°C of storage, the extension of the stem takes place and the blackening in the gills occurs. As per a finding, storage life of 11 days at 5°C and 7 days at 10°C was observed when the freshly harvested mushrooms were treated using ionizing radiations of 1.5–2.0 kGy. Organoleptic characteristics of the mushrooms remain unchanged when they are processed using ionizing radiations, as per most of the findings (IAEA, 2001). Higher doses of treatment, about 3–5 kGy were advised for the irradiation of dried vegetable powders (Dong-Ho et al., 2002). The significant effect of ionizing radiations on the consistency of the fruits decreased after they were pre-treated with calcium chloride (Prakash et al., 2007). A host of Salmonella strains in case of were having their decimal reduction values in the range of 0.6–0.8 kGy (Niemira and Lonczynski, 2006). As the ionizing radiations are least effective against viruses, the applicable doses of irradiation against the majority of the fresh and clean horticultural commodities will have limited effectiveness (Koopmans and Duizer, 2004).

4.7.2.2 Disadvantages Regarding Infrastructure and Economics of a New Process

Successful application of a new science needs to be cleared by a number of infrastructure checkpoints present inside a country. The process of food irradiation requires generally the same kind of infrastructure as different other methodologies like freezing, canning, drying, etc. Take for example the case of a production unit that utilizes any one of the above methodologies and places them at the center where enough quantity of food material is manufactured and conveyed to processing unit prior to its release into the market (Urbain, 1982).

4.7.2.3 Disadvantages with Regards to Consumer Concerns about Irradiated Products

While food irradiation is known to be associated with nuclear technology, an introduction of irradiated food can be enormously linked with radioactive products. Thus, there seems to be some widespread opinion among some countrywide authorities that buyers would be concerned about food being irradiated as part of its production. The majority of end users cannot understand why food must be treated with ionizing radiation when irradiated food is found on the market. Researchers fear that exposure to irradiation will result in the development of resistant microbes. No evidence exists to support such a threat; on the contrary, microorganisms that survive in the presence of irradiated water are mostly vulnerable to conditions adverse to microbial growth (i.e., low temperature) that are probably killed by cooking.

Various surveys on purchaser attitudes on usage of irradiated food had been conducted on one-of-a-kind targets companies in the last ten years. The survey varies extensively based on the patterns of questionnaires along with technical heritage of interviewees. Current family surveys are conducted in the USA to determine whether consumers will consume fresh food that has been irradiated (Malone Jr., 1990). Three-quarters of the food purchasers indicated a willingness to purchase items that have been irradiated. Three-fourths of shopkeepers were unaware that irradiation had been conducted. And 37% of those who were unaware refused to buy irradiated goods due to insufficient information.

Therefore, educating the purchaser is crucial for the success of food irradiation. The irradiated products should be clearly printed with the "Radura" logo (Figure 4.6) and the expression "ionization". The majority of customers show their faith in the government for determination of safety of the food irradiated items. National authorities, food enterprise, and changed associations and agencies therefore play a crucial role in imparting the information and advantages of irradiated food to the consumer (Moog, 1988).

Based on the results of the tests, it can be concluded that buyers will not only receive irradiated products once safety and benefits of the treatment are understood. Furthermore, they would be organized to buy them over and over again. This resulted to be the opposite of what companies opposed to food irradiation had hoped for. There is no doubt that these corporations fully understand the effects of market tests threatening picketing and public boycotts of stores that raise them (P Loaharanu, et al., 1991).

Figure 4.6   The "Radura" logo. (Image source: WHO, FAO, and the IAEA guidelines for food irradiation.)

4.8 FUTURE PROSPECTS OF IRRADIATION OF FRUITS AND VEGETABLES

By analyzing a few food-related issues in the world, let us look for ways by which we can use irradiation to ease out the processing.

4.8.1 Prospects of Irradiating Tropical Fruits and Vegetables

The main aim of using irradiation on commodities is to elevate their storage life by fulfilling the below characteristics:

(1) stoppage of sprouts
(2) delaying fruit ripening
(3) checking growth of fungi
(4) killing pests and stopping related growth.

After irradiation of some chosen fruits and vegetables which are produced in tropical regions, certain satisfying outcomes have been witnessed (KF MacQueen, 1969).

4.8.1.1 Prospects of Irradiating Papaya

When papayas at the Tropical Agriculture College at Hawaii University were treated with ionizing radiation, they ascertained positive effects of the process. If a locally made product can be availed of for export prior to irradiating, a host of wonderful possibilities could be assured. The developing nations can avail a host of benefits as they can export to other countries and earn forex reserves and utilize that for developing their agricultural economy. There is no need to overlook this possibility (KF MacQueen, 1969).

4.8.1.2 Prospects of Irradiating Bananas

In one rigorous research at the University of California by Maxie, the beneficial effects of ionizing radiation on bananas were ascertained. One particular commercial variety by the name "Gros Michel" which is cultivated in Central America, when irradiated at small doses before their ripening, their storage period was elongated by up to 480 hours before their ripening. Another benefit observed was that the fruit could remain in a ripened state and fit for consumption for up to 72 hours more than normal ripened fruits. The findings proved to be important for the reason that bananas ripened on shipping route are very much prone to mechanical injuries and irradiation delays the process making it easier to complete the shipping of fruit without much damage (KF MacQueen, 1969).

A separate approach of irradiating bananas in research is being done at the Atomic Energy Establishment, Trombay, India. Semi-dried bananas with about 40% moisture were stuffed inside polyethylene for radiating with less strong amounts of radiation could be stored well for a minimum of a quarter of a year.

4.8.1.3 Prospects of Irradiating other Tropical Fruits and Vegetables

Profitable outcomes have been witnessed from different research done by KF MacQueen (1965) with fruits and vegetables like mangoes, guavas, potatoes, and sapotas. Full research on potato irradiation has been carried out in Pakistan and obtained results supporting its storability at atmospheric conditions.

Researchers in Israel and the United States have done work on irradiation to point out its potential in the processing of citrus fruits and the reduction of post-harvest losses.

Different materials showcasing particular potential to their processing by irradiation consist of peaches, plums, onions, apricots, mushrooms, plums, and strawberries.

4.8.2 Importance of Radiation Sources in Irradiation Process and Facilities

The utilization of EBI in the near future is overlooked for a number of things. The demand for important uses of utility software will be high in the future due to:

- dose homogeneity and distribution
- incorporation into the handling process
- results that can be reproduced.

4.8.3 Importance of Wholesomeness Aspects of the Irradiated Products

Upon evaluating a large amount of information on wholesomeness aspect of foods treated by ionizing radiation: radiolytical effects of irradiation, ionizing radiation related toxicity of food, dietary aspects, and microbiological outcomes. Irradiating foods to radiation doses of up to 10 kGy resulted in healthy food for all investigations as supported by the study.

4.8.4 Importance of Toxicological Aspects of the Irradiated Products

Despite the assurance by the Joint FAO/WHO Expert Committee that consumption of irradiated foods is safe for health, and no need for toxicological studies ascertaining the same, major companies state that there is not enough data to cite whether or not long-term effects will happen.

4.8.5 Importance of Using Irradiation with Mixed Processing for More Efficiency

Irradiation in association with different other processing methods like ionizing radiations along with thermal treatment or chemical additives would be as good as cooked and convenience foods. We can say that mixed processing can help consumers and companies to provide a cheap alternative treatment and would result in product safety and component enhancement through the use of such methods of processing.

4.8.6 Importance of Public Information about Irradiation or Irradiated Products

For a method to gain popularity and wide acceptance, it must be appealing to the eyes of the end user. The acceptance or denial of a processing method by the consumer dictates the overall success or downfall of the process. And this will narrate the upcoming priorities in scientific findings, processors, and the toolmakers themselves in order to make the process successful. Thus, to achieve acceptability in the markets and utilization by people, the principles of contemporary marketing are taken into consideration.

4.9 CONCLUSION

Food irradiation can certainly kill bacteria and can be a beneficial tool in controlling sure pathogens in raw food. However, it took years, if not decades, for these applications to grow to be a reality and are presently not licensed via regulators. Further research on specific food and food irradiation is needed to answer many unanswered questions. What doses are advantageous towards pathogens of pastime in a specific food? Can irradiation successfully forestall recurrence whilst first lowering the pathogen load to a safe and suited level? Is the sensual first rate of the irradiated product acceptable to consumers? Irradiation can be blended with different food processing practices and/or "upstream" interventions in the food chain to enlarge the effectiveness of the irradiation phase and restrict unwanted facet effects. Which aggregate is high quality for which food? Is agricultural lookup economically and logistically feasible? Will the fruits and vegetables under investigation be conventional or rejected by means of the industry and the standard public? The lack of excellent scientific answers to these questions, rather than political opposition or timid bureaucracy, explains the slow use of irradiation to control problems caused by *E coli* in fresh leafy vegetables. As the need for concrete improvements in agricultural security increases, we will pursue different strategies such as GMP, GAP, and HACCP structures that no longer require substantial prior lookup expenses so that they can be carried out greater rapidly and at a decreased cost and a lot more controversial than irradiation. Ultimately, market electricity determines whether food investigations play a necessary function in efforts to improve the safety of agricultural products. Outbreaks of agricultural illnesses are prompted with due to *E. coli* and different pathogens, especially when the agricultural industry, regulators, and other stakeholders are conscious of the pathogen infection of raw ingredients being effectively handled. There might also be less lobbying in the usage of food research for agricultural products. Meanwhile, GMP, GAP, and HACCP "upstream" movements will be taken as motives due to economics or feasibility in the tournament of a highly publicized outbreak of food poisoning involving raw food in the near future. Contamination is not sufficiently recognized by the industry and may its recognition may be vital in order to efficiently enhance the safety of fruits and vegetables from microbial contaminants. The latter state of affairs may offer value-added irradiation in markets where there are undue issues about product protection and hence convince some food producers to use it. There may in addition be extended pressure on the FDA to approve raw food investigations. If there is enough scientific proof of efficacy and safety, the FDA will probably to approve it; however, it is tough to predict when such a choice will be made. FDA approval would without a doubt enhance the prospects of investigated products, but the business success of investigated fruits and vegetables is no longer guaranteed. It is not yet known whether or not the investigated merchandise meets patron expectations in terms of protection and quality, and whether a robust demand for these products will continue and fuel a boom in the market. Therefore, the most likely future scenario will see various stages from that of unchanged (not searching for fresh produce to kill pathogens) to that of an expanding introduction of fruit and vegetable market niches till slowly being investigated. However, even in the latter case, most products will stay unexplored for the time being. Vendors selling fruits and vegetables that are no longer inspected in the quality check could receive complaints from the various sections of the buying department. Claims of reliability and security apply to each class of products. Irradiation is an extra safety measure which is implemented gradually but does not by itself supply whole safety assurance. Yet a good, well-implemented inter-farm danger administration machine does the very same, hence barring the want for radiation that is needed in order to grant security. The fruits and vegetables surveyed will remain in a small specialized market that lobbies them; however they will constitute the bulk of the products provided to consumers.

REFERENCES

Afify, A. M. R., Rashed, M. M., Ebtesam, A. M., & El-Beltagi, H. S. (2013). Effect of gamma radiation on the lipid profiles of soybean, peanut and sesame seed oils. *Grasas y Aceites*, *64*(4), 356–368.

Akamine, E. K., & Moy, J. H. (2018). Delay in postharvest ripening and senescence of fruits. In E. K. Adamine & J. H. Moy (Eds), *Preservation of food by ionizing radiation* (pp. 129–158). Florida: CRC Press.

Akram, K., Ahn, J. J., & Kwon, J. H. (2012). Analytical methods for the identification of irradiated foods. In E. Belotserkovsky (Ed.), *Ionizing radiation: Applications, sources and biological effects* (pp. 1–36). New York: Nova Science Publishers.

Alothman, M., Bhat, R., & Karim, A. A. (2009). Effects of radiation processing on phytochemicals and antioxidants in plant produce. *Trends in Food Science and Technology*, *20*(5), 201–212.

Arvanitoyannis, I. S., Stratakos, A. C., & Tsarouhas, P. (2009). Irradiation applications in vegetables and fruits: A review. *Critical Reviews in Food Science and Nutrition*, *49*(5), 427–462.

Ashraf, S., Sood, M., Bandral, J. D., Trilokia, M., & Manzoor, M. (2019). Food irradiation: A review. *International Journal of Chemical Studies*, *7*(2), 131–136.

Ashwar, B. A., Shah, A., Gani, A., Rather, S. A., Wani, S. M., Wani, I. A., ... Gani, A. (2014). Effect of gamma irradiation on the physicochemical properties of alkali-extracted rice starch. *Radiation Physics and Chemistry*, *99*, 37–44.

Atomic Energy. (1996). (Control of irradiation food). *Rules*.

Audette-Stuart, M., Houée-Levin, C., & Potier, M. (2005). Radiation-induced protein fragmentation and inactivation in liquid and solid aqueous solutions. Role of OH and electrons. *Radiation Physics and Chemistry*, *72*(2–3), 301–306.

Bettaïeb, N. B., Jerbi, M. T., & Ghorbel, D. (2014). Gamma radiation influences pasting, thermal and structural properties of corn starch. *Radiation Physics and Chemistry*, *103*, 1–8.

Beyers, M., den Drijver, L., Holzapfel, C. W., Niemand, J. G., & Pretorius, I. (1983). Chemical consequences of irradiation of subtropical fruits. In P. S. Elias & A. J. Cohen (Eds.), *Recent advances in food irradiation*. Amsterdam, New York: Elsevier Biomedical Press.

Bidawid, S., Farber, J. M., & Sattar, S. A. (2000). Inactivation of hepatitis A virus (HAV) in fruits and vegetables by gamma irradiation. *International Journal of Food Microbiology*, *57*(1–2), 91–97.

Bradley, M. V., Hall, L. L., & Trebilcock, S. J. (1968). Low pH of irradiated sucrose in induction of chromosome aberrations. *Nature*, *217*(5134), 1182–1183.

Brennan, J. G., Grandison, A. S., & Lewis, M. J. (2006). Separations in food processing. In J. G. Brennan & A. S. Grandison (Eds.), *Food processing handbook* (vol. 429). Wiley.

Bruhn, C. M., & Noell, J. W. (1987). Consumer in-store response to irradiated papayas. *Food Technology*, *41*(9), 83–85.

Burditt, A. K. Jr. (1982). *Food irradiation as a quarantine treatment of fruits* (No. IAEA-TECDOC--271). International Atomic Energy Agency (IAEA).

Bustos-Griffin, E., Hallman, G. J., & Griffin, R. L. (2012). Current and potential trade in horticultural products irradiated for phytosanitary purposes. *Radiation Physics and Chemistry*, *81*(8), 1203–1207.

Castagna, A., Chiavaro, E., Dall'Asta, C., Rinaldi, M., Galaverna, G., & Ranieri, A. (2013). Effect of postharvest UV-B irradiation on nutraceutical quality and physical properties of tomato fruits. *Food Chemistry*, *137*(1–4), 151–158.

Ceci, L. N., Curzio, O. A., & Pomilio, A. B. (1991). Effects of irradiation and storage on the flavor of garlic bulbs cv "Red". *Journal of Food Science*, *56*(1), 44–46.

Chawla, S. P., Chander, R., & Sharma, A. (2007). Antioxidant formation by γ-irradiation of glucose–amino acid model systems. *Food Chemistry*, *103*(4), 1297–1304.

Cia, P., Pascholati, S. F., Benato, E. A., Camili, E. C., & Santos, C. A. (2007). Effects of gamma and UV-C irradiation on the postharvest control of papaya anthracnose. *Postharvest Biology and Technology*, *43*(3), 366–373.

Clemmons, H. E., Clemmons, E. J., & Brown, E. J. (2015). Electron beam processing technology for food processing. In Suresh D. Pillai & Shima Shayanfar (Eds), *Electron beam pasteurization and complementary food processing technologies* (pp. 11–25). Woodhead Publishing.

Dauphin, J. F., & Saint-Lebe, L. R. (1977). Radiation chemistry of lipids. In P. S. Elias & A. J. Cohen (Eds.), *Radiation chemistry of lipids*. North Holland: Elsevier Biomedical Press.

De Silva Aquino, A. (2012). Sterilization by gamma irradiation. In Feriz Adrovic (Ed.), *Gamma radiation* (pp. 171–206). London: InTechOpen.

Dionísio, A. P., Gomes, R. T., & Oetterer, M. (2009). Ionizing radiation effects on food vitamins – A review. *Brazilian Archives of Biology and Technology, 52*(5), 1267–1278.

Dong-Ho, K., Hyun-Pa, S., Hong-Sun, Y., Young-Jin, C., Yeung-Ji, K., & Myung-Woo, B. (2002). Distribution of microflora in powdered raw grains and vegetables and improvement of hygienic quality by gamma irradiation. *Journal of the Korean Society of Food Science and Nutrition, 31*(4), 589–593.

Elias, P. S. (1987). Food irradiation. In K. Miller (Ed.), *Toxicological aspects of food* (pp. 295–346). New York: Elsevier.

Erkan, N., & Özden, Ö. (2007). The changes of fatty acid and amino acid compositions in sea bream (Sparus aurata) during irradiation process. *Radiation Physics and Chemistry, 76*(10), 1636–1641.

Falguera, V., Pagán, J., & Ibarz, A. (2011). Effect of UV irradiation on enzymatic activities and physico-chemical properties of apple juices from different varieties. *LWT – Food Science and Technology, 44*(1), 115–119.

Fan, X., & Sokorai, K. J. B. (2008). Effect of ionizing radiation on furan formation in fresh-cut fruits and vegetables. *Journal of Food Science, 73*(2), C79–C83.

Fan, X., Sokorai, K. J. B., Sommers, C. H., Niemira, B. A., & Mattheis, J. P. (2005). Effects of calcium ascorbate and ionizing radiation on the survival of Listeria monocytogenes and product quality of fresh-cut 'Gala' apples. *Journal of Food Science, 70*(7), m352–m358.

Farkas, J., & Mohácsi-Farkas, C. (2011). History and future of food irradiation. *Trends in Food Science and Technology, 22*(2–3), 121–126.

Farkas, J., Saray, T., Mohacsi-Farkas, C., Horti, K., & Andrassy, E. (1997). Effects of low-dose gamma radiation on shelf-life and microbiological safety of pre-cut/prepared vegetables. *Advances in Food Sciences, 19*(3–4), 111–119.

Fellows, P. J. (2001). *Food processing technology* (2nd ed.). Cambridge: CRC Press. Woodhead Publishing, pp. 199.

Hajare, S. N., Saxena, S., Kumar, S., Wadhawan, S., More, V., Mishra, B. B., … Sharma, A. (2010). Quality profile of litchi (Litchi chinensis) cultivars from India and effect of radiation processing. *Radiation Physics and Chemistry, 79*(9), 994–1004.

Hallman, G. J. (2017). Process control in phytosanitary irradiation of fresh fruits and vegetables as a model for other phytosanitary treatment processes. *Food Control, 72*, 372–377.

Holsten, R. D., Sugii, M., & Steward, F. C. (1965). Direct and indirect effects of radiation on plant cells: Their relation to growth and growth induction. *Nature, 208*(5013), 850–856.

Hussain, P. R., Dar, M. A., Meena, R. S., Wani, A. M., Mir, M. A., & Shafi, F. (2008). Changes in quality of apple (Malus domestica) cultivars due to γ-irradiation and storage conditions. *Journal of Food Science and Technology, 45*(1), 44–49.

Hussain, P. R., Omeera, A., Suradkar, P. P., & Dar, M. A. (2014). Effect of combination treatment of gamma irradiation and ascorbic acid on physicochemical and microbial quality of minimally processed egg-plant (Solanum melongena L.). *Radiation Physics and Chemistry, 103*, 131–141.

Hussain, P. R., Suradkar, P. P., Wani, A. M., & Dar, M. A. (2015). Retention of storage quality and post-refrigeration shelf-life extension of plum (Prunus domestica L.) cv. Santa Rosa using combination of carboxymethyl cellulose (CMC) coating and gamma irradiation. *Radiation Physics and Chemistry, 107*, 136–148.

IAEA (International Atomic Energy Agency). (2001). *Consumer acceptance and market development of irradiated food in*. Vienna, Austria: Asia and the Pacific, p. 98.

Ihsanullah, I., & Rashid, A. (2017). Current activities in food irradiation as a sanitary and phytosanitary treatment in the Asia and the Pacific Region and a comparison with advanced countries. *Food Control, 72*, 345–359.

Jeong, R. D., Chu, E. H., Lee, G. W., Cho, C., & Park, H. J. (2016). Inhibitory effect of gamma irradiation and its application for control of postharvest green mold decay of Satsuma mandarins. *International Journal of Food Microbiology, 234*, 1–8.

Jessup, A. J., Rigney, C. J., & Wills, P. A. (1988). Effects of gamma irradiation combined with hot dipping on quality of "Kensington Pride" mangoes. *Journal of Food Science, 53*(5), 1486–1489.

Jiang, T., Luo, S., Chen, Q., Shen, L., & Ying, T. (2010). Effect of integrated application of gamma irradiation and modified atmosphere packaging on physicochemical and microbiological properties of shiitake mushroom (Lentinus edodes). *Food Chemistry, 122*(3), 761–767.

Jongen, W. (Ed.). (2005). *Improving the safety of fresh fruit and vegetables.* New York: Elsevier.

Joshi, P., Nathawat, N. S., Chhipa, B. G., Hajare, S. N., Goyal, M., Sahu, M. P., & Singh, G. (2011). Irradiation of sangari (Prosopis cineraria): Effect on composition and microbial counts during storage. *Radiation Physics and Chemistry, 80*(11), 1242–1246.

Jung, K., Yoon, M., Park, H. J., Lee, K. Y., Jeong, R. D., Song, B. S., & Lee, J. W. (2014). Application of combined treatment for control of Botrytis cinerea in phytosanitary irradiation processing. *Radiation Physics and Chemistry, 99*, 12–17.

Kodenchery, U. K., & Nair, M. P. (1972). Metabolic changes induced by sprout inhibiting dose of. gamma.-irradiation in potatoes. *Journal of Agricultural and Food Chemistry, 20*(2), 282–285.

Kong, Q., Wu, A., Qi, W., Qi, R., Carter, J. M., Rasooly, R., & He, X. (2014). Effects of electron-beam irradiation on blueberries inoculated with Escherichia coli and their nutritional quality and shelf life. *Postharvest Biology and Technology, 95*, 28–35.

Koopmans, M., & Duizer, E. (2004). Foodborne viruses: An emerging problem. *International Journal of Food Microbiology, 90*(1), 23–41.

Kraybill, H. F. (1982). *Effect of processing on nutritive value of food: Irradiation.* Florida: CRC Press.

Kuan, Y. H., Bhat, R., & Karim, A. A. (2011). Emulsifying and foaming properties of ultraviolet-irradiated egg white protein and sodium caseinate. *Journal of Agricultural and Food Chemistry, 59*(8), 4111–4118.

Kume, T., Furuta, M., Todoriki, S., Uenoyama, N., & Kobayashi, Y. (2009). Status of food irradiation in the world. *Radiation Physics and Chemistry, 78*(3), 222–226.

Kushwah, S. S., & Daheriya, A. K. (2012). Irradiation in post harvest management of vegetables. In S. S. Katiyar (Ed.), *Everyman's Science* (p. 90). Kolkata, India: Indian Science Congress Association.

Lacroix, M., Turgis, M., Borsa, J., Millette, M., Salmieri, S., Caillet, S., & Han, J. (2009). Applications of radiation processing in combination with conventional treatments to assure food safety: New development. *Radiation Physics and Chemistry, 78*(11), 1015–1017.

Lafortune, R., Caillet, S., & Lacroix, M. (2005). Combined effects of coating, modified atmosphere packaging, and gamma irradiation on quality maintenance of ready-to-use carrots (Daucus carota). *Journal of Food Protection, 68*(2), 353–359.

Latorre, M. E., Narvaiz, P., Rojas, A. M., & Gerschenson, L. N. (2010). Effects of gamma irradiation on biochemical and physico-chemical parameters of fresh-cut red beet (Beta vulgaris L. var. conditiva) root. *Journal of Food Engineering, 98*(2), 178–191.

Lester, G. E., Hallman, G. J., & Pérez, J. A. (2010). γ-irradiation dose: Effects on baby-leaf spinach ascorbic acid, carotenoids, folate, α-tocopherol, and phylloquinone concentrations. *Journal of Agricultural and Food Chemistry, 58*(8), 4901–4906.

Loaharanu, P., & Ahmed, M. (1991). Advantages and disadvantages of the use of irradiation for food preservation. *Journal of Agricultural and Environmental Ethics, 4*(1), 14–30.

Lu, Z., Yu, Z., Gao, X., Lu, F., & Zhang, L. (2005). Preservation effects of gamma irradiation on fresh-cut celery. *Journal of Food Engineering, 67*(3), 347–351.

Lynch, M. F., Tauxe, R. V., & Hedberg, C. W. (2009). The growing burden of foodborne outbreaks due to contaminated fresh produce: Risks and opportunities. *Epidemiology and Infection, 137*(3), 307–315.

Ma, L., Zhang, M., Bhandari, B., & Gao, Z. (2017). Recent developments in novel shelf life extension technologies of fresh-cut fruits and vegetables. *Trends in Food Science and Technology, 64*, 23–38.

MacQueen, K. F. (1969). Food irradiation and its future perspectives. *La Prensa Medica Mexicana, 34*(5), 156–163.

Mahto, R., & Das, M. (2013). Effect of gamma irradiation on the physico-chemical and visual properties of mango (Mangifera indica L.), cv.'dushehri' and 'Fazli' stored at 20 C. *Postharvest Biology and Technology, 86*, 447–455.

Malone Jr, J. W. (1990). Consumer willingness to purchase and to pay more for potential benefits of irradiated fresh food products. *Agribusiness, 6*(2), 163–178.

Mami, Y., Peyvast, G., Ziaie, F., Ghasemnezhad, M., & Salmanpour, V. (2013). Improvement of shelf-life and postharvest quality of white button mushroom by 60Co gamma-ray irradiation. *Plant Knowledge Journal*, *2*(1), 1–7.

Maraei, R. W., & Elsawy, K. M. (2017). Chemical quality and nutrient composition of strawberry fruits treated by γ-irradiation. *Journal of Radiation Research and Applied Sciences*, *10*(1), 80–87.

Matsuyama, A., & Umeda, K. (2018). Sprout inhibition in tubers and bulbs. In Edward S. Jisephson & Martin S. Peterson (Eds.), *Preservation of food by ionizing radiation* (pp. 159–213). Florida: CRC Press.

Meisberger, D., Brodie, A. D., Desai, A. A., Emge, D. G., Chen, Z. W., Simmons, R., … Munro, E. (1996). *U.S. patent no. 5,578,821*. Washington, DC: U.S. Patent and Trademark Office.

Mohacsi-Farkas, C. S., Farkas, J., Andrassy, E., Polyak-Feher, K., Brückner, A., Kisko, G., & Agoston, R. (2006). Improving the microbiological safety of some fresh pre-cut and prepackaged chilled produce by low dose gamma irradiation. In *Use of irradiation to ensure the hygienic quality of fresh, pre-cut fruits and vegetables and other minimally processed food of plant origin*. Proceedings of a final research coordination meeting organized by the Joint FAO/IAEA Programme of Nuclear Techniques in Food and Agriculture and held in Islamabad, Pakistan, 22–30 July 2005, 130.

Mondy, N. I., & Gosselin, B. (1989). Effect of irradiation on discoloration, phenols and lipids of potatoes. *Journal of Food Science*, *54*(4), 982–984.

Moog, P. (1988). *How to win consumer acceptance in the marketing of irradiated food. factors affecting practical application of food irradiation*. IAEA TECDOC-544.

Mostafavi, H. A., Mirmajlessi, S. M., & Fathollahi, H. (2012). The potential of food irradiation: Benefits and limitations. *Trends in Vital Food and Control Engineering*, *5*, 43–68.

Nawar, W. W. (1972). *Radiolytic changes in fats*. New York: Taylor & Francis, pp. 295–346.

Niemira, B. A. (2008). Irradiation compared with chlorination for elimination of Escherichia coli O157: H7 internalized in lettuce leaves: Influence of lettuce variety. *Journal of Food Science*, *73*(5), M208–M213.

Niemira, B. A., & Lonczynski, K. A. (2006). Nalidixic acid resistance influences sensitivity to ionizing radiation among Salmonella isolates. *Journal of Food Protection*, *69*(7), 1587–1593.

Niemira, B. A., Sommers, C. H., & Fan, X. (2002). Suspending lettuce type influences recoverability and radiation sensitivity of Escherichia coli O157:H7. *Journal of Food Protection*, *65*(9), 1388–1393.

Nunes, T. P., Martins, C. G., Behrens, J. H., Souza, K. L. O., Genovese, M. I., Destro, M. T., & Landgraf, M. (2008). Radio resistance of Salmonella species and Listeria monocytogenes on minimally processed arugula (Eruca sativa Mill.): Effect of irradiation on flavonoid content and acceptability of irradiated produce. *Journal of Agriculture and Food Chemistry*, *56*(4), 1264–1268.

Oufedjikh, H., Mahrouz, M., Amiot, M. J., & Lacroix, M. (2000). Effect of γ-irradiation on phenolic compounds and phenylalanine ammonia-lyase activity during storage in relation to peel injury from peel of Citrus clementina Hort. Ex. Tanaka. *Journal of Agricultural and Food Chemistry*, *48*(2), 559–565.

Pan, Y. G., & Zu, H. (2012). Effect of UV-C radiation on the quality of fresh-cut pineapples. *Procedia Engineering*, *37*, 113–119.

Patterson, M. (2005). *Food irradiation: Microbiological safety and disinfestation*. Thailand: International Symposium on New Frontier of Irradiated Food and Non-Food Products.

Pérez, M. B., Aveldaño, M. I., & Croci, C. A. (2007). Growth inhibition by gamma rays affects lipids and fatty acids in garlic sprouts during storage. *Postharvest Biology and Technology*, *44*(2), 122–130.

Pinela, J., Barreira, J. C., Barros, L., Verde, S. C., Antonio, A. L., Carvalho, A. M., … Ferreira, I. C. (2016). Suitability of gamma irradiation for preserving fresh-cut watercress quality during cold storage. *Food Chemistry*, *206*, 50–58.

Prakash, A., Chen, P. C., Pilling, R. L., Johnson, N., & Foley, D. (2007). 1% calcium chloride treatment in combination with gamma irradiation improves microbial and physicochemical properties of diced tomatoes. *Foodborne Pathogens and Disease*, *4*(1), 89–98.

Prakash, A., Inthajak, P., Huibregtse, H., Caporaso, F., & Foley, D. M. (2000). Effects of low-dose gamma irradiation and conventional treatments on shelf life and quality characteristics of diced celery. *Journal of Food Science*, *65*(6), 1070–1075.

Radomyski, T., Murano, E. A., Olson, D. G., & Murano, P. S. (1994). Elimination of pathogens of significance in food by low-dose irradiation: A review. *Journal of Food Protection*, *57*(1), 73–86.

Rajkowski, K. T., & Thayer, D. W. (2000). Reduction of Salmonella spp. and strains of Escherichia coli O157: H7 by gamma radiation of inoculated sprouts. *Journal of Food Protection*, *63*(7), 871–875.

Reddy, S. V. R., Sharma, R. R., & Gundewadi, G. (2018). Use of irradiation for postharvest disinfection of fruits and vegetables. In Mohammed Wasim Siddiqui (Ed.), *Postharvest disinfection of fruits and vegetables* (pp. 121–136). New York: Academic Press.

Rezaee, M., Almassi, M., Minaei, S., & Paknejad, F. (2013). Impact of post-harvest radiation treatment timing on shelf life and quality characteristics of potatoes. *Journal of Food Science and Technology*, *50*(2), 339–345.

Scott Smith, J., & Suresh, P. (2004). Irradiation and food safety. *Food Technology, Irradiation and Food Safety*, *58*(11), 48–55.

Shahbaz, H. M., Ahn, J. J., Akram, K., Kim, H. Y., Park, E. J., & Kwon, J. H. (2014). Chemical and sensory quality of fresh pomegranate fruits exposed to gamma radiation as quarantine treatment. *Food Chemistry*, *145*, 312–318.

Sharma, P., Sharma, S. R., Dhall, R. K., & Mittal, T. C. (2020). Effect of γ-radiation on post-harvest storage life and quality of onion bulb under ambient condition. *Journal of Food Science and Technology*, *57*(7), 2534–2544.

Shurong, L., Meixu, G., & Chuanyao, W. (2006). Use of irradiation to ensure hygienic quality of fresh pre-cut and blanched vegetables and Tofu. Use of irradiation to ensure the hygienic quality of fresh, pre-cut fruits and vegetables and other minimally processed food of plant origin, p. 87. Proceedings of a Final Research Coordination Meeting held in Islamabad, Pakistan, 22–30 July 2005.

Skala, J. H., Mcgown, E. L., & Waring, P. P. (1987). Wholesomeness of irradiated foods. *Journal of Food Protection*, *50*(2), 1150–1160.

Song, H. P., Byun, M. W., Jo, C., Lee, C. H., Kim, K. S., & Kim, D. H. (2007). Effects of gamma irradiation on the microbiological, nutritional, and sensory properties of fresh vegetable juice. *Food Control*, *18*(1), 5–10.

Stefanova, R., Vasilev, N. V., & Spassov, S. L. (2010). Irradiation of food, current legislation framework, and detection of irradiated foods. *Food Analytical Methods*, *3*(3), 225–252.

Swallow, A. J. (1991). Wholesomeness and safety of irradiated foods. In M. Friedman (Ed.), *Nutritional and toxicological consequences of food processing* (pp. 11–31). Springerlink.

Thayer, D. W. (1990). Food irradiation: Benefits and concerns. *Journal of Food Quality*, *13*(3), 147–169.

Thomas, P., & Moy, J. H. (1986). Radiation preservation of foods of plant origin. III. Tropical fruits: Bananas, mangoes, and papayas. *Critical Reviews in Food Science and Nutrition*, *23*(2), 147–205.

Trigo, M. J., Sousa, M. B., Sapata, M. M., Ferreira, A., Curado, T., Andrada, L., … Veloso, M. G. (2009). Radiation processing of minimally processed vegetables and aromatic plants. *Radiation Physics and Chemistry*, *78*(7–8), 659–663.

Tripathi, J., Variyar, A. P., Mishra, P. K., & Variyar, P. S. (2016). Impact of radiation processing on the stability of cucurbitacin glycosides in ready-to-cook (RTC) pumpkin during storage. *LWT*, *73*, 239–242.

Tripathi, P. C., Sankar, V., Mahajan, V. M., & Lawande, K. E. (2011). Response of gamma irradiation on postharvest losses in some onion varieties. *Indian Journal of Horticulture*, *68*, 556–560.

Urbain, W. M. (1982). *Considerations on economics and energy requirements of food irradiation applications in developing countries. Anticipated benefits of irradiation* (No. IAEA-TECDOC--271).

Vaishnav, J., Adiani, V., & Variyar, P. S. (2015). Radiation processing for enhancing shelf life and quality characteristics of minimally processed ready-to-cook (RTC) cauliflower (Brassica oleracea). *Food Packaging and Shelf Life*, *5*, 50–55.

Vanamala, J., Cobb, G., Loaiza, J., Yoo, K., Pike, L. M., & Patil, B. S. (2007). Ionizing radiation and marketing simulation on bioactive compounds and quality of grapefruit (Citrus paradisi cv Rio Red). *Food Chemistry*, *105*(4), 1404–1411.

Verde, S. C., Trigo, M. J., Sousa, M. B., Ferreira, A., Ramos, A. C., Nunes, I., … Botelho, M. L. (2013). Effects of gamma radiation on raspberries: Safety and quality issues. *Journal of Toxicology and Environmental Health, Part A*, *76*(4–5), 291–303.

Vojdani, J. D., Beuchat, L. R., & Tauxe, R. V. (2008). Juice-associated outbreaks of human illness in the United States, 1995 through 2005. *Journal of Food Protection*, *71*(2), 356–364.

Wang, J., & Chao, Y. (2002). Drying characteristics of irradiated apple slices. *Journal of Food Engineering*, *52*(1), 83–88.

Wang, J., & Du, Y. (2005). The effect of γ-ray irradiation on the drying characteristics and final quality of dried potato slices. *International Journal of Food Science and Technology, 40*(1), 75–82.

Wang, Z., Ma, Y., Zhao, G., Liao, X., Chen, F., Wu, J., … Hu, X. (2006). Influence of gamma irradiation on enzyme, microorganism, and flavor of cantaloupe (Cucumis melo L.) juice. *Journal of Food Science, 71*(6), M215–M220.

Wilkinson, V. M., & Gould, G. W. (1996). *Food irradiation: A reference guide.* London, UK: Woodhead Publishing.

Xu, F., Cao, S., Shi, L., Chen, W., Su, X., & Yang, Z. (2014). Blue light irradiation affects anthocyanin content and enzyme activities involved in postharvest strawberry fruit. *Journal of Agricultural and Food Chemistry, 62*(20), 4778–4783.

Yang, J. S., Perng, F. S., Liou, S. E., & Wu, J. J. (1993). Effects of gamma irradiation on chromatophores and volatile components of grass shrimp muscle. *Radiation Physics and Chemistry, 42*(1–3), 319–322.

Zhang, K., Deng, Y., Fu, H., & Weng, Q. (2014). Effects of Co-60 gamma-irradiation and refrigerated storage on the quality of Shatang mandarin. *Food Science and Human Wellness, 3*(1), 9–15.

Zhu, M. J., Mendonca, A., Ismail, H. A., & Ahn, D. U. (2009). Fate of Listeria monocytogenes in ready-to-eat turkey breast rolls formulated with antimicrobials following electron-beam irradiation. *Poultry Science, 88*(1), 205–213.

Chapter 5

Effect of Ozone Processing on Quality Characteristics of the Fruit Juice Processing Industry

Celale Kirkin and Esra Dogu-Baykut

CONTENTS

5.1 INTRODUCTION

In recent years, consumers' tendency towards healthy foods has increased. Hence, the demand for fruit juices has also increased due to their nutritional value and health-promoting effects. Fruit juices are rich sources of micronutrients (vitamins and minerals) and bioactive compounds (mainly polyphenols). Consumption of fruit juices can meet some of the daily vitamin and mineral needs of people (Mitchell et al., 2020). Moreover, bioactive compounds may contribute to the prevention of chronic diseases such as cardiovascular disease, cancer, diabetes, or age-related functional decline (Liu, 2013; Zurbau et al., 2020).

Fruit juices are susceptible to enzymatic and microbial spoilage because of their high water content. They also carry a risk of foodborne disease outbreaks if not processed properly. *Escherichia coli O157:H7* and *Salmonella* are the main foodborne pathogens associated with fruit juice outbreaks (Vojdani, Beuchat, & Tauxe, 2008). The United States Food and Drug Administration (FDA) published a final rule requiring manufacturers to achieve at least 5 log reduction of the pertinent pathogen in fruit juices to ensure microbiological safety (FDA, 2001a). Pasteurization is the most widely utilized thermal treatment for fruit juices to inactivate microorganisms and enzymes that catalyze spoilage. The major problem associated with the pasteurization of fruit juices is that heating mostly leads to detrimental changes in nutritional composition, sensory properties, and bioactive compounds (Petruzzi et al., 2017). The demand for safe, nutritious, and healthier fruit juices with absence of additives and preservatives has encouraged researchers to study nonthermal food preservation techniques in fruit juice processing (Noguera, Lima, Filho, Fonteles, & Rodrigues, 2021).

Ozone is a promising non-thermal technology for the processing of fruit juices, as it is a safe and highly effective disinfectant and does not leave a chemical residue in treated foods. It has widespread applications in the food industry, such as drinking water disinfection, wastewater treatment, and surface decontamination of fruits and vegetables (Guzel-Seydim, Greene, & Seydim, 2004). Ozone has received Generally Recognized as Safe (GRAS) status from FDA for use as a disinfectant for foods in 1997, and FDA officially approved ozone, in gas and aqueous phases, for use in the food industry, also for direct contact with foods in 2001 (FDA, 2001b). Ozone has a high germicidal effect against a broad spectrum of microorganisms, including bacteria, viruses, and protozoa and the efficacy of ozone in fruit juice processing to assure safety and quality has been evaluated in several recent studies (Cullen et al., 2010; Pandiselvam, Sunoj, Manikantan, Kothakota, & Hebbar, 2017). The objective of this chapter is to provide an overview of ozone processing of fruit juices, including its control parameters and efficacy on microbial safety. The effects of ozone processing on nutritional and bioactive compounds, food quality, enzyme inactivation, and pesticide degradation in fruit juices are also discussed.

5.2 GENERATION OF OZONE

Ozone (O_3) is a highly unstable triatomic oxygen molecule, formed by the addition of a free radical of oxygen to molecular diatomic oxygen (O_2). It is a naturally occurring gaseous molecule in the atmosphere that was first described in 1840 (Rubin, 2001). Ozone is a colorless gas at room temperature when generated from high purity oxygen, but it is blue gas when generated from dried air. Regardless of how it is generated, the color of ozone is unnoticeable for food processing applications because of the used concentrations (Pandiselvam et al., 2017). The density of ozone (2.14 g L^{-1}) is higher than air (1.28 g L^{-1}) at 0°C and 101.3 kPa (Kim, Yousef, & Khadre, 2003). The boiling point of ozone is –119.1°C at 1 atm, the melting point is –192.5°C at 1 atm and the molecular weight is 48 (Guzel-Seydim et al., 2004). Ozone has the ability to break down easily into dioxygen and atomic oxygen or to react with other compounds. Thus, it has a very strong oxidation potential (2.07 V) compared to chlorine (1.36 V) and hypochlorous acid (1.49 V) (Kim, Yousef, & Dave, 1999). Ozone has a characteristic and pungent odor, which is noticeable at concentrations above 0.01 ppm. It can be toxic or nontoxic depending on the concentration ($\geq$0.1 ppm) and length of exposure. Short-term exposure of 0.1–1.0 ppm may result in symptoms such as dry throat, headaches, eye irritation, and respiratory irritation and higher exposure levels (1–100 ppm) may result in symptoms such as asthma-like symptoms, tiredness, and loss of appetite (Pascual, Llorca, & Canut, 2007).

Ozone is generated artificially with different methods. Since ozone is highly unstable and degrades spontaneously, it is generated at the site of application and in closed systems. Therefore, it is not possible to store or transport ozone like other industrial gases (Kim et al., 1999). Corona discharge (electrical discharge), ultraviolet (UV) radiation, and electrolysis are the main three methods to generate ozone. All three of these methods rely on the same principle; they split the oxygen (O_2) molecule into two atoms of oxygen and free oxygen ($O\bullet$) atoms that have excess electrons and rapidly combine with molecular oxygen (O_2) to form the highly unstable ozone molecule (O_3). Because this reaction is endothermic, the generation of ozone requires a significant energy input (Guzel-Seydim et al., 2004; Karaca & Velioglu, 2007).

Corona discharge is a commonly used and most applicable method to generate a high concentration of ozone. Corona discharge generators are operated by passing dried, dust-free, and oil-free air or an oxygen-containing gas mixture, or oxygen itself. Oxygen passes through the space of a high-energy electrical field between two electrodes, one of which serves as the ground electrode, and the other as the dielectric medium. When electrons have sufficient energy to dissociate the oxygen molecule, the oxygen molecule splits apart and forms active atomic oxygen radicals that combine with other oxygen molecules to form ozone. Since ozone may be decomposed by heat (especially above 35°C), a temperature control and heat removal system is necessary. Therefore, ozone generators generally have a cooling system (Brodowska, Nowak, & Śmigielski, 2018; Karaca & Velioglu, 2007; Muthukumarappan, O'Donnell, & Cullen, 2009).

Ozone also can be generated commercially by UV ozone generators. In this method, ozone is produced photochemically from fresh air by passing across low-intensity, low-pressure mercury lamps producing 185 nm radiation. Oxygen molecules are split into oxygen atoms, which combine with surrounding oxygen molecules to form ozone. Due to the low efficiencies of these UV lamps, low concentrations of ozone (0.03 ppm) are produced. Therefore, although the installation and operating costs of this method are low, it cannot be used in processes that require high concentrations of ozone (Cullen et al., 2010; Guzel-Seydim et al., 2004; Kim et al., 1999).

Electrolysis is an electrochemical method in which ozone is generated between an anode and a cathode placed in an electrolytic solution containing water and highly electronegative anions. By electrolysis, water is split into oxygen and hydrogen atoms, then hydrogen atoms are separated and oxygen atoms are combined to form ozone. This system is small, portable, and can be designed to run continuously. Higher concentrations of ozone than the UV or corona discharge method can be achieved. However, the cost of the process is much higher than those methods (Kim et al., 1999; Muthukumarappan et al., 2009).

5.3 CONTROL PARAMETERS AND EFFECTIVENESS OF OZONE

The efficacy of ozone is influenced by intrinsic and extrinsic factors while processing liquids. Therefore, for optimum efficiency, it is important to understand how these factors influence ozone disinfection capability. These key factors are listed in Figure 5.1 and described below.

5.3.1 *pH*

An increase in ozone decomposition rate negatively affects microbial inactivation. Ozone efficacy decreases by increasing pH due to the acceleration of ozone degradation. Therefore, acidic environments can be considered more suitable for ozonation (Zuma, Lin, & Jonnalagadda, 2009). Jamil, Farooq, and Hashmi (2017) reported that the survival of *E. coli* was higher at pH 8 than the pH 6 and 7 in distilled water after ozonation. Patil, Valdramidis, Cullen, Frias, and Bourke (2010) also studied the effect of pH on the inactivation rate of *E. coli* in apple juice at different pH ranges (3.0, 3.5, 4.0, 4.5, and 5.0). They indicated that the *E. coli* inactivation rate of ozone in apple juice was much faster at low pH than at higher pH values. In another study, survival of *Staphylococcus aureus* in acetate buffer after ozonation was tested in different pH values (4.0, 5.5, and 6.0). Under more acidic conditions (pH 4), a higher reduction in *S. aureus* survival was found as compared with the other two pH values (Britton, Draper, & Talmadge, 2020).

5.3.2 Organic Matters in Food Matrix

Organic matters in the food matrix are known to adversely affect the effectiveness of ozone for the inactivation of microorganisms, due to the potential of ozone to react with microorganisms as well as with other particles (Kim et al., 1999). Fruit juices are rich in organic content such as sugars, pectic substances, and ascorbic acid. Williams, Sumner, and Golden (2004) found that the efficacy of ozonation for the inactivation of *E.coli* in orange juice is reduced in the presence of organic compounds compared with inactivation in pure water. Patil, Bourke, Frias, Tiwari, and Cullen (2009) observed that the rate for complete *E. coli* ATCC 25922 inactivation

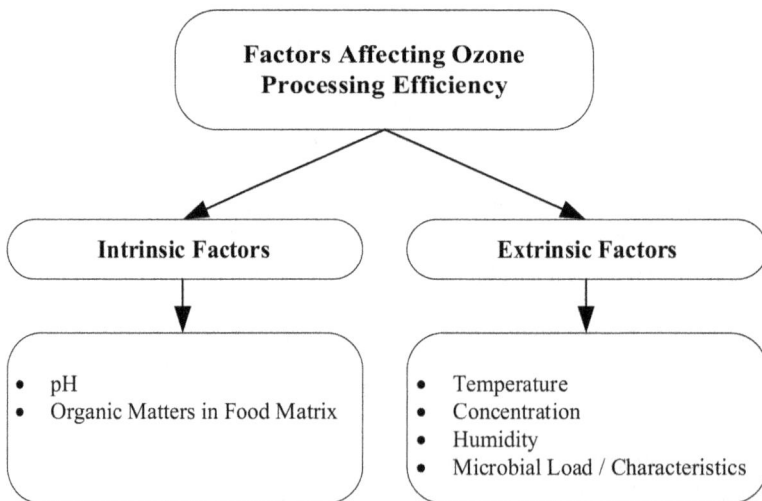

Figure 5.1   Factors affecting ozone processing efficiency.

for unfiltered orange juice (18 min) was slower than the inactivation in filtered one (5 min). In addition, Restaino, Frampton, Hemphill, and Palnikar (1995) reported that the type of organic matter present also highly affects the effectiveness of ozone more than the amount present. They observed a reduction in the effectiveness of ozone for inactivating *S. aureus, Listeria monocytogenes, E. coli*, and *Salmonella typhimurium* by adding 20 ppm of bovine serum albumin added to deionized water while 20 ppm of soluble starch had no effect on ozone activity. Furthermore, Labbe, Kinsley, and Wu (2001) reported the presence of 3% sucrose in buffer markedly reduced the effectiveness of ozone for two strains of *Candida* and *Pseudomonas fluorescens* isolated from maple sap compared with the buffer with no added sucrose.

5.3.3 *Temperature*

The effect of temperature on the efficacy of ozone is not clear according to the controversial results in the literature. The solubility of ozone in water at 0–30°C is 13 times higher than the solubility of oxygen in water (Rice, 1986). A reduction of the temperature of the water increases ozone solubility and stability, thus resulting in an increase in its efficacy. On the contrary, a rise in temperature increases the inactivating capabilities of disinfectants (Pascual et al., 2007).

Williams et al. (2004) reported that inactivation of *E. coli* and *Salmonella* in apple cider and orange juice treated with ozone at 4°C was generally higher than at 20°C, and ozone treatment at 50°C was higher than treatment at either 4°C or 20°C. In another study, inactivation of *Cryptosporidium parvum* with ozonation is found to be increased with increasing temperature (1–20°C) (Driedger, Rennecker, & Mariñas, 2001). However, an opposite trend was observed in the study of Patil, Cullen, Kelly, Frías, and Bourke (2009). They examined the effects of four different temperatures (ambient: 12–15°C, 20°C, 25°C, and 30°C) on the efficacy of ozone inactivation on *E. coli* ATCC 25922 in tryptic soy broth and ambient temperature gave the best inactivation levels. Thus, these results show that the influence of temperature on ozone efficacy can vary on particular applications due to experimental conditions.

5.3.4 Concentration

Higher ozone concentrations shorten the inactivation time. Patil, Cullen, et al. (2009) studied five different levels of ozone concentration to inactivate *E. coli* ATCC 25922 in tryptic soy broth. The authors stated that the time required for 5 log inactivation decreased with increasing concentration. Similarly, Zuma et al. (2009) found higher inactivation levels for *E. coli* in water at higher ozone concentrations. But higher ozone concentrations and long exposure times may cause alterations in food components. Therefore, ozonation should not exceed a certain threshold (Afsah-Hejri, Hajeb, & Ehsani, 2020; Miller, Silva, & Brandão, 2013).

It is important to point out that the measurement of the "residual ozone" which refers to the detectable concentration of ozone is also important to detect the effectiveness of ozone. Because the presence of ozone-demanding material in the medium affects the available ozone for microbial inactivation. Therefore, stating only the applied dose in studies may lead to incorrect determination of the effectiveness of ozone against microorganisms (Kim et al., 2003).

5.3.5 Humidity

High relative humidity (RH) contributes to increased microbial inactivation rates in ozone applications. It has been reported that the optimum level is 90–95% RH and below 45–50 RH the microbial inactivation effect of ozone was negligible (Kuprianoff, 1953; Pascual et al., 2007). Naitou and Takahara (2008) studied the sporicidal effect of ozone with a concentration of 50 ppm at five different RH (11, 33, 52, 80, and 95%). They found that spores at higher RH (80 and 95% RH) were more sensitive to ozone than spores at lower RH (11–33% RH). Kim et al. (1999) used gaseous ozone to inactivate natural contaminants in a powdered food having different water activities (a_w). The authors indicate that microbial inactivation occurred when a_w of food was 0,95, but ozone was not effective on products with a_w less than 0.85.

5.3.6 Microbial Load/Characteristics

Microorganisms have different sensitivity to ozone. Thus, one important factor affecting the efficacy of ozone in food treatment is the sensitivity of the targeted microorganism (Pascual et al., 2007). In addition, the physiological state and initial load of microorganisms also significantly affect the antimicrobial effect of ozone (Cullen et al., 2010). These topics are covered in Section 5.4.

5.4 EFFECTS OF OZONE ON MICROBIAL SAFETY

Ozone is a strong antimicrobial agent that is effective against a broad range of pathogenic microorganisms, including bacteria, yeasts, molds, and viruses (Kim et al., 2003).

Each microorganism has a sensitivity to ozone. However, since the efficacy of ozone can be significantly affected by changes in experimental variables, comparing sensitivity results of microorganisms in different studies may not give accurate results. In general, ozone is more effective for bacteria than yeast and fungi, and it is more effective for gram-positive bacteria than gram-negative bacteria. Also, spores are more resistant than vegetative cells (Pascual et al., 2007).

The antimicrobial action of ozone has been investigated for different types of fruit juices. Up to 3.5 log reductions of total mesophilic bacteria and yeast and mold counts have been reported in watermelon juice treated with ozone (Lee, Ting, & Thoo, 2021). Ozone treatment was also found effective in reducing the number of pathogens such as *E. coli* O157:H7, *Salmonella,* and *L. monocytogenes* in several studies (Patil, Torres, et al., 2010; Patil, Valdramidis, et al., 2010; Sung, Song, Kim, Ryu, & Kang, 2014; Williams et al., 2004). *Listeria innocua*, a nonpathogenic surrogate for *L. monocytogenes,* was studied in peach juice (Garcia Loredo, Guerrero, & Alzamora, 2015) and cantaloupe melon juice (Sroy, Fundo, Miller, Brandão, & Silva, 2019) as target microorganism and 5 log reductions of *L. innocua* was reported in both studies. Torlak (2014) and Fundo et al. (2018) also reported that ozone has potential for the control of *Alicyclobacillus acidoterrestris* spores which is a cause of spoilage in hot-fill fruit and fruit juices. A summary of the studies evaluating the antimicrobial effect of ozone on fruit juices is given in Table 5.1.

5.4.1 Microbial Inactivation Mechanism

The inactivation of microorganisms by ozone can be triggered by a direct reaction of the molecular ozone or by the action of free radicals (hydroxyl, hydroperoxyl, and superoxide) produced by its breakdown (Guzel-Seydim et al., 2004). Although some researchers suggest that the molecular ozone is more effective for the inactivation of microorganisms (Hunt & Mariñas, 1997; Zuma & Jonnalagadda, 2010), there is no clear consensus on this issue.

TABLE 5.1   ANTIMICROBIAL EFFECT OF OZONE ON FRUIT JUICES

Product	Treatment	Inoculated Microorganisms	Findings	Reference
Apple juice	2.0–3.0 g m^{-3} 20 s, 40 s, 1 min 3.0 L min^{-1} 25, 45, 50, 55°C	• E. coli O157:H7, S. typhimurium, and L. monocytogenes	• 1.84 and 2.20 log reduction for S. typhimurium at 25 and 45°C, respectively • 1.50 and 1.60 log reduction for E. coli O157:H7 at 25 and 45°C, respectively • 0.79 and 0.93 log reduction for L. monocytogenes at 25 and 45°C, respectively • Below the detection limit at 50 and 55°C for all 3 microorganisms	(Sung et al., 2014)
	2.8, 5.3 mg L^{-1} 10, 20, 30, 40 min 0.4 L min^{-1} 4 and 22°C	• Alicyclobacillus acidoterrestris spores	• 1.8 and 2.4 log reductions after 40 min treatment at 22°C with 2.8 mg/L and 5.3 mg/L ozone, respectively • 2.2 and 2.8 log reductions after 40 min treatment at 4°C with 2.8 mg/L and 5.3 mg/L ozone, respectively	(Torlak, 2014)
	0.048 mg min^{-1} mL^{-1} 5 min 0.12 L min^{-1} pH: 3.0, 3.5, 4.0, 4.5, 5.0	• E. coli ATCC 25922 • E. coli NCTC 12900	• 5 log reduction of the 2 E. coli strains at pH 5.0 in 18 min, at pH 4.5 in 11 min, at pH 4.0 in 9 min, at pH 3.5 in 6 min, at pH 3.5 in 5 min	(Patil, Valdramidis, Cullen, Frias, & Bourke, 2010)
	0.048 mg min^{-1} mL^{-1} 5 min 0.12 L min^{-1} pH: 3.8 20 ± 1.5°C	• E. coli ATCC 25922 • E. coli NCTC 12900	• 5 log reduction of the 2 E. coli strains	(Patil, Torres, et al., 2010)
Apple cider	2.4 L/min 14, 45, 75, 240 min 0.9 g h^{-1} 4 and 50°C pH: 3.8	• E. coli O157:H7 (five strain mixture) • Salmonella (five serovar mixture)	• 6 log reduction of the E. coli O157:H7 after 45 min at 50°C • 4.8 log reduction of the E. coli O157:H7 after 240 min at 4°C • 4.8 log reduction of the Salmonella after 15 min at 50°C • 4.5 log reduction of the Salmonella after 240 min at 4°C	(Williams, Sumner, & Golden, 2004)

(Continued)

TABLE 5.1 (CONTINUED) ANTIMICROBIAL EFFECT OF OZONE ON FRUIT JUICES

Product	Treatment	Inoculated Microorganisms	Findings	Reference
Cantaloupe melon juice	7.7 ± 2.4 g L⁻¹ 2–10 min 5 g h⁻¹	• *Listeria innocua*	• 5.3 log reduction after 6 min • Below the detection limit after 10 min	(Sroy et al., 2019)
	7.0 ± 2.4 g L⁻¹ 30, 60 min 5 g h⁻¹	• *Alicyclobacillus acidoterrestris* spores	• 2.22 ± 0.04 log reduction after 60 min	(Fundo et al., 2018)
Orange juice	75–78 mg min⁻¹ mL⁻¹ 0–18 min 0.12 L min⁻¹ 12–15°C pH: 3.5–4.0	• *E. coli* ATCC 25922 • *E. coli* NCTC 12900	• 5 log reduction of the 2 *E. coli* strains in unfiltered orange juice at ~12 min • 5 log reduction of the 2 *E. coli* strains in filtered orange juice at ~5 min	(Patil, Bourke, Frias, Tiwari, & Cullen, 2009)
	2.4 L/min 14, 45, 75, 240 min 0.9 g h⁻¹ 4 and 50°C pH: 3.8	• *E. coli* O157:H7 (five strain mixture) • *Salmonella* (five serovar mixture)	• 6 log reduction of the *E. coli* O157:H7 after 75 min at 50°C • 5.4 log reduction of the *E. coli* O157:H7 after 240 min at 4°C • 6 log reduction of the *Salmonella* after 15 min at 50°C • 4.2 log reduction of the *Salmonella* after 240 min at 4°C	(Williams, Sumner, & Golden, 2004)
Peach juice	10 and 18 ppm 12 min 5 L/min 20 ± 1°C	• *E. coli* ATCC 11229, *L. innocua* ATCC 33090 and *S. cerevisiae* KE162	• At least 3.9 up to 4.9 log reductions after 12 min using 10 or 18 ppm ozone for *E. coli* ATCC 11229, *L. innocua* ATCC 33090 • *L. innocua* ATCC 33090 was found more sensitive than *E. coli* ATCC 11229 at both ozone concentrations • Only 1 log reduction after 12 min using both ozone concentrations for *S. cerevisiae*	(Loredo, Guerrero, & Alzamora, 2015)
Watermelon juice	5, 10, 15, 20, 25 min 1 L min⁻¹	• Natural flora (total mesophilic bacteria, yeasts, and molds)	• Inactivation in unclarified and clarified watermelon juices improved with increasing processing time of the ozone treatment • 3.15 and 3.466 log reductions after 25 min for total mesophilic bacteria in unclarified and clarified watermelon juices, respectively • 3.046 and 3.411 log reductions after 25 min for yeasts and molds in unclarified and clarified watermelon juices, respectively	(Lee et al., 2021)

Ozone application causes a change in the permeability of the cell membrane of the microorganism, which leads to cell death due to leakage of cell contents (Guzel-Seydim et al., 2004; Khadre, Yousef, & Kim, 2001). Because of the lysis of the cell membrane, intracellular components are easily disrupted including proteins, unsaturated lipids, and enzymes in cell membranes (Cullen et al., 2010). Enzymes and nucleic acids in the cytoplasm are also attacked (Girgin Ersoy, Barisci, & Dinc, 2019; Khadre et al., 2001). The inactivation mechanism of ozone treatment is faster than the disinfectants because disinfectant agents need to permeate through the cell membrane in order to be effective (Pascual et al., 2007).

5.4.2 Mycotoxin Degradation

Mycotoxins are the secondary metabolites of filamentous fungi that are very hazardous to human health. The most common mycotoxins found in fruit juices are aflatoxin B1, aflatoxin G1, ochratoxin A, and patulin (Drusch & Ragab, 2003).

Studies investigating the effect of ozone on mycotoxins started in the 1960s and it was stated in many different studies that ozone degrades aflatoxins and leaves fewer degradation products (Freitas-Silva & Venâncio, 2010). Although it is generally believed that ozone reacts with the functional groups in mycotoxin molecules to change their molecular structure and form lower molecular weight, fewer double bonds, and less toxic products, the detoxification mechanism of all mycotoxins is still not fully known (Afsah-Hejri et al., 2020).

Patulin is a very hazardous toxin commonly found in apple juices and sometimes can be found in pear juice. FDA has established the maximum permitted level for patulin in apple juice of 50 µg/kg (FDA, 2001a). Diao et al. (2019) found that ozone treatment with an ozone concentration of 12 mg L^{-1} and flow rate of 3 L min^{-1} significantly decreased patulin in apple juice from 201.06 to below 50 µg L^{-1} within 15 min. Cataldo (2008) added patulin to apple juice and injected it into an HPLC column. They observed the disappearance of patulin on the chromatogram after bubbling ozone into the solution. The authors stated that ozone is a promising application for apple juice to reduce or eliminate patulin and can be a standard industry practice for fruit juices.

5.5 EFFECTS OF OZONE ON NUTRITIONAL COMPONENTS AND FOOD QUALITY

Even though the main objective of ozonation in the processing of several food products is microbial decontamination, the applications of the treatment for the purpose of enhancing food quality are also increasing. For instance, it can be used for the removal of off-odor, controlling respiration rate, color control, and depolymerization in various food products (Brodowska et al., 2018; Buluc & Koyuncu, 2020; Seo, King, & Prinyawiwatkul, 2007). The effects of the ozone treatments on the nutrients and bioactive composition and quality of foods can be positive or negative depending on the treatment conditions and the properties of the product (Brodowska et al., 2018).

A summary of the studies evaluating the effects of ozonation on the quality properties of fruit juices is given in Table 5.2 and explained in more detail in the following subsections.

5.5.1 Color

As ozone is a powerful oxidizer, it can lead to discoloration in many products including fruit juices (Cullen et al., 2010). The change in color of fruit juices can be explained by the degradation of

TABLE 5.2   EFFECTS OF OZONE ON QUALITY PROPERTIES OF FRUIT JUICES

Product	Treatment	Findings	Reference
Apple juice	12 mg L^{-1} 5–30 min 3 L min^{-1}	• Reduction in color, malic acid, ascorbic acid, and total phenol contents with treatment time • No effects on pH, soluble solids, and total acids	(Diao, Wang, Li, Wang, & Gao, 2018)
	2.0–3.0 g m^{-3} 20 s, 40 s, 1 min 3.0 L min^{-1} 25, 45, 50, 55°C	• Color values (L, a, b) were not affected by the treatments	(Sung et al., 2014)
	2.8, 5.3 mg L^{-1} 10, 20, 30, 40 min 0.4 L min^{-1}	• Decrease in total phenolic content after 5.3 mg L^{-1} when longer treatment times were applied.	(Torlak, 2014)
	1–4.8% w/w 0–10 min 0.125 L min^{-1}	• Color degradation (decrease in a value, increase in L and b values) as affected by the concentration and treatment time • Reduction in consistency index due to ozone concentration • Decrease in the contents of chlorogenic acid, caffeic acid, cinnamic acid, and total phenol	(Torres et al., 2011)
	0.048 mg min^{-1} mL^{-1} 10 min 0.12 L min^{-1}	• Increase in L* and b* values, decrease in a* value • Significant reduction in phenolic contents	(Patil, Torres, et al., 2010)
Apple cider	860 ppm 28 min 0.5 L min^{-1}	• Significant decrease in Hunter L, a, and b values after treatment, but the differences were not perceivable by the eye • No change in pH • The turbidity and soluble solids of the ozone-treated samples were lower, and the sedimentation was higher compared to the control but decreased after 14- and 21-d storage	(Choi & Nielsen, 2005)
Blackberry juice	0–7.8%w/w 0–10 min 0.0625 L min^{-1}	• Increase in lightness and decrease in redness and yellowness due to concentration and treatment time • Response surface methodology can be used to model the color change	(Tiwari, O'Donnell, Muthukumarappan, et al., 2009)

(Continued)

TABLE 5.2 (CONTINUED)   EFFECTS OF OZONE ON QUALITY PROPERTIES OF FRUIT JUICES

Product	Treatment	Findings	Reference
Cantaloupe melon juice	7.7 ± 2.4 g L^{-1} 2–10 min 5 g h^{-1}	• The total color difference increased during 13-d refrigerated storage. • No significant changes in total phenolic content, soluble solids, and pH • Reduction in antioxidant activity and vitamin C content • Retention of vitamin C was higher compared to heat pasteurization	(Sroy et al., 2019)
	7.0 ± 2.4 g L^{-1} 30, 60 min 5 g h^{-1}	• Significant changes in color values (increase in L* value, decrease in a* and b* values and chroma) after ozonation, but the change due to treatment time was insignificant • As treatment time increased hue angle decreased significantly • Decrease in pH with treatment time • No change in soluble solids content • Reduction in vitamin C and carotenoid contents and total antioxidant activity	(Fundo et al., 2018)
Coconut water	0.075–0.37 mg mL^{-1} 125 mL min^{-1} 10, 20, 30°C	• Small differences in color values (TCD <2) due to treatment • No change in pH and acidity • Total soluble solids content was slightly affected by the treatment conditions	(Porto et al., 2020)
Grape juice	1.6–7.8% w/w 0–10 min 0.06 L min^{-1}	• Significant color change • Changes in L value and total color difference fitted well with zero-order kinetics, and changes in a and b values were described by first-order kinetics • Decrease in anthocyanin content • No change in pH, total soluble solids, and titratable acidity	(Tiwari, O'Donnell, Patras, Brunton, & Cullen, 2009a)
Orange juice	0.6–3.0, 1.0–4.8, 1.6–7.8, 2.3–10.0% w/w 0, 2, 4, 6, 8, 10 min 0.25, 0.125, 0.0625, 0.0312 L min^{-1}	• Increase in L value and decrease in a and b values due to concentration, flow rate, and treatment time • Changes in L value and total color difference were expressed with zero-order kinetics, whereas changes in a and b values fitted to first-order kinetics • Ascorbic acid content decreased after 10 min treatment and followed first-order kinetics • pH, titratable acidity, soluble solids, nonenzymatic browning, and cloud value did not change	(Tiwari, Muthukumarappan, O'Donnell, & Cullen, 2008a)
	0, 30, 60, 90, 120 min	• Discoloration (decrease in a value as treatment time increased) • Degradation in ascorbic acid	(Angelino, 2002)

(Continued)

TABLE 5.2 (CONTINUED)   EFFECTS OF OZONE ON QUALITY PROPERTIES OF FRUIT JUICES

Product	Treatment	Findings	Reference
Orange juice (prebiotic)	0.057, 0.128, 0.230 mg mL[-1]	• Decrease in hue angle, no changes in L* and chroma values • Decrease in oligosaccharide concentration • No change in pH, total phenolic content • Decrease in antioxidant activity when 0.230 mg mL[-1] was applied	(Almeida et al., 2015)
Peach juice	0.06–2.48 g L[-1] 0.3–12 min 5 L min[-1]	• Significant changes in color values (increase in a* and C*; decrease in L*, b*, and h*) after treatment • Decrease in soluble solids contents at doses at and above 1.03 g L[-1] • Increase in titratable acidity 2.48 g L[-1] • Decrease in apparent viscosity with treatment time • Decrease in consistency index and increase in flow behavior index as affected by treatment time • No change in pH	(Jaramillo-Sánchez et al., 2018)
Pummelo fruit juice	600 mg h[-1] 0–50 min	• The effect of treatment time on color was not significant • Reduction in ascorbic acid and total phenolic content	(Shah, Supian, et al., 2019)
Roselle fruit juice	600 mg h[-1] 0, 10, 20, 30 min 0.2 L min[-1]	• The total color difference increased with increasing treatment time • Reduction in ascorbic acid, total phenolic, and total anthocyanin contents • No change in pH and titratable acidity	(Shah, Ong, Supian, & Sulaiman, 2020)
Strawberry juice	0–7.8% w/w 0–10 min 0.0625 L min[-1]	• Increase in L* value and decrease in a* and b* values with concentration and treatment time • Changes in L* value and total color difference were expressed with zero-order kinetics, whereas changes in a* and b* values fitted to first-order kinetics • Degradation of anthocyanins and ascorbic acid contents that follow fraction conversion model was observed	(Tiwari, O'Donnell, Patras, et al., 2009b)
Tomato juice	0–7.8 %w/w 0–10 min 0.0625 L min[-1]	• L* value increased, whereas in a* and b* values decreased after ozonation • Changes in L* value and total color difference were expressed with zero-order kinetics, and changes in a* and b* values fitted to first-order kinetics • No changes in pH, soluble solids, titratable acidity, cloud value, and nonenzymatic browning after the treatment • Decrease in ascorbic acid content up to 96%	(Tiwari, O'Donnell, Brunton, et al., 2009)

pigments, such as anthocyanins and carotenoids, depending on the fruit type (Fundo et al., 2018; Shah, Sulaiman, Sidek, & Supian, 2019; Sroy et al., 2019; Tiwari, O'Donnell, Muthukumarappan, & Cullen, 2009). Some researchers argue that the color degradation can be explained by the oxidizing power of the newborn oxygen atom (Cullen et al., 2010; Tiwari, Muthukumarappan, O'Donnell, & Cullen, 2008a). The treatment can cause cleavage of the conjugated double bonds of the pigments, and ozone and hydroxyl radicals can rupture their aromatic rings and partially oxidize their products (Tiwari, Muthukumarappan, O'Donnell, & Cullen, 2008a; Tiwari, O'Donnell, Muthukumarappan, et al., 2009). The oxidation of the color molecules can follow the Criegee mechanism (Sartori, Angolini, Eberlin, & de Aguiar, 2017). According to this mechanism, ozone forms ozonides with the double bonds of the compounds, and the ozonides can decompose into carbonyl groups and then into carboxyl acids and ketones (Criegee, 1975; Tiwari, O'Donnell, Brunton, & Cullen, 2009).

It was stated that flow rate, treatment time, and concentration of ozone significantly affected the color degradation in orange juice, and the degradation can be estimated using response surface methodology with a Box-Behnken design (Tiwari, Muthukumarappan, O'Donnell, & Cullen, 2008b).

Ozone has been widely applied in the processing of fruits for various purposes (Karaca, 2010; Karaca & Velioglu, 2007). The application of ozone on a fruit can also influence the quality properties of its juice. For instance, the L*, a, and C* values were higher and h* value was lower in açai beverage that was produced from aqueous ozone ($4\,mg\,L^{-1}$ for 10 min) treated fruits compared to those prepared with chloride washing (Bezerra, Freitas-Silva, Damasceno, Mamede, & Cabral, 2017).

5.5.2 Changes in Nutrients and Bioactive Compounds

Ozonation can influence the concentrations of the components (i.e., vitamin C, anthocyanins, carotenoids, phenolics, flavonoids, antioxidant capacity, etc.) of fruits and vegetables depending on the cultivar and the ozone treatment conditions (Alothman, Kaur, Fazilah, Bhat, & Karim, 2010; Horvitz & Cantalejo, 2014; Karaca & Velioglu, 2007). The nutritional properties and bioactive contents of fruit juices can also be affected by ozone (Cullen et al., 2010).

The sucrose content of ozone-treated ($0.5\,L\,min^{-1}$ at 860 ppm for 28 min) apple cider was lower at the end of the 21-day storage period (Choi & Nielsen, 2005). Also, a reduction in sucrose, fructose, citric acid, and malic acid after ozone treatment ($0.057-0.230\,mg\,mL^{-1}$ for 1–6 min) was reported in orange juice (Alves Filho et al., 2016). However, the concentration of β-glucose was increased by ozonation. Similarly, according to Alves Filho et al. (2020) the effect of non-thermal methods, including ozonation on the composition (i.e., anacardic acids, sucrose, tyrosine, and odor profile) of cashew apple juice was lower compared to high temperature short time and ultra-high temperature treatments. Furthermore, It was reported that ozonation ($20\,mg\,L^{-1}$ at $1\,L\,min^{-1}$ for 5 min) in combination with nisin (5000 IU) caused a decrease in ascorbic acid and total phenolic content but maintained the total sugar content (Rajashri, Roopa, Negi, & Rastogi, 2020). However, the total sugar content was decreased during storage. The authors also noted that ozone treatment with nisin did not change the total flavonoid content and the total antioxidant capacity. In contrast, ozone treatment ($0.074-0.296\,mg\,mL^{-1}$ for 2–6 min) was reported to increase the phenolic and vitamin C contents, and the antioxidant capacity in guava, orange, tangerine juices depending on the applied dose (Noguera et al., 2021).

Polyphenols and ascorbic acid in fruit juices can be oxidized and degraded due to the direct and indirect reactions of ozone (Cullen et al., 2010; Patil, Torres, et al., 2010; Tiwari, O'Donnell,

Patras, Brunton, & Cullen, 2009b). The former include the reactions of ozone, and the indirect reactions include reactions of the hydroxyl radicals that are produced due to the decomposition of ozone (Criegee, 1975; Lee et al., 2021; Patil, Torres, et al., 2010; Tiwari, O'Donnell, Brunton, et al., 2009).

5.5.3 Rheological Properties

Rheological behavior and properties are important quality parameters of fruit juices. The ozonation process has been reported to cause changes in apparent viscosity, consistency index, and flow behavior index of fruit juices in several studies (Jaramillo-Sánchez, Garcia Loredo, Gómez, & Alzamora, 2018; Torres et al., 2011). The decrease in apparent viscosity can be associated with the depolymerization of the macromolecules in the juice, degradation of the β-D-glucoside linkages, oxidative damage by the formation of free radicals (such as hydrogen peroxide), and acid hydrolysis (Kabal'Nova et al., 2001; Tiwari, Muthukumarappan, O'Donnell, Chenchaiah, & Cullen, 2008). It was also stated that the trend of rheological behavior of fruit juices, such as apple juice and peach juice, turned from non-Newtonian toward Newtonian after ozonation (Jaramillo-Sánchez et al., 2018; Torres et al., 2011).

5.5.4 Sensory Properties

The sensory properties of food products are amongst the most crucial factors that influence the perception and preference of consumers. It has been stated that ozone can reduce an undesired flavor, color, and odor by causing the degradation of the organic and inorganic compounds (Rakness, 2011). The sensory properties of fruits can be compromised if the applied ozone treatment conditions are not appropriate (Karaca & Velioglu, 2007). Thus, the effects of the treatment on the sensory properties are evaluated and controlled before the implementation of the system.

For instance, the sensory properties of açai beverage prepared with aqueous ozone-treated fruits were not significantly different from those prepared with blanched (80°C, 90°C for 10 s) or chlorinated fruits (Bezerra et al., 2017). The tender coconut water samples that were treated with ozone (20 mg L^{-1} at 1 L min^{-1} for 5 min) and nisin (5000 IU) were sensorily acceptable three weeks after the treatment (Rajashri et al., 2020). Ozone treatment (at 20.2 g m^{-3} for 3 min) demonstrated a negative effect on the sensory color properties of sour cherry juice, but no effects on the flavor, sweetness, sourness, and aftertaste were observed (Akdemir Evrendilek, Keskin, & Golge, 2020).

5.6 ENZYME INACTIVATION BY OZONE

Enzyme activity plays an important role in the quality of fruit juices. For instance, polyphenol oxidase causes brown coloration and degradation of phenols as well as the sensory and nutritional properties of fruits (Queiroz, Mendes Lopes, Fialho, & Valente-Mesquita, 2008; Silva & Gibbs, 2004). In addition, the peroxidase enzyme is linked with lipid oxidation, off-flavor, browning, and degradation of vitamins in fruits (Silva & Gibbs, 2004). It is also highly heat resistant (Silva & Gibbs, 2004). On the other hand, pectin methylesterase activity is required for the degradation of pectins and the decrease of viscosity and cloudiness (Lee et al., 2021; Rodrigo et al., 2006; Tiwari, O'Donnell, Brunton, et al., 2009).

Jaramillo Sánchez, Garcia Loredo, Contigiani, Gómez, and Alzamora (2018) studied the kinetics of enzyme inactivation in peach juice using ozonation. They found the Weibull model useful in explaining the inactivation of peroxidase and polyphenol oxidase enzymes. The authors also reported that the inactivation effect was greater at higher ozone concentration (0.20 mg O_3 min^{-1} mL^{-1}) with longer treatment time (12 min), and peroxidase was more susceptible to ozone than polyphenol oxidase. In addition, Rajashri, Roopa, Negi, and Rastogi (2020) reported that ozonation in combination with nisin treatment decreased the browning index of tender coconut water and inactivated polyphenol oxidase and peroxidase enzymes. The inactivation of the enzymes was higher in samples treated with ozone and nisin compared to the samples treated with ultrasound or conventional heating. A similar finding was also reported by Panigrahi, Mishra, and De, (2020). The authors declared that the ozone treatment (10–30%, 5–20 min, 3–10 L min^{-1}) resulted in the inactivation of polyphenol oxidase (67.8%) and peroxidase (75.3%) enzymes in sugarcane juice. In addition, Porto et al. (2020) claimed that 0.075 mg mL^{-1} ozone treatment was sufficient for the inactivation of peroxidase in coconut water. Moreover, peroxidase and polyphenol oxidase activities were lower in ozone-treated (2–6 min, 0.074–0.296 mg mL^{-1}) guava juice compared to non-processed control samples (Noguera et al., 2021).

Ozonation can be utilized in the processing of sugarcane juice, as it can facilitate the inactivation of the enzymes and control of browning (Garud, Priyanka, Rastogi, Prakash, & Negi, 2018; Panigrahi et al., 2020). Decreases in enzyme activities of polyphenol oxidase (up to 83.24%) and peroxidase (up to 86.03%) in sugarcane juice due to ozone treatment (0.45 mg min^{-1} mL^{-1}, 50 min) were reported (Panigrahi, Vishwakarma, Mishra, & De, 2021). It was also added that the Weibull model was found as the best model that explains the inactivation kinetics of the enzymes.

The effect of ozone treatment (at 600 mg h^{-1} for up to 50 min) on the pectin methylesterase activity in pummelo juice was investigated by Shah, Supian, and Hussein (2019). The authors found that the activity of the enzyme did not change in filtered fruit juice, but it was increased by the ozone treatment time in the unfiltered juice samples. In another study, it was reported that pectin methylesterase activity was increased after ozonation (at 1 L min^{-1} for up to 25 min) of clarified and unclarified watermelon juice (Lee et al., 2021).

5.7 PESTICIDE DEGRADATION BY OZONE

Fruits are treated with pesticides before harvesting in order to protect them from pests and diseases. The residues of the pesticides can remain in the fruit after the harvest and be transferred into the fruit juice (Picó, Font, José Ruiz, & Fernández, 2006). As pesticides are linked to several health issues (Alavanja, Hoppin, & Kamel, 2004; Ntzani, Chondrogiorgi, Ntritsos, Evangelou, & Tzoulaki, 2013), the maximum residue levels in the food are limited (EC, 2006). However, there have been studies that report that the pesticide residues in fruit juices are still higher than the maximum allowed levels, or unallowed pesticides were used (Torović, Vuković, & Dimitrov, 2021). Ozonation is a potential solution for the degradation of pesticide residues when it is applied alone or in combination with other nonthermal methods, such as UV irradiation (Pandiselvam et al., 2020; Wang, Wang, Wang, Li, & Wu, 2019). However, it should be noted that the byproducts of ozonation can be more toxic depending on the original pesticide (Velioglu, Fikirdeşici Ergen, Aksu, & Altındağ, 2018; Verstraeten, Thurman, Lindsey, Lee, & Smith, 2002; Wang et al., 2019). The effectiveness of ozone treatment on the elimination of pesticides from produces can be affected by the concentration, temperature, and treatment time (Pandiselvam et al., 2020).

It was stated that 15 min ozone treatment at 0.4 ppm resulted in a 100% reduction in the carbosulfan concentration in tomato juice that was spiked with the pesticide up to 10 ppm (Al-Antary, Al-Dabbas, & Shaderma, 2015). Akdemir Evrendilek, Keskin, and Golge (2020) investigated the effects of several food processing methods, including ozone, pulsed electric fields, ultrasound, and their combinations, on the pesticide degradation in sour cherry juice. The authors reported that the treatments provided a significant reduction of chlorpyrifos ethyl, τ-fluvalinate, cyprodinil, pyraclostrobin, and malathion. They also added that the ozone-included treatments compromised the color values of the samples.

5.8 COMBINATION OF OZONE TREATMENT WITH OTHER NONTHERMAL METHODS

As stated above, ozone can successfully be applied in fruit juice processing for microbial decontamination, and it can also exhibit desired and undesired effects on other quality parameters. In order to control or improve the influences of the ozone technology on fruit juice quality, possible synergistic effects with other nonthermal technologies have been evaluated.

Del Rosario García-Mateos, Quiroz-González, Corrales-García, Ybarra-Moncada, and Leyva-Ruelas (2019) investigated the combined effects of ozonation (24 mg L^{-1} min^{-1} for 7 min) and high pressure (316 MPa for 5 min) on the quality of pitaya juice. They found that the ozone and high pressure treated juice demonstrated higher sensory acceptability but lower lightness and chroma values, antioxidant activity, total betalains, and phenolic contents after the treatments. The authors also observed 5.1 log reduction in the *Listeria innocua* population.

Ozone technology can also be applied in combination with ultrasound in fruit juice processing. For instance, Oliveira et al. (2018) applied ultrasound-assisted (19 kHz, 350–700 J mL^{-1}) and ozone treatments (1.5 ppm for 5–10 min) in the processing of açai juice. They reported that cloud value was increased by ultrasound treatment and decreased by ozonation. The treatments decreased the antioxidant activity, anthocyanins content, viscosity, and microbial load of the juice. The activity of peroxidase was decreased by the combined and single treatments; however, the polyphenol activity was increased after the combined treatment of ultrasound and ozone. The pH and titratable acidity of the samples were not affected.

In another study, the effects of the combination of ozone (20.2 g m^{-3} dose for 3 min) with pulsed electric fields (500 pps, 24.7 kV cm^{-1}, 60 mL min^{-1}, 327 µs and 655 µs) and ultrasound (35 kHz for 3 min) on sour cherry juice was evaluated (Akdemir Evrendilek et al., 2020). The authors stated that the sensory properties of the samples were not changed due to the treatments except for color. A significant decrease in the antioxidant properties and in the instrumental and sensory color values was observed after the treatments with ozone. The authors also observed significant microbial inactivation and pesticide reduction after the treatments.

5.9 COMMERCIALIZATION OF OZONE

Electrochemical, ultraviolet, or corona discharge systems can be used for ozone generation, but the initial costs for electrochemical and ultraviolet systems are higher compared to corona discharge, and they are not as efficient as the corona system (Miller et al., 2013). The purchasing and installation cost of an ozonation system can be compensated in a short time due to the possible increase in the production yield and reduced cost compared to the systems that require higher operating and maintenance costs while providing products with lower quality (Miller et al., 2013; Rice, 2005). The operating cost for an ozonation system is generally low, and the system also requires less cost for utilities and wastewater treatment (Pascual et al., 2007).

Ozonation can be used for pasteurization, removal of undesired compounds, such as pesticides and mycotoxins, and enzyme inactivation in fruit juice treatment. It can also be applied for the washing and decontamination of the fruits before processing them into the juice. Ozone treatment has been commercially used in the food industry for several purposes, such as disinfection of barrels in wineries, in CIP systems, treatment of drinking water and the water that will be used in food production, odor removal, fruit and vegetable washing and sanitizing, and disinfection of food packaging material and food and food-contact surfaces, etc. (Brodowska et al., 2018; Guzel-Seydim et al., 2004; Loeb, Thompson, Drago, Takahara, & Baig, 2012; Pandiselvam et al., 2019; Pascual et al., 2007; Rice, Graham, & Lowe, 2002). It was also reported that an increasing trend was followed in the installation of ozonation systems (Loeb et al., 2012). Even though there are several commercial alternatives of ozonation equipment in the market for fruit treatment, the commercial application of ozone treatment in fruit juice pasteurization is still emerging and being used by some companies in the US and Europe (Cullen et al., 2010).

The costs for the ozone generating system, ozone contactors, ozone destruction system, discharge ozone quenching system, piping, pumping, valves, electrical and instrumentation system, and storage are the main elements of the direct costs to be considered when estimating the capital cost for an in-plant ozone system according to the United States Environmental Protection Agency (Mundy et al., 2018). A multiplier is then applied in order for the total direct cost to include the engineering and construction costs. The indirect costs (i.e., building, land, operator training, piloting, etc.) should also be taken into account. The authors also provided multipliers, percentages, and unit costs for several elements that were used in calculating the capital and operating costs of an ozone system for municipal water treatment.

5.10 CONCLUSIONS

The demand for non-heat-treated, natural fruit juices is increasing. Ozone is a potential non-thermal alternative for the heat treatment of fruit juices. The application of ozone for microbial safety of fruit juices has given promising results with negligible losses of nutrients, bioactive compounds, or sensory qualities. It is also found effective for mycotoxin and pesticide degradation and enzyme inactivation in fruit juices.

Ozone is a safe, low-cost, and environmentally friendly process. Although the potential of ozone has been investigated in many studies, the adaptation of the ozone technology in the fruit juice processing industry has some challenges. Because every ozone application is unique, all treatment conditions must be optimized for any type of juice for effective and safe use before starting a commercial application.

REFERENCES

Afsah-Hejri, L., Hajeb, P., & Ehsani, R. J. (2020). Application of ozone for degradation of mycotoxins in food: A review. *Comprehensive Reviews in Food Science and Food Safety, 19*(4), 1777–1808. https://doi.org/10.1111/1541-4337.12594.

Akdemir Evrendilek, G., Keskin, E., & Golge, O. (2020). Interaction and multi-objective effects of multiple non-thermal treatments of sour cherry juice: Pesticide removal, microbial inactivation, and quality preservation. *Journal of the Science of Food and Agriculture, 100*(4), 1653–1661. https://doi.org/10.1002/jsfa.10178.

Al-Antary, T. M., Al-Dabbas, M., & Shaderma, A. (2015). Evaluation of three treatments on carbosulfan removal in tomato juice. *Fresenius Environmental Bulletin, 24*(3), 733–739.

Alavanja, M. C. R., Hoppin, J. A., & Kamel, F. (2004). Health effects of chronic pesticide exposure: Cancer and neurotoxicity. *Annual Review of Public Health, 25*(1), 155–197. https://doi.org/10.1146/annurev .publhealth.25.101802.123020.

Almeida, F. D. L., Cavalcante, R. S., Cullen, P. J., Frias, J. M., Bourke, P., Fernandes, F. A. N., & Rodrigues, S. (2015). Effects of atmospheric cold plasma and ozone on prebiotic orange juice. *Innovative Food Science and Emerging Technologies, 32*, 127–135. https://doi.org/10.1016/j.ifset.2015.09.001.

Alothman, M., Kaur, B., Fazilah, A., Bhat, R., & Karim, A. A. (2010). Ozone-induced changes of antioxidant capacity of fresh-cut tropical fruits. *Innovative Food Science and Emerging Technologies, 11*(4), 666–671. https://doi.org/10.1016/j.ifset.2010.08.008.

Alves Filho, E. G., Almeida, F. D. L., Cavalcante, R. S., de Brito, E. S., Cullen, P. J., Frias, J. M., … Rodrigues, S. (2016). 1H NMR spectroscopy and chemometrics evaluation of non-thermal processing of orange juice. *Food Chemistry, 204*, 102–107. https://doi.org/10.1016/j.foodchem.2016.02.121.

Alves Filho, E. G., Silva, L. M. A., Wurlitzer, N. J., Fernandes, F. A. N., Fonteles, T. V., Rodrigues, S., & de Brito, E. S. (2020). An integrated analytical approach based on NMR, LC–MS and GC–MS to evaluate thermal and non-thermal processing of cashew apple juice. *Food Chemistry, 309*, 125761. https://doi.org /10.1016/j.foodchem.2019.125761.

Angelino, P. D. (2002). *Effect of ozone treatment on quality of orange juice.* Knoxville, TN: The University of Tennessee.

Bezerra, V. S., Freitas-Silva, O., Damasceno, L. F., Mamede, A. M. G. N., & Cabral, L. M. C. (2017). Sensory analysis and consumers studies of açaí beverage after thermal, chlorine and ozone treatments of the fruits. *Journal of Food Processing and Preservation, 41*(3), e12961. https://doi.org/10.1111/jfpp.12961.

Britton, H. C., Draper, M., & Talmadge, J. E. (2020). Antimicrobial efficacy of aqueous ozone in combination with short chain fatty acid buffers. *Infection Prevention in Practice, 2*(1), 100032. https://doi.org/10 .1016/j.infpip.2019.100032.

Brodowska, A. J., Nowak, A., & Śmigielski, K. (2018). Ozone in the food industry: Principles of ozone treatment, mechanisms of action, and applications: An overview. *Critical Reviews in Food Science and Nutrition, 58*(13), 2176–2201. https://doi.org/10.1080/10408398.2017.1308313.

Buluc, O., & Koyuncu, M. A. (2020). Effects of intermittent ozone treatment on postharvest quality and storage life of pomegranate. *Ozone: Science and Engineering*, 1–9. https://doi.org/10.1080/01919512 .2020.1816449.

Cataldo, F. (2008). Ozone decomposition of patulin—A micotoxin and food contaminant. *Ozone: Science and Engineering, 30*(3), 197–201. https://doi.org/10.1080/01919510801925930.

Choi, L. H., & Nielsen, S. S. (2005). The effects of thermal and nonthermal processing methods on apple cider quality and consumer acceptability. *Journal of Food Quality, 28*(1), 13–29. https://doi.org/10.1111/j .1745-4557.2005.00002.x.

Criegee, R. (1975). Mechanism of ozonolysis. *Angewandte Chemie International Edition in English, 14*(11), 745–752. https://doi.org/10.1002/anie.197507451.

Cullen, P. J., Valdramidis, V. P., Tiwari, B. K., Patil, S., Bourke, P., & O'Donnell, C. P. (2010). Ozone processing for food preservation: An overview on fruit juice treatments. *Ozone: Science and Engineering, 32*(3), 166–179. https://doi.org/10.1080/01919511003785361.

del Rosario García-Mateos, M., Quiroz-González, B., Corrales-García, J., Ybarra-Moncada, M. C., & Leyva-Ruelas, G. (2019). Ozone-high hydrostatic pressure synergy for the stabilization of refrigerated pitaya (Stenocereus pruinosus) juice. *Innovative Food Science and Emerging Technologies, 56*, 102187. https:// doi.org/10.1016/j.ifset.2019.102187.

Diao, E., Wang, J., Li, X., Wang, X., & Gao, D. (2018). Patulin degradation in apple juice using ozone detoxification equipment and its effects on quality. *Journal of Food Processing and Preservation, 42*(7), e13645. https://doi.org/10.1111/jfpp.13645.

Diao, E., Wang, J., Li, X., Wang, X., Song, H., & Gao, D. (2019). Effects of ozone processing on patulin, phenolic compounds and organic acids in apple juice. *Journal of Food Science and Technology, 56*(2), 957–965. https://doi.org/10.1007/s13197-018-03561-0.

Driedger, A. M., Rennecker, J. L., & Mariñas, B. J. (2001). Inactivation of Cryptosporidium parvum oocysts with ozone and monochloramine at low temperature. *Water Research, 35*(1), 41–48. https://doi.org/10 .1016/S0043-1354(00)00260-8.

Drusch, S., & Ragab, W. (2003). Mycotoxins in fruits, fruit juices, and dried fruits. *Journal of Food Protection*, *66*(8), 1514–1527. https://doi.org/10.4315/0362-028X-66.8.1514.

EC. (2006). Commission regulation (EC) no 178/2006 of 1 February 2006 amending regulation (EC) No 396/2005 of the European Parliament and of the council to establish annex I listing the food and feed products to which maximum levels for pesticide residues apply (1). *Official Journal of the European Union*, *49*, 3–25.

FDA. (2001a). Hazard analysis and critical control point (HACCP); procedures for the safe and sanitary processing and importing of juice. *Federal Register Final Rule*, *66*(13), 6137–6202.

FDA. (2001b). Secondary direct food additives permitted in food for human consumption. *Federal Register*, *66*(123), 153–154.

Freitas-Silva, O., & Venâncio, A. (2010). Ozone applications to prevent and degrade mycotoxins: A review. *Drug Metabolism Reviews*, *42*(4), 612–620. https://doi.org/10.3109/03602532.2010.484461.

Fundo, J. F., Miller, F. A., Tremarin, A., Garcia, E., Brandão, T. R. S., & Silva, C. L. M. (2018). Quality assessment of cantaloupe melon juice under ozone processing. *Innovative Food Science and Emerging Technologies*, *47*, 461–466. https://doi.org/10.1016/j.ifset.2018.04.016.

Garcia Loredo, A. B., Guerrero, S. N., & Alzamora, S. M. (2015). Inactivation kinetics and growth dynamics during cold storage of Escherichia coli ATCC 11229, Listeria innocua ATCC 33090 and Saccharomyces cerevisiae KE162 in peach juice using aqueous ozone. *Innovative Food Science and Emerging Technologies*, *29*, 271–279. https://doi.org/10.1016/j.ifset.2015.02.007.

Garud, S. R., Priyanka, B. S., Rastogi, N. K., Prakash, M., & Negi, P. S. (2018). Efficacy of ozone and lactic acid as nonthermal hurdles for preservation of sugarcane juice. *Ozone: Science and Engineering*, *40*(3), 198–208. https://doi.org/10.1080/01919512.2017.1415802.

Girgin Ersoy, Z., Barisci, S., & Dinc, O. (2019). Mechanisms of the Escherichia coli and Enterococcus faecalis inactivation by ozone. *LWT*, *100*, 306–313. https://doi.org/10.1016/j.lwt.2018.10.095.

Guzel-Seydim, Z. B., Greene, A. K., & Seydim, A. C. (2004). Use of ozone in the food industry. *LWT - Food Science and Technology*, *37*(4), 453–460. https://doi.org/10.1016/j.lwt.2003.10.014.

Horvitz, S., & Cantalejo, M. J. (2014). Application of ozone for the postharvest treatment of fruits and vegetables. *Critical Reviews in Food Science and Nutrition*, *54*(3), 312–339. https://doi.org/10.1080/10408398.2011.584353.

Hunt, N. K., & Mariñas, B. J. (1997). Kinetics of Escherichia coli inactivation with ozone. *Water Research*, *31*(6), 1355–1362. https://doi.org/10.1016/S0043-1354(96)00394-6.

Jamil, A., Farooq, S., & Hashmi, I. (2017). Ozone disinfection efficiency for indicator microorganisms at different pH values and temperatures. *Ozone: Science and Engineering*, *39*(6), 407–416. https://doi.org/10.1080/01919512.2017.1322489.

Jaramillo Sánchez, G. M., Garcia Loredo, A. B., Contigiani, E. V., Gómez, P. L., & Alzamora, S. M. (2018). Inactivation kinetics of peroxidase and polyphenol oxidase in peach juice treated with gaseous ozone. *International Journal of Food Science and Technology*, *53*(2), 347–355. https://doi.org/10.1111/ijfs.13591.

Jaramillo-Sánchez, G. M., Garcia Loredo, A. B., Gómez, P. L., & Alzamora, S. M. (2018). Ozone processing of peach juice: Impact on physicochemical parameters, color, and viscosity. *Ozone: Science and Engineering*, *40*(4), 305–312. https://doi.org/10.1080/01919512.2017.1417111.

Kabal'Nova, N. N., Murinov, K. Y., Mullagaliev, I. R., Krasnogorskaya, N. N., Shereshovets, V. V., Monakov, Y. B., & Zaikov, G. E. (2001). Oxidative destruction of chitosan under the effect of ozone and hydrogen peroxide. *Journal of Applied Polymer Science*, *81*(4), 875–881. https://doi.org/10.1002/app.1506.

Karaca, H. (2010). Use of ozone in the citrus industry. *Ozone: Science and Engineering*, *32*(2), 122–129. https://doi.org/10.1080/01919510903520605.

Karaca, H., & Velioglu, Y. S. (2007). Ozone applications in fruit and vegetable processing. *Food Reviews International*, *23*(1), 91–106. https://doi.org/10.1080/87559120600998221.

Khadre, M. A., Yousef, A. E., & Kim, J.-G. (2001). Microbiological aspects of ozone applications in food: A review. *Journal of Food Science*, *66*(9), 1242–1252. https://doi.org/10.1111/j.1365-2621.2001.tb15196.x.

Kim, J.-G., Yousef, A. E., & Dave, S. (1999). Application of ozone for enhancing the microbiological safety and quality of foods: A review. *Journal of Food Protection*, *62*(9), 1071–1087. https://doi.org/10.4315/0362-028X-62.9.1071.

Kim, J.-G., Yousef, A. E., & Khadre, M. A. (2003). Ozone and its current and future application in the food industry. *Advances in Food and Nutrition Research, 45,* 167–218. https://doi.org/10.1016/S1043-4526(03)45005-5.

Kuprianoff, J. (1953). The use of ozone for the cold storage of fruit. *Z. Kalentechnik, 10,* 1–4.

Labbe, R. G., Kinsley, M., & Wu, J. (2001). Limitations in the use of ozone to disinfect maple sap. *Journal of Food Protection, 64*(1), 104–107. https://doi.org/10.4315/0362-028X-64.1.104.

Lee, B. J., Ting, A. S. Y., & Thoo, Y. Y. (2021). Impact of ozone treatment on the physico-chemical properties, bioactive compounds, pectin methylesterase activity and microbiological properties of watermelon juice. *Journal of Food Science and Technology.* https://doi.org/10.1007/s13197-021-05102-8.

Liu, R. H. (2013). Dietary bioactive compounds and their health implications. *Journal of Food Science, 78*(s1), A18–A25. https://doi.org/10.1111/1750-3841.12101.

Loeb, B. L., Thompson, C. M., Drago, J., Takahara, H., & Baig, S. (2012). Worldwide ozone capacity for treatment of drinking water and wastewater: A review. *Ozone: Science and Engineering, 34*(1), 64–77. https://doi.org/10.1080/01919512.2012.640251.

Miller, F. A., Silva, C. L. M., & Brandão, T. R. S. (2013). A review on ozone-based treatments for fruit and vegetables preservation. *Food Engineering Reviews, 5*(2), 77–106. https://doi.org/10.1007/s12393-013-9064-5.

Mitchell, E. S., Musa-Veloso, K., Fallah, S., Lee, H. Y., Chavez, P. J. De, & Gibson, S. (2020). Contribution of 100% fruit juice to micronutrient intakes in the United States, United Kingdom and Brazil. *Nutrients, 12*(5). https://doi.org/10.3390/nu12051258.

Mundy, B., Kuhnel, B., Hunter, G., Jarnis, R., Funk, D., Walker, S., … Schulz, C. (2018). A review of ozone systems costs for municipal applications. Report by the municipal committee – IOA pan American group. *Ozone: Science and Engineering, 40*(4), 266–274. https://doi.org/10.1080/01919512.2018.1467187.

Muthukumarappan, K., O'Donnell, C. P., & Cullen, P. J. (2009). Ozone treatment of food materials. In S. Jun & J. M. Irudayaraj (Eds.), *Food processing operations modeling design and analysis* (pp. 266–267). Boca Raton, FL: CRC Press.

Naitou, S., & Takahara, H. (2008). Recent developments in food and agricultural uses of ozone as an antimicrobial agent-food packaging film sterilizing machine using ozone. *Ozone: Science and Engineering, 30*(1), 81–87. https://doi.org/10.1080/01919510701813376.

Noguera, N. H., Lima, D. C., Filho, E. G. A., Fonteles, T. V., & Rodrigues, S. (2021). Influence of different non-thermal processing on guava, orange, and tangerine juices and the food matrix effects. *Food and Bioprocess Technology.* https://doi.org/10.1007/s11947-021-02663-6.

Ntzani, E. E., Chondrogiorgi, M., Ntritsos, G., Evangelou, E., & Tzoulaki, I. (2013). *Literature review on epidemiological studies linking exposure to pesticides and health effects.* EFSA Supporting Publication, 2013, EN-497. https://doi.org/10.2903/sp.efsa.2013.EN-497.

Oliveira, A. F. A., Mar, J. M., Santos, S. F., da Silva Júnior, J. L., Kluczkovski, A. M., Bakry, A. M., … Campelo, P. H. (2018). Non-thermal combined treatments in the processing of açai (Euterpe oleracea) juice. *Food Chemistry, 265,* 57–63. https://doi.org/10.1016/j.foodchem.2018.05.081.

Pandiselvam, R., Kaavya, R., Jayanath, Y., Veenuttranon, K., Lueprasitsakul, P., Divya, V., … Ramesh, S. V. (2020). Ozone as a novel emerging technology for the dissipation of pesticide residues in foods – A review. *Trends in Food Science and Technology, 97,* 38–54. https://doi.org/10.1016/j.tifs.2019.12.017.

Pandiselvam, R., Subhashini, S., Banuu Priya, E. P., Kothakota, A., Ramesh, S. V., & Shahir, S. (2019). Ozone based food preservation: A promising green technology for enhanced food safety. *Ozone: Science and Engineering, 41*(1), 17–34. https://doi.org/10.1080/01919512.2018.1490636.

Pandiselvam, R., Sunoj, S., Manikantan, M. R., Kothakota, A., & Hebbar, K. B. (2017). Application and kinetics of ozone in food preservation. *Ozone: Science and Engineering, 39*(2), 115–126. https://doi.org/10.1080/01919512.2016.1268947.

Panigrahi, C., Mishra, H. N., & De, S. (2020). Effect of ozonation parameters on nutritional and microbiological quality of sugarcane juice. *Journal of Food Process Engineering, 43*(11), e13542. https://doi.org/10.1111/jfpe.13542.

Panigrahi, C., Vishwakarma, S., Mishra, H. N., & De, S. (2021). Kinetic modeling for inactivation of polyphenoloxidase and peroxidase enzymes during ozonation of sugarcane juice. *Journal of Food Processing and Preservation, 45*(1), e15094. https://doi.org/10.1111/jfpp.15094.

Pascual, A., Llorca, I., & Canut, A. (2007). Use of ozone in food industries for reducing the environmental impact of cleaning and disinfection activities. *Trends in Food Science and Technology*, *18*, S29–S35. https://doi.org/10.1016/j.tifs.2006.10.006.

Patil, S., Bourke, P., Frias, J. M., Tiwari, B. K., & Cullen, P. J. (2009). Inactivation of Escherichia coli in orange juice using ozone. *Innovative Food Science and Emerging Technologies*, *10*(4), 551–557. https://doi.org /10.1016/j.ifset.2009.05.011.

Patil, S., Cullen, P. J., Kelly, B., Frías, J. M., & Bourke, P. (2009). Extrinsic control parameters for ozone inactivation of Escherichia coli using a bubble column. *Journal of Applied Microbiology*, *107*(3), 830–837. https://doi.org/10.1111/j.1365-2672.2009.04255.x.

Patil, S., Torres, B., Tiwari, B. K., Wijngaard, H. H., Bourke, P., Cullen, P. J., ... Valdramidis, V. P. (2010). Safety and quality assessment during the ozonation of cloudy apple juice. *Journal of Food Science*, *75*(7), M437–M443. https://doi.org/10.1111/j.1750-3841.2010.01750.x.

Patil, S., Valdramidis, V. P., Cullen, P. J., Frias, J., & Bourke, P. (2010). Inactivation of Escherichia coli by ozone treatment of apple juice at different pH levels. *Food Microbiology*, *27*(6), 835–840. https://doi .org/10.1016/j.fm.2010.05.002.

Petruzzi, L., Campaniello, D., Speranza, B., Corbo, M. R., Sinigaglia, M., & Bevilacqua, A. (2017). Thermal treatments for fruit and vegetable juices and beverages: A literature overview. *Comprehensive Reviews in Food Science and Food Safety*, *16*(4), 668–691. https://doi.org/10.1111/1541-4337.12270.

Picó, Y., Font, G., José Ruiz, M., & Fernández, M. (2006). Control of pesticide residues by liquid chromatography-mass spectrometry to ensure food safety. *Mass Spectrometry Reviews*, *25*(6), 917–960. https:// doi.org/10.1002/mas.20096.

Porto, E., Alves Filho, E. G., Silva, L. M. A., Fonteles, T. V., do Nascimento, R. B. R., Fernandes, F. A. N., ... Rodrigues, S. (2020). Ozone and plasma processing effect on green coconut water. *Food Research International*, *131*, 109000. https://doi.org/10.1016/j.foodres.2020.109000.

Queiroz, C., Mendes Lopes, M. L., Fialho, E., & Valente-Mesquita, V. L. (2008). Polyphenol oxidase: Characteristics and mechanisms of browning control. *Food Reviews International*, *24*(4), 361–375. https://doi.org/10.1080/87559120802089332.

Rajashri, K., Roopa, B. S., Negi, P. S., & Rastogi, N. K. (2020). Effect of ozone and ultrasound treatments on polyphenol content, browning enzyme activities, and shelf life of tender coconut water. *Journal of Food Processing and Preservation*, *44*(3), e14363. https://doi.org/10.1111/jfpp.14363.

Rakness, K. L. (2011). *Ozone in drinking water treatment: Process design, operation, and optimization*. Denver, CO: American Water Works Association.

Restaino, L., Frampton, E. W., Hemphill, J. B., & Palnikar, P. (1995). Efficacy of ozonated water against various food-related microorganisms. *Applied and Environmental Microbiology*, *61*(9), 3471–3475. https:// doi.org/10.1128/aem.61.9.3471-3475.1995.

Rice, R. G. (1986). Application of ozone in water and waste water treatment. In R. G. Rice & M. J. Browning (Eds.), *Analytical aspects of ozone treatment of water and waste water* (pp. 7–26). Syracuse, NY: The Institute.

Rice, R. G. (2005). IOA-PAG user success reports—Commercial applications of ozone in agri-foods. *International ozone association pan American group conference*, Georgia, USA, 9–12.

Rice, R. G., Graham, D. M., & Lowe, M. T. (2002). Recent ozone applications in food processing and sanitation. *Food Safety Magazine*, *8*(5), 10–17.

Rodrigo, D., Cortés, C., Clynen, E., Schoofs, L., Loey, A. Van, & Hendrickx, M. (2006). Thermal and high-pressure stability of purified polygalacturonase and pectinmethylesterase from four different tomato processing varieties. *Food Research International*, *39*(4), 440–448. https://doi.org/10.1016/j.foodres .2005.09.007.

Rubin, M. B. (2001). The history of ozone: The Schönbein period, 1839–1868. *Bulletin for the History of Chemistry*, *26*(1), 40–56.

Sartori, J. A. D., Angolini, C. F. F., Eberlin, M. N., & de Aguiar, C. L. (2017). Criegee mechanism as a safe pathway of color reduction in sugarcane juice by ozonation. *Food Chemistry*, *225*, 181–187. https://doi.org /10.1016/j.foodchem.2017.01.028.

Seo, S., King, J. M., & Prinyawiwatkul, W. (2007). Simultaneous depolymerization and decolorization of chitosan by ozone treatment. *Journal of Food Science*, *72*(9), C522–C526. https://doi.org/10.1111/j.1750 -3841.2007.00563.x.

Shah, N. N. A. K., Ong, M. C., Supian, N. M. A., & Sulaiman, A. (2020). Ultrasound-assisted extraction and gaseous ozone as preservation method for Roselle (Hibiscus sabdariffa L.) fruit juice. *Food Research*, *4*(S6), 85–95.

Shah, N. N. A. K., Sulaiman, A., Sidek, N. S. M., & Supian, N. A. M. (2019). Quality assessment of ozone-treated citrus fruit juices. *International Food Research Journal*, *26*(5), 1405–1415.

Shah, N. N. A. K., Supian, N. A. M., & Hussein, N. A. (2019). Disinfectant of pummelo (Citrus grandis L. Osbeck) fruit juice using gaseous ozone. *Journal of Food Science and Technology*, *56*(1), 262–272. https://doi.org/10.1007/s13197-018-3486-2.

Silva, F. V. M., & Gibbs, P. (2004). Target selection in designing pasteurization processes for shelf-stable high-acid fruit products. *Critical Reviews in Food Science and Nutrition*, *44*(5), 353–360. https://doi .org/10.1080/10408690490489251.

Sroy, S., Fundo, J. F., Miller, F. A., Brandão, T. R. S., & Silva, C. L. M. (2019). Impact of ozone processing on microbiological, physicochemical, and bioactive characteristics of refrigerated stored Cantaloupe melon juice. *Journal of Food Processing and Preservation*, *43*(12), e14276. https://doi.org/10.1111/jfpp.14276.

Sung, H.-J., Song, W.-J., Kim, K.-P., Ryu, S., & Kang, D.-H. (2014). Combination effect of ozone and heat treatments for the inactivation of Escherichia coli O157:H7, Salmonella Typhimurium, and Listeria monocytogenes in apple juice. *International Journal of Food Microbiology*, *171*, 147–153. https://doi.org/10 .1016/j.ijfoodmicro.2013.11.001.

Tiwari, B. K., Muthukumarappan, K., O'Donnell, C. P., Chenchaiah, M., & Cullen, P. J. (2008). Effect of ozonation on the rheological and colour characteristics of hydrocolloid dispersions. *Food Research International*, *41*(10), 1035–1043. https://doi.org/10.1016/j.foodres.2008.07.011.

Tiwari, B. K., Muthukumarappan, K., O'Donnell, C. P., & Cullen, P. J. (2008a). Kinetics of freshly squeezed orange juice quality changes during ozone processing. *Journal of Agricultural and Food Chemistry*, *56*(15), 6416–6422. https://doi.org/10.1021/jf800515e.

Tiwari, B. K., Muthukumarappan, K., O'Donnell, C. P., & Cullen, P. J. (2008b). Modelling colour degradation of orange juice by ozone treatment using response surface methodology. *Journal of Food Engineering*, *88*(4), 553–560. https://doi.org/10.1016/j.jfoodeng.2008.03.021.

Tiwari, B. K., O'Donnell, C. P., Brunton, N., & Cullen, P. J. (2009). Degradation kinetics of tomato juice quality parameters by ozonation. *International Journal of Food Science and Technology*, *44*(6), 1199–1205. https://doi.org/10.1111/j.1365-2621.2009.01946.x.

Tiwari, B. K., O'Donnell, C. P., Muthukumarappan, K., & Cullen, P. J. (2009). Anthocyanin and colour degradation in ozone treated blackberry juice. *Innovative Food Science and Emerging Technologies*, *10*(1), 70–75. https://doi.org/10.1016/j.ifset.2008.08.002.

Tiwari, B. K., O'Donnell, C. P., Patras, A., Brunton, N., & Cullen, P. J. (2009a). Anthocyanins and color degradation in ozonated grape juice. *Food and Chemical Toxicology*, *47*(11), 2824–2829. https://doi.org/10 .1016/j.fct.2009.09.001.

Tiwari, B. K., O'Donnell, C. P., Patras, A., Brunton, N., & Cullen, P. J. (2009b). Effect of ozone processing on anthocyanins and ascorbic acid degradation of strawberry juice. *Food Chemistry*, *113*(4), 1119–1126. https://doi.org/10.1016/j.foodchem.2008.08.085.

Torlak, E. (2014). Efficacy of ozone against Alicyclobacillus acidoterrestris spores in apple juice. *International Journal of Food Microbiology*, *172*, 1–4. https://doi.org/10.1016/j.ijfoodmicro.2013.11.035.

Torović, L., Vuković, G., & Dimitrov, N. (2021). Pesticide residues in fruit juice in Serbia: Occurrence and health risk estimates. *Journal of Food Composition and Analysis*, *99*, 103889. https://doi.org/10.1016/j .jfca.2021.103889.

Torres, B., Tiwari, B. K., Patras, A., Wijngaard, H. H., Brunton, N., Cullen, P. J., & O'Donnell, C. P. (2011). Effect of ozone processing on the colour, rheological properties and phenolic content of apple juice. *Food Chemistry*, *124*(3), 721–726. https://doi.org/10.1016/j.foodchem.2010.06.050.

Velioglu, Y., Fikirdeşici Ergen, Ş., Aksu, P., & Altındağ, A. (2018). Effects of ozone treatment on the degradation and toxicity of several pesticides in different groups. *Journal of Agricultural Sciences*, *24*, 245–255. https://doi.org/10.15832/ankutbd.446448.

Verstraeten, I. M., Thurman, E. M., Lindsey, M. E., Lee, E. C., & Smith, R. D. (2002). Changes in concentrations of triazine and acetamide herbicides by bank filtration, ozonation, and chlorination in a public water supply. *Journal of Hydrology*, *266*(3), 190–208. https://doi.org/10.1016/S0022-1694(02)00163-4.

Vojdani, J. D., Beuchat, L. R., & Tauxe, R. V. (2008). Juice-associated outbreaks of human illness in the United States, 1995 through 2005. *Journal of Food Protection, 71*(2), 356–364. https://doi.org/10.4315/0362-028X-71.2.356.

Wang, S., Wang, J., Wang, T., Li, C., & Wu, Z. (2019). Effects of ozone treatment on pesticide residues in food: A review. *International Journal of Food Science and Technology, 54*(2), 301–312. https://doi.org/10.1111/ijfs.13938.

Williams, R. C., Sumner, S. S., & Golden, D. A. (2004). Survival of Escherichia coli O157:H7 and Salmonella in apple cider and orange juice as affected by ozone and treatment temperature. *Journal of Food Protection, 67*(11), 2381–2386. https://doi.org/10.4315/0362-028X-67.11.2381.

Zuma, F., & Jonnalagadda, S. B. (2010). Studies on the O3-initiated disinfection from Gram-positive bacteria Bacillus subtilis in aquatic systems. *Journal of Environmental Science and Health, Part A, 45*(2), 224–232. https://doi.org/10.1080/10934520903429931.

Zuma, F., Lin, J., & Jonnalagadda, S. B. (2009). Ozone-initiated disinfection kinetics of Escherichia coli in water. *Journal of Environmental Science and Health, Part A, 44*(1), 48–56. https://doi.org/10.1080/10934520802515335.

Zurbau, A., Au-Yeung, F., Blanco Mejia, S., Khan, T. A., Vuksan, V., Jovanovski, E., … Sievenpiper, J. L. (2020). Relation of different fruit and vegetable sources with incident cardiovascular outcomes: A systematic review and meta-analysis of prospective cohort studies. *Journal of the American Heart Association, 9*(19), e017728. https://doi.org/10.1161/JAHA.120.017728.

Chapter 6

Supercritical CO$_2$ Extraction of Phytoconstituents from Fruits and Vegetables

M. Selvamuthukumaran

CONTENTS

6.1 INTRODUCTION

Supercritical fluid extraction is one form of green technology used in order to effectively extract several phytoconstituents, especially from fruit- and vegetable-based products. This technology is very simple and does not involve such a tedious operation. The major advantages of using this technique are that it does not involve higher temperatures for extraction and it is eco-friendly i.e. it will not pollute the environment (Baldino et al., 2017 (Figure 6.1). This technique can be merged with a mechanical extraction process in order to achieve maximum extraction efficiency (Misra et al., 2017). This technology can be successfully utilized to extract antioxidants from the by-products of fruit processing (Kelly et al., 2019).

6.2 ABOUT THE SUPERCRITICAL FLUID EXTRACTION (SFE)

Supercritical fluid can exhibit characteristics of liquid cum solid above the critical temperature and pressure. Because of this nature, it possesses greater diffusivity with lesser viscosity than ordinary solvents, which enhances the solvent penetration power via the solid matrix, which can further increase the yield of bioactive constituent extraction (Ameer et al., 2017; Gallego et al., 2019). The important notable properties of this fluid are a density that can be modified by altering either temperature or pressure, which in turn can help in the efficient extraction of phytoconstituents to a greater extent (Pimentel-Moral et al., 2019). The unique solvent selection

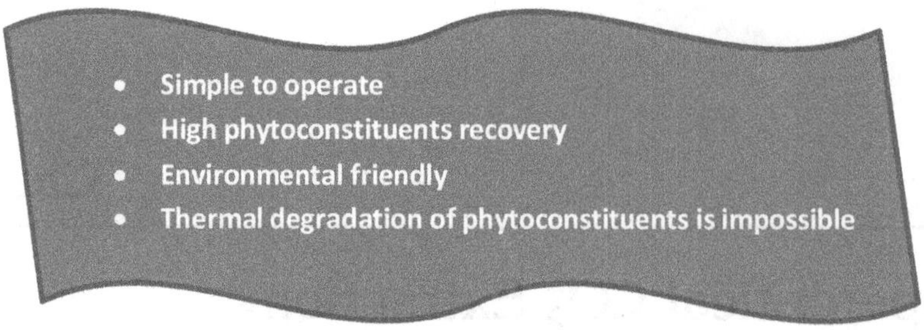

- Simple to operate
- High phytoconstituents recovery
- Environmental friendly
- Thermal degradation of phytoconstituents is impossible

Figure 6.1   Benefits of using the supercritical fluid extraction process.

for specific extraction is very much essential for its use in supercritical fluid and therefore one should place emphasis on parameters like toxicity level, cost, power of solvation, and critical conditions of temperature and pressure (Pereira and Meireles, 2010). There are several advantages that are associated with the use of CO_2 in supercritical fluid extraction, which include non-corrosive properties, low cost, easy availability, recyclability, and its ecofriendly nature (Wrona et al., 2017). In addition to the above properties, it is non-polar in nature, it is tasteless as well as being an odorless gas, which is non-combustible with a lesser critical value i.e. a pressure of around 73.8 bars and a temperature of around 31.1°C. Therefore, these good properties make it convenient for extraction, especially of thermolabile bioactive components in a safe mode (Pimentel-Moral et al., 2019; Da Silva et al., 2016).

Herrero et al. (2006) reported that the use of CO_2 has immense benefits, as it is gas even at room temperature and pressure; its removal after extraction is relatively easy by decompressing the entire system and it further helps to achieve chemical solvent free extracts from the sample. The only drawback observed in this extraction is it is very difficult to extract polar-based phytoconstituents, but this can be rectified by incorporating some modifiers like ethanol during extraction, which can enhance the solubilizing power of CO_2 thereby enhancing the extraction efficiency of several polar components (Pimentel-Moral et al., 2019; Wrona et al., 2017; Ameer et al., 2017). The operation parameters finally determine the efficiency of extraction of phytoconstituents from fruit and vegetable products. The various processing parameters, which include pressure, temperature, time of extraction, composition of modifier used during extraction, mass transfer rate, size of the particle, solid solvent ratio, flow rate of solvent etc. (Tita et al., 2021). Therefore, in order to achieve maximum extraction efficiency one has to standardize the above critical parameters, so that the yield of bioactive constituents obtained from fruits and vegetables can be enhanced to a greater extent.

6.3 EXTRACTION OF PHYTOCONSTITUENTS FROM FRUIT BY-PRODUCTS

The supercritical fluid extraction technique is one opt extraction process used for extracting phytoconstituents from fruit-based by-products (Table 6.1). The compound is successfully extracted by using this technique, which includes essential oil and phenolic components. Casas et al. (2010) adopted this method in order to extract resveratrol from the grape pomace – they found that resveratrol solubility was enhanced as a result of co-solvent addition and they obtained good extraction at a temperature of 35–55°C with a pressure of 10–40 MPa.

TABLE 6.1   EXTRACTION OF PHYTOCONSTITUENTS FROM FRUIT-BASED BY-PRODUCTS BY USING THE SUPERCRITICAL FLUID EXTRACTION PROCESS

Name of the Material	Name of the Phytoconstituents Extracted	References
Grape pomace	Resveratrol	Casas et al. (2010)
Grape seed	Gallic acid, catechin, epicatechin	Marqués, et al. (2013)
Grape bagasse	Vanillic acid, quercetin, p-coumaric acid, p-hydroxybenzene	Farías-Campomanes et al. (2013)
Grape marc	Polyphenol	Da Porto et al. (2015)
Orange pomace	Polyphenol	Espinosa-pardo et al. (2016)
Apple pomace-Freeze dried	Polyphenol	Ferrentino et al. (2018)
Mango peel	Carotene	Sánchez-camargo et al. (2019)
Elderberry	Polyunsaturated linoleic acid, α-linolenic fatty acid	Kitryt et al. (2020)

Certain antioxidants like epicatechin, gallic acid, resveratrol, and catechin were successfully extracted from grape seed by using this technique (Marqués et al., 2013). The recovered phytoconstituents, which include catechin of 49 mg, epicatechin of 53 mg, resveratrol of 667 mg, gallic acid of 51 mg per kg of grape seed with a temperature of 40°C, pressure range of 15 MPa, with CO$_2$ molar fractions of 0.98. The residue left after extraction possesses a high amount of phenols ranging from 15 to 22g equivalent gallic acid per kg of seed and this residue can be used for the development of various functional food products, which contain higher dietary fiber.

From grape bagasse several phenolic components were extracted with the help of the SFE technique, which includes vanillic acid, gallic acid, syringic acid, quercetin, formic acid, p-coumaric acid, p-hydroxybenzene (Farías-Campomanes et al., 2013). They demonstrated that the target compounds were extracted using the SFE technique with a pressure of 20 MPa and a temperature of 40°C. The supercritical fluid extraction methods were merged with another extraction technique, which can further enhance the yield of phytoconstituents. Passos et al. (2009) extracted the components by using enzymatic treatment coupled with the SFE process – they found that a combination of both processes enhanced the extraction rate. Similar research was carried out by Da Porto et al. (2015) in the extraction of polyphenols from grape marc, they found that a combination of ultrasound-assisted extraction-cum-SFE processes can further enhance the polyphenol recovery and antioxidant activity by 28 and 62%.

The SFE technique was also employed to extract biological active constituents from by-products of orange processing. The biologically active constituents were obtained from orange fruit pomace by using this SFE technique and further the extraction efficiency was compared with soxhlet-, hydrodistillation- and ultrasonic-assisted extraction processes (Benelli et al., 2010). Their work confirmed that the phytoconstituents extracted by using SFE technique exhibited higher antioxidant-cum-antibacterial properties. Similar research was also confirmed by Espinosa-Pardo et al. (2016) in phenolic component extraction from orange fruit pomace. They proved that SFE helps to save time, energy, and as well as lowering the consumption of solvent it may require less time compared to traditional methods.

Antioxidants like polyphenols, carotene, and tocopherols were extracted by using the SFE technique on fruit by-products like apple, mango, and passion fruit (Sánchez-Camargo et al.,

2011; Ferrentino et al., 2018; Hatami et al., 2020). Ferrentino et al. (2018) implemented this SFE technique for phenolic constituent extraction from freeze-dried apple pomace. They used extraction parameter conditions i.e. a pressure of 30 MPa and a temperature of 45°C, which lead to the extraction of higher antioxidant activity compared to traditional solvent methods of extraction: the extraction efficiency achieved was around three times higher than the traditional method. Carotenoids were successfully isolated from mango peel by using the SFE technique (Sánchez-Camargo et al., 2019). Kitryt et al. (2020) used a pressure of 35 MPa and a temperature and time of 53°C and 45 min in order to extract various lipophilic components from pomace of elderberry. They recovered around 14 g of such constituents from 100 g of pomace. Polyunsaturated linoleic acid of around 42% and α-linolenic fatty acids of around 34.1% were extracted from this pomace. They found that the SFE technique leads to a higher yield in less time when compared to the traditional soxhlet extraction process. The yield enhancement of around 4% with an eight-fold reduced extraction time was obtained when compared to the traditional extraction process.

6.4 EXTRACTION OF PHYTOCONSTITUENTS FROM VEGETABLE BY-PRODUCTS

Lycopene, one of the antioxidant phytoconstituents obtained from tomato, can be successfully extracted by using the supercritical fluid extraction technique (Table 6.2). The processing parameter conditions viz. pressure and temperature in the range of 30–50 MPa and 50–80°C with a flow rate of around 3–5 g CO_2/min for a time period of 105 min was successfully utilized for this SFE technique (Kehili et al., 2017). The authors also used a conventional method of extraction of lycopene with various solvents, which includes ethanol, ethyl acetate, and hexane; they finally concluded that maximum lycopene yield recovery was obtained by using the SFE technique under processing parameter conditions viz. pressure of 40 MPa, a temperature of 80°C, and CO_2 of 4g/min. Under these conditions the lycopene yield obtained was 729 mg/kg of dry tomato peels, while the convention method of extraction gave only 285 to 608 mg lycopene/kg of dry tomato peels. There was a higher concentration of lycopene: 90% was obtained with a processing condition with a temperature of around 80°C and a pressure of 46 MPa (Vagi et al., 2007)

Devani et al. (2020) implemented CCRD (central composite rotatable design) to extract the oleoresin, pyruvate, and sulfur content from the rotten waste of onion, the temperature used

TABLE 6.2   EXTRACTION OF PHYTOCONSTITUENTS FROM VEGETABLE-BASED BY-PRODUCTS BY USING THE SUPERCRITICAL FLUID EXTRACTION PROCESS

Name of the Material	Name of the Phytoconstituents Extracted	References
Tomato	Lycopene	Kehili et al. (2017); Vagi et al. (2007)
Onion rotten waste	Oleoresin, pyruvate, and sulfur content	Devani et al. (2020)
Brown onion	Flavonoids	Campone et al. (2018)
Soybean oilcake	Isoflavones	Kao et al. (2008)
Soybean oilcake	Tocopherols	Fang et al. (2007)
Soybean oilcake	Flavonoids and polyphenols	Alvarez et al. (2019)

was 50–90°C, a pressure of 15 to 45 MPa, and extraction time ranges from 30 to 150 min. The authors obtained maximum extraction efficiency by using this technique i.e. oleoresin content of 1.01%, sulfur content of 31 g per kg of oleoresin, in addition to 10.4 micromolar pyruvate per gram of onion fresh weight that was obtained and this extraction was further compared with the traditional extraction process. Their results confirmed that the yield of extractants had been increased to around twice that of the traditional extraction process. Campone et al. (2018) extracted flavonoids from peels of brown onion by using ethanol as an extraction solvent with a pressure of 10 MPa and a temperature of 40°C, and achieved higher quercetin derivatives.

This SFE technique seems to be the most preferred technique for the recovery of bioactive constituents from soybean-based by-products. Isoflavones were successfully extracted from a pressed oil cake of soybean (Kao et al., 2008). The authors demonstrated that the SFE technique leads to extraction of higher amounts of aglycone as well as acetylglucoside with an operating pressure of 350 bar and a temperature of 80°C.

Fang et al. (2007) concentrated tocopherols from the oil deodorizer distillate, they achieved higher tocopherol recovery at 20 MPa, and therefore the SFE technique can be widely employed for the concentration of tocopherols from by-products obtained from the soybean oil refining industry. Alvarez et al. (2019) obtained higher flavonoids and polyphenols from by-products of soybean oil processing by using ethanol at 25% as a co-solvent, and they achieved good extraction efficiency with a processing condition i.e. a pressure of 40 Mpa, a temperature of 35°C, and by using CO_2 solvent of 5 kg CO_2/kg expeller. The conclusion was that SFE is one of the safest as well as one of the greenest extraction techniques in order to obtain several phytoconstituents from soybean expellers.

6.5 CONCLUSION

Supercritical fluid extraction technique helps to extract several phytoconstituents from various by-products of fruits and vegetables. This technique helps to further extract enhanced phytoconstituents recovery compared to traditional extraction process. The technique needs only an opt process parameter standardization when it is being applied on a wide scale.

REFERENCES

Alvarez, M.V.; Cabred, S.; Ramirez, C.L.; Fanovich, M.A. Valorization of an agroindustrial soybean residue by supercritical fluid extraction of phytochemical compounds. *J. Supercrit. Fluids* 2019, 143, 90–96.

Ameer, K.; Shahbaz, H.M.; Kwon, J.H. Green extraction methods for polyphenols from plant matrices and their byproducts: A review. *Compr. Rev. Food Sci. Food Saf.* 2017, 16(2), 295–315.

Baldino, L.; Della Porta, G.; Reverchon, E. Supercritical CO₂ processing strategies for pyrethrins selective extraction. *J. CO2 Util.* 2017, 20, 14–19.

Benelli, P.; Riehl, C.A.S.; Smânia, A.; Smânia, E.F.A.; Ferreira, S.R.S. Bioactive extracts of orange (Citrus sinensis L. Osbeck) pomace obtained by SFE and low pressure techniques: Mathematical modeling and extract composition. *J. Supercrit. Fluids* 2010, 55(1), 132–141.

Campone, L.; Celano, R.; Lisa, A.; Pagano, I.; Carabetta, S.; Di, R.; Russo, M.; Ibañez, E.; Cifuentes, A.; Rastrelli, L. Response surface methodology to optimize supercritical carbon dioxide/co-solvent extraction of brown onion skin by-product as source of nutraceutical compounds. *Food Chem.* 2018, 269, 495–502.

Casas, L.; Mantell, C.; Rodríguez, M.; De Ossa, E.J.M.; Roldán, A.; De Ory, I.; Caro, I.; Blandino, A. Extraction of resveratrol from the pomace of Palomino fino grapes by supercritical carbon dioxide. *J. Food Eng.* 2010, 96(2), 304–308.

Da Porto, C.; Natolino, A.; Decorti, D. The combined extraction of polyphenols from grape marc: Ultrasound assisted extraction followed by supercritical CO_2 extraction of ultrasound-raffinate. *LWT Food Sci. Technol.* 2015, 61(1), 98–104.

Da Silva, R.P.F.F.; Rocha-Santos, T.A.P.; Duarte, A.C. Supercritical fluid extraction of bioactive compounds. *TrAC Trends Anal. Chem.* 2016, 76, 40–51.

Devani, B.M.; Jani, B.L.; Balani, P.C.; Akbari, S.H. Optimization of supercritical CO_2 extraction process for oleoresin from rotten onion waste. *Food Bioprod. Process.* 2020, 119, 287–295.

Espinosa-pardo, F.A.; Mayumi, V.; Alves, G.; Alves, J.; Martínez, J. Extraction of phenolic compounds from dry and fermented orange pomace using supercritical CO_2 and cosolvents. *Food Bioprod. Process.* 2016, 101, 1–10.

Fang, T.; Goto, M.; Wang, X.; Ding, X.; Geng, J.; Sasaki, M.; Hirose, T. Separation of natural tocopherols from soybean oil byproduct with supercritical carbon dioxide. *J. Supercrit. Fluids* 2007, 40(1), 50–58.

Farías-Campomanes, A.M.; Rostagno, M.A.; Meireles, M.A.A. Production of polyphenol extracts from grape bagasse using supercritical fluids: Yield, extract composition and economic evaluation. *J. Supercrit. Fluids* 2013, 77, 70–78.

Ferrentino, G.; Morozova, K.; Mosibo, O.K.; Ramezani, M.; Scampicchio, M. Biorecovery of antioxidants from apple pomace by supercritical fluid extraction. *J. Clean. Prod.* 2018, 186, 253–261.

Gallego, R.; Bueno, M.; Herrero, M. Sub- and supercritical fluid extraction of bioactive compounds from plants, food-by-products, seaweeds and microalgae – An update. *TrAC Trends Anal. Chem.* 2019, 116, 198–213.

Hatami, T.; Cristina, L.; Leone, G.; Virginia, P.; Pontes, D.A.; Augusto, E.; Batista, C.; Helena, L.; Mei, I.; Martínez, J. Integrated supercritical extraction and supercritical adsorption processes from passion fruit by-product: Experimental and economic analyses. *J. Supercrit. Fluids* 2020, 162, 162.

Herrero, M.; Cifuentes, A.; Ibañez, E. Sub- and supercritical fluid extraction of functional ingredients from different natural sources: Plants, food-by-products, algae and microalgae-A review. *Food Chem.* 2006, 98(1), 136–148.

Kao, T.; Chien, J.; Chen, B. Extraction yield of isoflavones from soybean cake as affected by solvent and supercritical carbon dioxide. *Food Chem.* 2008, 107(4), 1728–1736.

Kehili, M.; Kammlott, M.; Choura, S.; Zammel, A.; Zetzl, C.; Smirnova, I.; Allouche, N.; Sayadi, S. Supercritical CO_2 extraction and antioxidant activity of lycopene and β-carotene-enriched oleoresin from tomato (Lycopersicum esculentum L.) peels by-product of a Tunisian industry. *Food Bioprod. Process.* 2017, 102, 340–349.

Kelly, N.P.; Kelly, A.L.; O'Mahony, J.A. Strategies for enrichment and purification of polyphenols from fruit-based materials. *Trends Food Sci. Technol.* 2019, 83, 248–258.

Kitryt, V.; Syrpas, M.; Pukalskas, A.; Pukalskas, A.; Venskutonis, P.R. Modeling and optimization of super-critical carbon dioxide extraction for isolation of valuable lipophilic constituents from elderberry (Sambucus nigra L.) pomace. *J. CO2 Util.* 2020, 35, 225–235.

Marqués, J.L.; Della Porta, G.; Reverchon, E.; Renuncio, J.A.R.; Mainar, A.M. Supercritical antisolvent extraction of antioxidants from grape seeds after vinification. *J. Supercrit. Fluids* 2013, 82, 238–243.

Misra, N.N.; Koubaa, M.; Roohinejad, S.; Juliano, P.; Alpas, H.; Inàcio, R.S.; Saraiva, J.A.; Barba, F.J. Landmarks in the historical development of twenty first century food processing technologies. *Food Res. Int.* 2017, 97, 318–339.

Passos, C.P.; Silva, R.M.; Da, F.A.; Coimbra, M.A.; Silva, C.M. Enhancement of the supercritical fluid extrac-tion of grape seed oil by using enzymatically pre-treated seed. *J. Supercrit. Fluids* 2009, 48(3), 225–229.

Pereira, C.G.; Meireles, M.A.A. Supercritical fluid extraction of bioactive compounds: Fundamentals, appli-cations and economic perspectives. *Food Bioprocess Technol.* 2010, 3(3), 340–372.

Pimentel-Moral, S.; Borrás-Linares, I.; Lozano-Sánchez, J.; Arráez-Román, D.; Martínez-Férez, A.; Segura-Carretero, A. Supercritical CO_2 extraction of bioactive compounds from Hibiscus sabdariffa. *J. Supercrit. Fluids* 2019, 147, 213–221.

Sánchez-Camargo, A.P.; Martinez-Correa, H.A.; Paviani, L.C.; Cabral, F.A. Supercritical CO_2 extraction of lipids and astaxanthin from Brazilian redspotted shrimp waste (Farfantepenaeus paulensis). *J. Supercrit. Fluids* 2011, 56, 164–173.

Sánchez-Camargo, P.; Gutiérrez, L.; Milena, S.; Martinez-correa, H.A.; Parada-alfonso, F.; Narváez-cuenca, C. Valorisation of mango peel: Proximate composition, supercritical fluid extraction of carotenoids, and application as an antioxidant additive for an edible oil. *J. Supercrit. Fluids* 2019, 152, 104574.

Tita, G.J.; Navarrete, A.; Martín, Á.; Cocero, M.J. Model assisted supercritical fluid extraction and fractionation of added-value products from tobacco scrap. *J. Supercrit. Fluids* 2021, 167, 105046.

Vági, E.; Simándi, B.; Vásárhelyiné, K.P.; Daood, H.; Kéry, Á.; Doleschall, F.; Nagy, B. Supercritical carbon dioxide extraction of carotenoids, tocopherols and sitosterols from industrial tomato by-products. *J. Supercrit. Fluids* 2007, 40(2), 218–226.

Wrona, O.; Rafińska, K.; Możeński, C.; Buszewski, B. Supercritical fluid extraction of bioactive compounds from plant materials. *J. AOAC Int.* 2017, 100(6), 1624–1635.

Chapter 7

Ultrasound Application for the Fruit and Vegetable Processing Industry

M. Selvamuthukumaran

CONTENTS

7.1 INTRODUCTION

Ultrasound is a non-thermal processing technique that helps to reduce the effects of processing, increasing quality as well as food product safety (Knorr et al., 2011). This technology is usually based on the frequency of mechanical waves, which may be more than the human hearing threshold i.e. 16 kHz. The frequencies can be classified into lower and higher energy frequencies. A lower energy ultrasound has a frequency of more than 100 kHz with intensities of less than 1 Wcm^{-2}, while the higher energy frequency may have an intensity less than 1 Wcm^{-2} with a frequency range between 20 kHz and 500 kHz (Mason et al., 2011). Generally the frequency used in this technique may range from 20 kHz–500 MHz (Yusaf & Al-Juboori, 2014). A higher frequency can assure knowledge of fruit firmness, acidity, ripeness, and sugar content. The ultrasound of lower frequencies can alter the physic-chemical properties of food (Soria & Villamiel, 2010). This ultrasound technique can be widely used in the quality control of fresh fruits and vegetables, which can be applied both for pre- as well as post-harvesting of fruits and vegetables. Liquid foods' structural characteristics can also be studied by using this technique (McClements, 1997).

7.2 WORKING MECHANISM OF THE ULTRASOUND PROCESSING TECHNIQUE

The acoustic cavitation effect is achieved when the ultrasound technique is applied to liquid foods like beverages, resulting in the growth and collapse of the bubbles. The propagation of

DOI: 10.1201/9781003222170-7

ultrasound waves leads to the oscillation of bubbles, finally in their collapse, which can produce mechanical, thermal, and chemical effects. The generation of free radicals leads to chemical effects, while shear stress, turbulences, and collapse pressure can lead to mechanical effects (Lateef et al., 2007; Yu Yusaf & Al-Juboori, 2014).

The cavitation zone effect generates higher temperatures i.e. 5000 K and a pressure of 1000 atm (Soria & Villamiel, 2010). Based on ultrasound frequencies, the generated negative and positive pressures can lead to the compression and expansion of material, which results in a rupturing of cells. The ultrasound effect can cause water hydrolysis inside the bubbles, which are oscillating, which can further lead to the formation of H+ and OH– free radicals that may be captured in some chemical reactions – for example, the scavenging of free radicals by amino acids of the enzymes, which are involved in the binding of substrates, structural stability, and these effects are highly resisted by homogenous liquids (Ercan & Soysal, 2011). Bubbles are produced, which are classified into two kinds based on their structure; the first type is non-linear, which forms larger bubble clouds that are named stable cavitation bubbles; and the second is non-stable, which rapidly collapses and divides into a small bubble. Such bubbles can be internal or transient cavitation bubbles.

The tiny bubbles can dissolve rapidly, but during the stretching of bubbles, the interfacial area is higher and the mass transfer boundary layer, which is thinner, projects more air which is transferred into the bubbles during the stretching phase – after that, they can be leaked out during the collapse phase (Tiwari & Mason, 2012).

7.3 PROS AND CONS OF ULTRASOUND

7.3.1 Pros

- The waves generated by ultrasound are not toxic and are ecofriendly (Kentish and Ashok Kumar, 2011)
- It does not involve many processing costs
- It is easy to operate
- It provides excellent power output
- Microorganisms can be inactivated by merging this technique with other non-thermal processing methods (Vercet et al., 2002)
- It does not require sophisticated facilities (Gallego-Juarez et al., 2010)
- It provides higher yields when we use it for extracting components compared to traditional extraction methods (Balachandran et al., 2006)
- It cuts the cost of energy and expenditure; it leads to a smaller loss of flavor and retains consistency (Chouliara et al., 2010)
- It can be used in the food industry for extraction, homogenization, preservation, and emulsification (Chemat et al., 2011)

7.3.2 Cons

- Inactivation of released products can happen due to shear stress generated by swirls forming shock waves (mechanical effects) (Lateef et al., 2007).
   When used on a commercial scale one has to think about the need for greater input of energy (Yusaf & Al-Jauboori, 2014).

Sometimes issues in the quality parameter of processed foods with this technique may occur.

- It can help to form radicals ascribed to critical pressure and temperature that can bring changes in food components. Product quality defect is achieved because of production of radicals (H and OH) in the medium, which stimulates reaction of radicals, which can further degrade compounds and ultimately leading to reduction in product quality (Czechowska-Biskup et al., 2005).
- Ultrasound wave's frequency imposes mass transfer resistance (Esclapez et al., 2011).

    Based on medium characteristics ultrasound power can exhibit changes in the material. So power has to be reduced in the food industry in order to get good results (Feng et al., 2011).

7.4 APPLICATION OF ULTRASOUND IN FRUIT AND VEGETABLE PROCESSING

The application of ultrasound in fruit and vegetable processing were described in Table 7.1 and Figure 7.1. This technique can be used for retaining especially post- and pre-harvest quality-related attributes in several fruits and vegetables (Gallego-Juárez et al., 2010). It can be replaced as a substitute for washing steps while processing fruits and vegetables in the food industries (Alexandre et al., 2013). This technique can be successfully used to retain the freshness quality, nutritional constituents, and safety of guava juice (Cheng et al., 2007), tomato juice (Wu et al., 2008), and orange juice. Jabbar et al. (2014) reported that the ultrasound treatment can retain nutrient losses that occur during blanching. Ultrasonication cleaners of around 20–400 kHz can be used to produce fruits and vegetables which are free of contamination (Lin & Erel, 1992). It was observed that the quality and safety of strawberry fruit can be achieved by using this technique with a frequency range of 40 kHz (Cao et al., 2010).

The ultrasound technique was used to inactivate the peroxidase enzyme in tomato fruit, the complete inactivation of this enzyme was achieved when the ultrasound technique was used at 50% (Ercan & Soysal, 2011). Enhancement of cavitation by using the ultrasound technique can help to augment the inactivation of pectinic methyl esterase activity (PME) (Raviyan et al., 2005).

TABLE 7.1   APPLICATION OF ULTRASOUND TECHNIQUES FOR FRUIT AND VEGETABLE PROCESSING

Name of the fruit/vegetable or its products	Effect achieved	References
Strawberry fruit	Color retention	Alexandre et al. (2012)
Strawberry fruit	Destruction of mold growth	Aday et al. (2013)
Watercress	Color retention	Cruz et al., 2006
Strawberry fruit	Fruit firmness improved	Alexandre et al. (2012)
Carrot disc	Vitamin C retained	Rawson et al. (2011)
Tomato fruit	Inactivation of peroxidase enzyme	Ercan and Soysal, 2011
Tomato fruit	Inactivation of pectinic methyl esterase activity (PME)	Raviyan et al. 2005
Orange juice	Retention of vitamin C	Lee and Feng (2011)

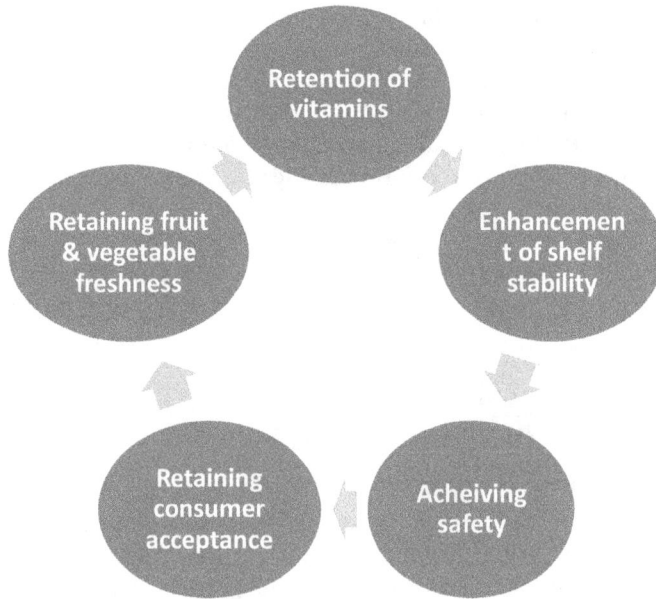

Figure 7.1   Application of ultrasound in fruits and vegetable processing.

Lee and Feng (2011) justified that orange juice treated with this ultrasound technique can help to retain vitamin C effectively when compared to a thermal method of processing.

Rawson et al. (2011) observed the vitamin C retention after applying ultrasound treatment in the carrot discs sample; they found that this treatment was found to be highly effective in vitamin C retention when ultrasound of a frequency of 20 kHz is being used. Likewise the phenolic content was increased as a result of subjecting kasturi fruit juice to ultrasound treatment of 25 kHz (Bhat et al., 2011). Application of ultrasound can bring changes in functional properties like solubility and viscosity (Arzeni et al., 2012). Strawberry fruit firmness was retained and improved when ultrasound was applied at 35 kHz (Alexandre et al., 2012).

Application of this technique during watercress blanching resulted in a lesser yellow color appearance compared to fresh watercress (Cruz et al., 2006). The ultrasound-treated grape fruit exhibited a significant color difference when compared to untreated fruit (Fava et al., 2011). Alexandre et al. (2012) reported that treatment of strawberry fruit with ultrasound of 35 kHz for a time period of 2 min exhibited good color retention. Strawberry fruit shelf stability can also be enhanced by subjecting the fruit to ultrasound treatment with a power level ranging from 30 to 60 W, it also destroyed the mold growth to a higher extent (Aday et al., 2013).

7.5 CONCLUSION

Ultrasound techniques can be effectively used in fruit and vegetable processing, which can not only enhance the quality of fruits and vegetables, but at the same time can increase the safety aspects to a greater extent. This technique can be an option for replacing washing steps during the preliminary processing of fruits and vegetables. This can further extend the stability of fresh fruits and vegetables including processed products.

REFERENCES

Aday, M. S., Temizkan, R., Büyükcan, M. B., & Caner, C. (2013). An innovative technique for extending shelf life of strawberry: Ultrasound. *LWT - Food Science and Technology, 52*(2), 93–101.

Alexandre, E. M. C., Brandão, T. R. S., & Silva, C. L. M. (2012). Efficacy of non-thermal technologies and sanitizer solutions on microbial load reduction and quality retention of strawberries. *Journal of Food Engineering, 108*(3), 417–426.

Alexandre, E. M. C., Brandao, T. R. S., & Silva, C. L. M. (2013). Impact of non-thermal technologies and sanitizer solutions on microbial load reduction and quality factor retention of frozen red bell peppers. *Innovative Food Science and Emerging Technologies, 17*, 199–205.

Arzeni, C., Martínez, K., Zema, P., Arias, A., Pérez, O. E., & Pilosof, A. M. R. (2012). Comparative study of high intensity ultrasound effects on food proteins functionality. *Journal of Food Engineering, 108*(3), 463–472.

Balachandran, S., Kentish, S. E., Mawson, R., & Ashokkumar, M. (2006). Ultrasonic enhancement of the supercritical extraction from ginger. *Ultrasonics Sonochemistry, 13*(6), 471–479.

Bhat, R., Kamaruddin, N. S. B. C., Min-Tze, L., & Karim, A. A. (2011). Sonication improves kasturi lime (Citrus microcarpa) juice quality. *Ultrasonics Sonochemistry, 18*(6), 1295–1300.

Cao, S., Hu, Z., Pang, B., Wang, H., Xie, H., & Wu, F. (2010). Effect of ultrasound treatment on fruit decay and quality maintenance in strawberry after harvest. *Food Control, 21*(4), 529–532.

Chemat, F., Zill-e-Huma, & Khan, M. K. (2011). Applications of ultrasound in food technology: Processing, preservation and extraction. *Ultrasonics Sonochemistry, 18*(4), 813–835.

Cheng, L. H., Soh, C. Y., Liew, S. C., & Teh, F. F. (2007). Effects of sonication and carbonation on guava juice quality. *Food Chemistry, 104*(4), 1396–1401.

Chouliara, E., Georgogianni, K. G., Kanellopoulou, N., & Kontominas, M. G. (2010). Effect of ultrasonication on microbiological, chemical and sensory properties of raw, thermized and pasteurized milk. *International Dairy Journal, 20*(5), 307–313.

Cruz, R. M. S., Vieira, M. C., & Silva, C. L. M. (2006). Effect of heat and thermosonication treatments on peroxidase inactivation kinetics in watercress (Nasturtium officinale). *Journal of Food Engineering, 72*(1), 8–15.

Czechowska-Biskup, R., Rokita, B., Lotfy, S., Ulanski, P., & Rosiak, J. M. (2005). Degradation of chitosan and starch by 360-kHz ultrasound. *Carbohydrate Polymers, 60*(2), 175–184. http://doi.org/10.1016/j.carbpol.2004.12.001.

Ercan, S. S., & Soysal, Ç. (2011). Effect of ultrasound and temperature on tomato peroxidase. *Ultrasonics Sonochemistry, 18*(2), 689–695.

Esclapez, M. D., García-Pérez, J. V., Mulet, A., & Cárcel, J. A. (2011). Ultrasound-assisted extraction of natural products. *Food Engineering Reviews, 3*(2), 108–120. https://doi.org/10.1007/s12393-011-9036-6.

Fava, J., Hodara, K., Nieto, A., Guerrero, S., Alzamora, S. M., & Castro, M. A. (2011). Structure (micro, ultra, Nano), color and mechanical properties of Vitis labrusca L. (grape Berry) fruits treated by hydrogen peroxide, UV-C irradiation and ultrasound. *Food Research International, 44*(9), 594–595.

Feng, H., Barbosa-Canovas, G., & Weiss, J. (Eds.). (2011). *Ultrasound technologies for food and bioprocessing.* New York: Springer.

Gallego-Juárez, J., Rodriguez, G., Acosta, V., & Riera, E. (2010). Power ultrasonic transducers with extensive radiators for industrial processing. *Ultrasonics Sonochemistry, 17*(6), 953–964.

Jabbar, S., Abid, M., Hu, B., Wu, T., Hashim, M. A., Lei, S., … Zeng, X. (2014). Quality of carrot juice as influenced by blanching and sonication treatments. *LWT - Food Science and Technology, 55*(1), 16–21.

Kentish, S., & Ashokkumar, M. (2011). The physical and chemical effects of ultrasound. In H. Feng, G. V. BarbosaCanovas, & J. Weiss (Eds.), *Ultrasound technologies for food and bioprocessing* (pp. 1–12). London: Springer.

Knorr, D., Froehling, A., Jaeger, H., Reineke, K., Schlueter, O., & Schoessler, K. (2011). Emerging technologies in food processing. *Annual Review of Food Science and Technology, 2*, 203–235.

Lateef, A., Oloke, J. K., & Prapulla, S. G. (2007). The effect of ultrasonication on the release of fructosyltransferase from Aureobasidium pullulans CFR 77. *Enzyme and Microbial Technology, 40*(5), 1067–1070.

Lee, H., & Feng, H. (2011). Effect of power ultrasound on food quality. In H. Feng, G. V. Barbosa-Cánovas, & J. Weiss (Eds.), *Ultrasound technologies for food and bioprocessing* (pp. 559–582). New York: Springer.

Lin, I., & Erel, D. (1992). Dynamic ultrasonic cleaning and disinfecting device and method. US Patent No. 5113881A. Washington, DC: U.S. Patent and Trademark Office.

Mason, T. J., Chemat, F., & Vinatoru, M. (2011). The extraction of natural products using ultrasound or microwaves. *Current Organic Chemistry, 15*(2), 237–247.

McClements, D. J. (1997). Ultrasonic characterization of foods and drinks: Principles, methods, and applications. *Critical Reviews in Food Science and Nutrition, 37*(1), 1–46.

Raviyan, P., Zhang, Z., & Feng, H. (2005). Ultrasonication for tomato pectinmethylesterase inactivation: Effect of cavitation intensity and temperature on inactivation. *Journal of Food Engineering, 70*(2), 189–196.

Rawson, A., Tiwari, B. K., Tuohy, M. G., O'Donnel, C. P., & Brunton, N. (2011). Effect of ultrasound and blanching pretreatments on polyacetylene and carotenoid content of hot air and freeze dried carrots discs. *Ultrasonics Sonochemistry, 18*(5), 1172–1179.

Soria, A. C., & Villamiel, M. (2010). Effect of ultrasound on the technological properties and bioactivity of food: A review. *Trends in Food Science and Technology, 21*(7), 323–331.

Tiwari, B. K., & Mason, T. J. (2012). Ultrasound processing of fluid foods. In P. J. Cullen, B. K. Tiwari, & V. P. Valdramidis (Eds.), *Novel thermal and non-thermal technologies for fluid foods* (pp. 135–165). London: Academic Press.

Vercet, A., Sánchez, C., Burgos, J., Montañés, L., & Lopez Buesa, P. L. (2002). The effects of manothermo-sonication on tomato pectic enzymes and tomato paste rheological properties. *Journal of Food Engineering, 53*(3), 273–278.

Wu, J., Gamage, T. V., Vilkhu, K. S., Simons, L. K., & Mawson, R. (2008). Effect of thermosonication on quality improvement of tomato juice. *Innovative Food Science and Emerging Technologies, 9*(2), 186–195.

Yusaf, T., & Al-Juboori, R. A. (2014). Alternative methods of microorganism disruption for agricultural applications. *Applied Energy, 114*, 909–923.

Chapter 8

Pulsed Light Technology for the Fruit and Vegetable Processing Industry

Pervin Basaran and Ayşenur Betül Bilgin

CONTENTS

8.1 INTRODUCTION

Fruits and vegetables are an essential group of food with enormous nutritional and commercial benefits. They provide vital antioxidants, vitamins (vitamin C, vitamin A, and vitamin E), trace elements (selenium, copper, zinc), and rich sources of functional bioactive compounds including dietary fibers, carotenoids, polyphenols, phenolics, glucosinolates, phytoestrogens, alpha-lipoic acid, and lectins (Vincente et al., 2014). The consumption of a rich diet with fresh fruits and vegetables has been associated with lowered risk of all-cause, cancer, and cardiovascular diseases

DOI: 10.1201/9781003222170-8

(Aune et al., 2017). Fresh fruits are important in the diet not only because of their numerous health benefits but also because of their delicious taste and pleasant feeling. Even though quantities vary from country to country depending on the cultural habits, World Health Organization (WHO) recommends fruit and vegetable consumption per capita should exceed 400g daily (World Health Organization, 2003).

Fresh produce is commonly consumed after minimal or no processing and is greatly vulnerable to microbial contamination during pre- and postharvest stages. Mostly, they are washed, cleaned, cut, or sliced, packaged, and stored at low temperatures. Fresh produce is highly perishable with a short shelf life, which depends on various factors such as initial quality, processing technology applied, and the level and composition of microbial contamination (Watada & Qi, 1999). The open nature of the fresh produce chain can cause contamination in production, harvesting, and during processing. Microbial decay is one of the main factors that compromise the quality of fruits and vegetables in the field and later during the postharvest period (Sanzani et al., 2009). The population of background microbiota on the surface of the fruit or vegetable can vary widely depending on produce types, source, and sampling time. The main spoilage microbial groups are *Pseudomonas* spp., Enterobacteriaceae, yeasts, and molds. The natural microflora of raw fruits and vegetables is usually non-pathogenic for humans. However, during the harvest, and further processing and handling, the produce can be contaminated with hazardous pathogens such as *Salmonella*, *Listeria*, *Escherichia coli*, and *Campylobacter jejuni* (CDC, 2020; Iwu & Okoh, 2019). Although food pathogen contamination occurs at any point from field to fork, plant growing (bird or other animal feces, composts, or manures), irrigation, harvesting, washing, processing equipment, tools, and handling workers contaminated with pathogens (Hanning et al., 2009), post-processing contamination is the major contributing factor to foodborne illness outbreaks (Jung et al., 2014). According to the Centers for Disease Control and Prevention (CDC) in the USA, the number of reported foodborne illness outbreaks associated with fresh produce is around 57 annually (Wadamori et al., 2016). In fact, nearly 85% of food poisoning outbreaks reported in the USA were from lettuce, spinach, cabbage, strawberry, tomato, etc. especially when eaten raw (CDC, 2020).

Nowadays, a wide variety of minimal applications of non-thermal technologies are being developed as a response to consumers' desire to eat fresh produce which is safe and convenient for an urban lifestyle and further fully loaded with high nutritional and sensory properties. Since microbial contamination occurs only on the surface, external decontamination applications such as ultraviolet (UV) and PL are especially suitable for minimally processed ready-to-eat fresh fruits and vegetables.

8.2 LIGHT APPLICATIONS FORMS AND TYPES OF LIGHT PROCESSING EQUIPMENT

Continuous ultraviolet C (UV-C) light used as a disinfectant on food surfaces is conventionally generated by low-pressure mercury lamps that work at a monochromatic wavelength of 254 nm (Ma et al., 2017). UV light is produced when the atoms are exited and returned to their ground state as an electrical current is passed through (Gray, 2014). Commercial lamps are bulky (with a maximum lifespan of 10000 h), overheat (require air or water cooling for the operation), subject to deterioration and failure because of electrode erosion, and can cause hazardous mercury waste contamination in the environment (Gray, 2014; Lee & Bahneth, 2013; Ma et al., 2017).

Pulsed light (PL) is an improved environmentally friendly version and a next-generation approach to continuous UV light treatment (Rowan et al., 2015). PL is also referred to in publications as intense pulsed light, pulsed white light, high-intensity pulsed light, high-intensity broad-spectrum pulsed light, and pulsed UV light (Martín-Sómer et al., 2017; Rowan, 2019). PL technology is based upon the accumulation of high-discharge voltage in a capacitor where this deposited energy is discharged in ultra-short pulses through a rare gas (xenon or krypton) that results in intense light pulses. Even though the peak power of each pulse is high (nearly 20,000 times more intense than sunlight at sea level), the total pulse energy is relatively low because of its short duration (1 µs–0.1 seconds), releasing a powerful energy intensity of 0.01–50 J/cm^2 (Dunn et al., 1995; Gómez-Lopez et al., 2005). In a low or high-pressure mercury lamp, light is mainly emitted at wavelengths close to 254 nm and 365 nm respectively, whereas PL is composed of broad-spectrum light (100 to 1100 nm), and the percentage of the emitted energy changes depending on the infrastructure (Martín-Sómer et al., 2017). Wavelengths below 200 nm are classified as vacuum ultraviolet. It is estimated that at least 50% of the PL electromagnetic energy falls within the wavelength range from 170 to 260 nm UV light that is normally filtered by the atmosphere, 20% infrared, and 26% visible light radiation (Gómez-Lopez et al., 2005; Oms Oliu et al., 2010). The three UV sub-divisions as used in the literature are 200–280 nm (UV-C; about 20%), 280–320 nm (UV-B, about 8%), 320–400 nm (UV-A, about 12%), and 400–470 nm (blue light) are all being explored individually for their ability to inactivate microorganisms (Surjadinata et al., 2017; Wang et al., 2017). The PL provides some practical advantages over continuous UV sources with faster disinfection and reduced energy cost as the duration of samples for the treatment ranges in minutes instead of hours as in the case of continuous UVC treatment (Oms-Oliu et al., 2010; Rice & Ewell, 2001). Recent comparative studies reported that the antimicrobial efficacy of PL was 4–6 fold higher than continuous UV light for their respective wavelengths (Deng et al., 2020; Nyangaresi et al., 2019). Additionally, PL, because of its higher power with short duration requires no warm-up time, penetrates opaque or thick materials better than continuous UVC, and achieves a more effective microbial inactivation with significantly less thermal damage to the product.

The major elements of a PL system are flash lamps filled with xenon gas, a high-voltage power supply, a capacitor, a pulse forming setup, and a trigger that initiates discharging of the electrical energy turn to a flash of light (Mahendran et al., 2019; Uesugi & Moraru, 2009). Five major units that are commonly used to characterize PL treatment are: fluence rate (W/m^2), exposure (seconds), pulse width (fractions of seconds), pulse-repetition rate (pulses per second), and peak power (W). In recent years, light-emitting diodes (LED) based light transmission through two terminals have appeared as an alternative to conventional PL illumination systems and are tested to reduce pathogenic bacteria, especially on fresh-cut fruits and juices (Cossu et al., 2021). LEDs are semiconductor devices in which free electrons from a high-energy conduction band recombine holes in a valence band, releasing energy in the form of photons, and emitting monochromatic light by electroluminescence (Ghate et al., 2015). LEDs can emit light at any specific wavelength based on the composition and condition of the semiconducting materials used (Jo & Tayade, 2014). Mostly, LEDs tested for food applications have a visible wavelength range from 400 to 760 nm. Light extraction and internal quantum efficiency enable LEDs to generate white light to surpass the efficiency of conventional shining lighting by 15–20 times. Today, LED-based light systems are applied in various forms and with all types of commercial applications, the global LED lighting market in 2016 was estimated to be USD 26 billion.

Compared to conventional UVC and PL systems, UV-LEDs have numerous advantages due to the special manner of light generation, such as being of small compact size, non-hazardous (not containing mercury), having long lifespans (50,000–100,000 h), low energy consumption (>90% more efficient), low starting voltage (5 V), relatively less thermal effect on the product, higher luminous efficiency (>200 lm/W), are easy to install, flexible operating at multiple wavelengths, have a more constant light intensity, and application can be continued as needed, and having instantaneous on-off cycles (Hinds et al., 2019, 2020; Muramoto et al., 2014; Song et al., 2018). Even more importantly, LEDs can be designed to emit light at any specific wavelength by select-ing different semiconductor materials as needed for the food treatment (Jo & Tayade, 2014; Song et al., 2016). LEDs reduce undesirable effects such as sensory deterioration in food due to low emission of radiant heat in the form of infrared light (Mitchell et al., 2012). Sholtes and Linden (2019) designed an experiment to compare the inactivation efficacy of the UV mercury-arc lamps to UV-LEDs in various combinations. Sequencing of wavelengths and various pulsing strategies for the inactivation of *E. coli*, *Pseudomonas aeruginosa*, and bacteriophage MS-2 were tested (Sholtes & Linden, 2019). The inactivation efficiency of pulsed UV-LED with continuous UV-LED irradiation modes was investigated, and no statistically significant difference in log inactivation of *E. coli* and *Staphylococcus epidermidis* was observed between LED operating systems under various cycles (Lu et al., 2021). These studies concluded that the biological impacts of a specific light are directly related to the total dose, rather than the application methods.

Depending on the light wavelength, the germicidal action of PL has been attributed to three mechanisms, which may coexist: the photochemical effect (DNA cleavage) (Gomez-Lopez et al., 2007), the photo thermal effect (cell overheating) (Ferrario et al., 2013; Wekhof, 2000), and the photo physical effect (structural damage to cell wall and membrane) (Pataro et al., 2011; Schottroff et al., 2018; Vicente et al., 2005). The efficiency of shorter wavelengths within the vis-ible light spectrum and its results are more contradictory (Ghate et al., 2015; Schmid et al., 2019). The lethal action of PL has been primarily attributed to a photochemical mechanism, by which UV-C and a small portion of UV-B wavelength absorption induce irreparable alterations as cyclobutane pyrimidine dimers and pyrimidine (6–4) pyrimidone photoproducts in the genetic material (DNA) of target microorganisms (Aguirre et al., 2018). These photochemical changes cause severe damage, preventing the replication and gene transcription, eventually causing cell death (Giese & Darby, 2000; Krishnamurthy et al., 2008; Surjadinata et al., 2017; Wang et al., 2017). UV-A (315–400 nm) radiation mainly inactivates microorganisms by generating reac-tive oxygen species (e.g., H_2O_2, O_2, and OH) causing irreversible oxidative disturbance to bio-molecules, including DNA guanine bases, proteins, and lipids (Brem, et al., 2017; Nelson et al., 2018). The antimicrobial activity of blue light (400–470 nm) is associated with the presence of endogenous photosensitizer compounds (e.g., porphyrins, flavins, nucleic acids, vitamins, and nicotinamide adenine dinucleotides) broadly found in microbial cells. These molecules in their light-excited states can promote photodynamic reactions and generate reactive oxygen species that cause oxidative damage to cellular components (Gwynne & Gallagher, 2018; Nelson et al., 2018). Photo thermal effect is frequently associated with the temporary overheating caused by infrared radiation (Gómez Lopez et al., 2007).

The photo physical nature of light on cell constituents includes morphological changes (depo-larization, disruption, flattening out of the cells from the edges, damage or loss of cell mem-brane permeability, destruction of cell wall, damage and rupture of the plasma membrane), leakage of intracellular components, the loss of vital cell components, cytoplasmic shrinkage, DNA damage, and denaturation of some essential proteins and enzymes in the microbial cells (Garvey & Rowan, 2019; Ghate et al., 2019; Hyun & Lee, 2020; Kim et al. 2016, 2017; Kramer et al. 2017b; Preetha et al., 2021; Xu & Wu, 2016). In yeast cells, specifically increased intracellular reactive oxygen species result in lipid peroxidation (Ferrario et al., 2014; Takeshita et al., 2003). Additionally, in yeast cells, loss of esterase activity, vacuole expansion, membrane distortion, and change of cellular shape were observed (Farrell et al., 2011; Ferrario et al., 2014; Takeshita et al., 2003; Wu et al., 2018). Microbial cells accumulate the number of injuries until they exceed the capability of repair mechanism, consequently, the number of survivors declines exponentially (Gouma et al., 2015). The effectiveness of PL technology critically depends on various intrinsic and extrinsic factors as listed in Table 8.1. Depending on the treatment conditions (Table 8.1), PL treatment has been reported to achieve reduction of anywhere from 0.5 to 8 log CFU.

Recent proteomic analyses investigated gene expression and protein biosynthesis of stressed microorganism which was not identified in early PL studies. According to proteomic analysis, PL induced the expression of the general stress protein Ctc in *Listeria innocua* (Aguirre et al., 2018). Ctc is an ATP synthase β-subunit, and related with ribosome assembly, which participates in a multisubunit enzyme complex in ATP processing and intracellular pH homeostasis, and contributes to survival and adaptation in different environments (Aguirre et al., 2018; Stack et al., 2008; Uesugi et al., 2016). However, flagellin and some glucose metabolism proteins were down-regulated in the same PL-treated cells (Aguirre et al., 2018). Uesugi et al. (2016) reported 1.8-fold changes in the expression of genes related to flagella production in *Listeria monocytogenes* treated with a PL fluence of 33 mJ/cm^2. The antimicrobial effect of LED light was also found to be related to the up- or down-regulation of metabolic pathways (Kumar et al., 2017). In future studies, more evidence of the genetic effects of PL in microorganisms is expected to be researched.

8.3 APPLICATION OF PULSED LIGHT

8.3.1 Single PL Applications for Whole or Fresh-cut Fruits and Vegetables and Juices

The reduction of microbial load in most fruits and vegetables is currently carried out mainly by sanitation with chlorine solutions, which provide only limited (1–2 log) reductions of pathogens (Banach et al., 2017). A previous study showed that the decontaminating chemicals blocked by the hydrophobicity of the leaf surface on some vegetables and furthermore chlorine-halogenated side-products present health risks (Whipps et al., 2008). Therefore, there is a constant effort to find chlorine alternatives and new decontamination technologies for fresh produce. Since microbial contamination occurs only on the peel, surface decontamination applications such as PL have great potential and are tested as an alternative form of non-thermal application for the inactivating microorganism on the surface of fresh fruits and vegetables (Table 8.2).

Fresh berry fruits on the market are not washed for fear of fungal development due to damage to surface skin, and in general, no additional processing technique is adopted besides sufficient pre-cooling and cold storage (Romero Bernal et al., 2019). Only a few studies have addressed the efficacy of PL against mold decay in berry fruits after harvest (Duarte et al., 2016; Marquenie et al., 2003; Romero Bernal et al., 2019). Romero Bernal et al. (2019) reported a 2-day delay to the onset of the *Botrytis cinerea* infection on artificially contaminated strawberries exposed to PL (11.9 and 23.9 J/cm^2) for 10 s or 20 s compared to the control. A maximal inactivation of 3 and 4 log units for *B. cinerea* and *Monilinia fructigena* was reported, respectively, and the conidia of both fungi showed similar behavior to PL (Marquenie et al., 2003).

Ready to consume minimally processed fresh fruits and vegetables do not undergo intensive inactivation or preservation treatments during processing. Fruit and vegetables are cleaned, cut, appropriately packaged, and commercialized as ready-to-eat products preserved using mild techniques. They are expected to have a shelf life of 5–7 days at 4°C while ensuring food safety and maintaining nutritional and sensory quality (Cliffe-Byrnes & O'Beirne, 2005). Fresh-cut

TABLE 8.1   INTRINSIC AND EXTRINSIC FACTORS IMPACTING THE ANTIMICROBIAL EFFICACY OF PL

Food Matrix Characteristics	Process Parameters	Microbial Properties
• Opaqueness, turbidity • Viscosity, thickness of liquid foods • Food chemical composition (optically active food ingredients) • Suspended solids' particle size and material • Insoluble suspended matter (e.g., fibers) • Surface characteristics (morphology, topography, smoothness, or roughness) • Light reflection coefficient of solid fruit and vegetables • Color • Physical state/properties (solid or liquid) • Presence of packaging • Transparency of the liquid material • Sample volume • Presence of protective substances • Total soluble solids • Presence of particles shielding the microorganisms • The shape of the fruit or vegetable • Temperature • Water activity	• Light wavelengths • Energy dose (or fluence) • Composition of the emitted light spectrum (wavelength) • Intensity of illumination • The physical distance of the sample from the light source • Number and design of lamps • The geometry of the treatment cell • Power applied on the food item • Treatment duration • Environmental relative humidity • Light penetration depth	• Microbial species • Type of microorganism • Size of organism • Physiological condition of the inoculums used • Presence of nuclear membrane • Structural/compositional differences in the cell walls • Location of microbial contamination • Microbial population density • Microbial inoculum size • Microbial resistance against light • Ability to repair the injury on DNA • Physiological state (spore/vegetative form)

TABLE 8.2   USE OF PULSED LIGHT APPLICATION ON MINIMALLY PROCESSED FRUIT AND VEGETABLES

Food Products	Light System	Working Conditions Frequency, Voltage, Treatment Time	Target Microorganisms	Log Reduction ($\log_{10}$ CFU/g)	Reference
Apple and plum juice, carbonated beverages	Intense Pulsed Light	3000 pulses at intervals of 200 ms, light intensity 0.97–29.21 J/cm²	Pseudomonas aeruginosa	7.0 (12.17–24.35 J/cm²)	Hwang et al., 2015*
Apple juice	Pulsed Light	Batch mode, xenon lamp, 3 pulses/s, 0–71.6 J/cm²	Alicyclobacillus acidoterrestris spores	3.0–3.5	Ferrario & Guerrero, 2018
Apple juice	Pulsed Light	0.2–0.4 J/cm² per pulse, number of pulses (5–40) depth of the juice layer (6–10 mm)	Penicillium expansium	3.76	Maftei et al., 2014
Apple juice	UVC LED	280/365 nm, 707.2 mJ /cm², 40 min, 4 UV-LEDs	Escherichia coli K12	2 (cloudy juice), 4.4 (clear juice)	Akgün & Ünlütürk, 2017
Apple juice	UVC	254 nm, 38 W/m², 0–40 min, 4 mm of deepness juice	Neosartorya fischeri ascospores	4	Menezes et al., 2019
Apple juice	UV LED	275 nm, distance between light source and sample surface; 15 cm, 800, and 1200 mJ/cm²	Zygosaccharomyces rouxii	4.86- 5.46	Xiang et al., 2020*
Apple juice Orange juice	Pulsed Light	100–1100 nm (high intensity light), continuous flow system, xenon flash-lamp, 3Hz, 3 pulses/s (pulsed widht 360 µs), From 1.8 to 5.5 J/m² intensity	Listeria innocua and Escherichia coli DH5-α	2.98 and 4 (4 J/m²) 2.90 and 0.93 (4 J/m²)	Pataro et al., 2011
Apple juice Orange juice Milk	High-Intensity Pulsed Light	Batch system (0–8 s), short (100–400 µs) light pulses (200–1100 nm), 3 Hz frequency	Escherichia coli and Listeria innocua	5.66 and 3.88 (20 min) 4.7 and 1.93 (8 s) 1 (8 s) (both mos.) 1.06 and 0.84 (8 s)	Palgan et al., 2011*

(Continued)

TABLE 8.2 (CONTINUED)	USE OF PULSED LIGHT APPLICATION ON MINIMALLY PROCESSED FRUIT AND VEGETABLES				
Food Products	Light System	Working Conditions Frequency, Voltage, Treatment Time	Target Microorganisms	Log Reduction (log₁₀ CFU/g)	Reference
Carrot juice	Pulsed Light	500 J of a single pulse, two xenon lamps, 0.5 Hz and 20 pulses, distance between sample and lamp; 10 cm	*Escherichia coli* O157:H7	4.54	Zhu et al., 2019
Chokanan Mango juice	UVC	254 nm, 15, 30, and 60 min at 25°C, 3,525 J/m², the distance between sample and lamp; 35 cm	Aerobic bacteria	2.74	Santhirasegaram et al., 2015
			Coliform	1	
			Yeast and mould	2.42	
Coconut water	Pulsed Light	A continuous system, the capacity of 100 ml/min, a xenon flash lamp, 254 nm 95.2 J/cm², pulsed light treated juices at a fluence rate of 0.18, 2, and 5.6 W/cm². 3 pulses per second with a pulse width of 360 μs	*Escherichia coli*	5.33	Preetha et al., 2021*
Orange juice				4	
Pineapple juice				4.5	
Fresh-cut apple	Pulsed Light	4 xenon lamps, 0 to 157.5 kJ/m², pulse duration was 50 μs, 0.5 Hz	Viable counts	1	Igmat et al., 2014
			Lactobacillus brevis	3.0 (17.5 kJ/m²)	
			Listeria monocytogenes	2.7 (17.5 kJ/m2)	
Fresh-cut avocado	Pulsed Light	Two xenon lamps (200 to 1100 nm with 15% to 20% of the light in the UV region), 3.6– 14 J/cm²	Aerobic mesophilic microorganisms	1.20	Aguiló-Aguayo et al., 2014
Fresh-cut avocado	Intense Light Pulses	15 or 30 pulses at fluencies of 0.4 J/cm² per pulse	*Listeria innocua*	2.61 and 2.97 (15 and 30 pulses)	Ramos-Villarroel et al., 2011
			Escherichia coli	2.90 and 3.33 (15 and 30 pulses)	
Fresh-cut cantaloupe	Repetitive Pulsed Light	0.3–1.2 J/cm², exposing in polypropylene bag	Total microbial count	4.64 (0.9 J/cm²)	Koh et al., 2016
			Yeast and mold count	4.56 (0.9 J/cm²)	

(Continued)

TABLE 8.2 (CONTINUED)   USE OF PULSED LIGHT APPLICATION ON MINIMALLY PROCESSED FRUIT AND VEGETABLES

Food Products	Light System	Working Conditions Frequency, Voltage, Treatment Time	Target Microorganisms	Log Reduction (log_{10} CFU/g)	Reference
Fresh-cut lettuce	Pulsed Light	One xenon flash lamp, duration of one pulse emitted; 300 µs, 200 J/pulse (0.33 J/cm²), fluence; 4.0, 8.2, 12.5 and 16.8 J/cm²	Staphylococcus aureus, Escherichia coli, Salmonella enteritidis, Listeria monocytogenes	2.03–6.56	Tao et al., 2019*
Fresh-cut lettuce	UVC	10 UV-C lamp, 254 nm, 1.36–6.80 mW/cm², distance of 10–50 cm, two-sided exposure, exposure times (0.5 to 10 min)	Escherichia coli O157:H7 / Salmonella Typhimurium / Listeria monocytogenes	1.45 (1 min.) / 1.35 (1 min.) / 2.12 (1 min.)	Kim et al., 2013
Fresh-cut mango	LED	405 nm, total dose; 2.6–3.5 kJ/cm² / 20 mW/cm², distance sample and system; 3.5 cm	Escherichia coli O157:H7, Listeria monocytogenes, Salmonella	1–1.6	Kim et al., 2017b
Fresh-cut melon	UVC	Two UV-C lamp, 253.7 nm, irradiation 20 W/m², 0–10 min treatment time, 1200, 6000 and 12,000 J/m²	Total viable count and Enterobacteriaceae	2	Manzocco et al., 2011
Fresh-cut mushroom	Pulsed Light	180–1100 nm (15–20% of the light in the UV region), a fluence of 12 J/cm², duration of each pulse was 0.3 ms with a fluence of 0.4 J/cm²	Escherichia coli / Listeria innocua	3 / 2	Ramos-Villarroel et al., 2012b
Fresh-cut papaya	LED	405 nm, 10 mW/cm², 1700 J/cm²	Salmonella enterica	1.2	Kim et al., 2017a
Fresh-cut pineapples	LED	460 nm, 10 W, 92.0–254.7 mW/cm², 7950 J/cm²	Salmonella enterica	1.72 (92.0 mW/cm²)	
Fresh-cut watermelon	Intense Light Pulses	Duration of each pulse was 0.3 ms with a fluence of 0.4 J/cm², exposed 15 and 30 pulses (6 and 12 J/cm²)	Escherichia coli / Listeria innocua	>3 (12 J/cm²) / 2.79 (12 J/cm²)	Ramos-Villarroel et al., 2012a

(Continued)

TABLE 8.2 (CONTINUED)   USE OF PULSED LIGHT APPLICATION ON MINIMALLY PROCESSED FRUIT AND VEGETABLES

Food Products	Light System	Working Conditions Frequency, Voltage, Treatment Time	Target Microorganisms	Log Reduction ($\log_{10}$ CFU/g)	Reference
Grape juice	Pulsed Light	A spiral flow continuous system, 2 linear xenon flash lamps (200–1100 nm), 0.5 Hz or 1 Hz,	Escherichia coli		Xu et al., 2019
		100 pulses, 30 mL/min flow rate, intensities of 0.13, 0.40, and 0.66 J/cm² /pulse		2.1, 2.6, and 3.7 (intensity, respectively)	
		100 pulses, 0.66 J/cm² /pulse, flow rates of 0, 30, and 60 mL/min		1.0, 3.1, and 1.7 (flow rates, respectively)	
		the optimal levels of 80 pulses, the intensity of 0.66 J/cm² /pulse, and flow rate of 40 mL/min		4.89	
Grape juice	UVC	Continuous flow UV system, 0 -282.24 mJ/cm², 6 low-pressure mercury UV lamps (254 nm) Max. operational capacity;333 l/min for clear liquids, 83 l/min for low transmittance liquids	Saccharomyces cerevisiae	3.39 (clear juice) (65.50 mJ/cm²)	Kaya & Unluturk, 2016
			Yeasts	1.54 (turbid juice)	
			Lactic acid bacteria	1.64 (turbid juice)	
Grape tomato	UVC	Low-pressure mercury-vapor fluorescent lamps, 254 nm, duration of 0–100 s, 0–6.0 kJ/m² doses	Escherichia coli O157:H7	2.3–3.5	Mukhopadhyay et al., 2014
			Salmonella enterica	2.15–3.1	
Lettuce and strawberry	UVC	254 nm, 2 mW/cm² at dosages 1.2–7.2 J/cm², distance of sample and lamps; 8 cm, treatment time; 10, 20, 30, 45, and 60 min	Escherichia coli	1.75	Birmpa et al., 2013
			Listeria innocua	1.27	
			Salmonella Enteritidis	1.39	
			Staphylococcus aureus	1.21	

(Continued)

TABLE 8.2 (CONTINUED)   USE OF PULSED LIGHT APPLICATION ON MINIMALLY PROCESSED FRUIT AND VEGETABLES

Food Products	Light System	Working Conditions Frequency, Voltage, Treatment Time	Target Microorganisms	Log Reduction (log$_{10}$ CFU/g)	Reference
Lettuce (Romaine) (In-package)	Pulsed Light	180–1000 nm of which 40% belongs to the UV region, 1 min (63 J/cm²) // 0.35 J/cm²/pulse, 3 pulses of light per second with a width of 500 µs	Total aerobic bacteria and mold and yeast Pathogen Escherichia coli O157:H7	>1 1 (1 s – .1.05 J/cm²) >2	Mukhopadhyay et al., 2021
Mung bean sprouts and salad	Pulsed Light	3 xenon lamps, 200 and 1100 nm, the distance between the reflector and sample table; 0–20 cm, 320, and 580 mJ/cm² per light pulse	Total microbial count	2.5 (60 s)	Kramer et al., 2017a
Orange juice	UV LED	460 nm, 105 W, 92.0–254.7 mW/cm², 4500 J/cm²	Salmonella enterica	3.3 -3.6	Ghate et al., 2016
Raspberries	Pulsed UV-Light	1.27 J/cm² per pulse for input of 3,800 V, 100–1100 nm, 6 and 72 J/cm², 3 pulses per second	Escherichia coli Salmonella	~4.5 log ~3.5 log	Bialka et al., 2008
Raspberry	Pulsed Light	Quartz PL lamp (200–1100 nm), 30 s (with fluence of 28.2 J/cm²), 3 pulses/sec pulse rate	Salmonella Escherichia coli O157:H7	4.5 3.9	Xu & Wu, 2016
Spinach	Intense Pulsed Light	Xenon flash lamps (180-1100 nm), duration of each pulse was 0.3 ms with a fluence of 4 kJ m⁻², 0 - 120 kJ m⁻² (0–30 pulses)	Listeria innocua Escherichia coli Natural microbial count	2.6 (120 kJ m⁻²) 2.3 (120 kJ m⁻²) 0.4–2.2	Agüero et al., 2016
Strawberries	Pulsed UV-Light	1.27 J/cm² per pulse for the input of 3,800 V, 100–1100 nm, 5.4 and 64.4 J/cm², 3 pulses per second	Escherichia coli Salmonella	~2 ~2	Bialka et al., 2008

(Continued)

TABLE 8.2 (CONTINUED)   USE OF PULSED LIGHT APPLICATION ON MINIMALLY PROCESSED FRUIT AND VEGETABLES

Food Products	Light System	Working Conditions Frequency, Voltage, Treatment Time	Target Microorganisms	Log Reduction (log$_{10}$ CFU/g)	Reference
Strawberry	High-Intensity Pulsed Light	1.27 J/cm^2 per pulse for the input of 3,800 V, 100–1100 nm, pulse rate of 3 pulses per second, and a pulse width of 360 µs	Molds	16–42% reduction (2.4–47.8 J/cm^2)	Duarte-Molina et al., 2016
Strawberry	Pulsed Light	Xenon flash lamp (200–1000 nm), duration of light pulse was 112 µs, and frequency was 5 Hz, power of each pulse ranged from 0.07 to 0.9 MW	Naturally mesophilic bacteria	2.2 (3.9 J/cm^2)	Luksiene et al., 2013
			Bacillus cereus	1.5 (3.9 J/cm^2)	
			Listeria monocytogenes	1.1 (3.9 J/cm^2)	
			Yeast and fungi	1.0 (3.9 J/cm^2)	
Strawberry	Pulsed Light	1 to 40 s (fluences: 1.2 to 47.8 J/cm^2, respectively)	Botrytis cinerea	2 and 3.5 (11.9 and 23.9 J/cm^2)	Romero Bernal et al., 2019
Strawberry	Pulsed Light	3–5 J/cm^2 (10, 16, and 20 s), the distance between sample and lamp; 35 and 55 cm	Salmonella	0.4–0.8 (approximately)	Cao et al., 2019
Tomatoes	UVC	Two lamps, 254 nm, distance between lamp and sample; 43.8, doses; 0 to 223.1 mJ/cm^2	Salmonella enterica	3.22 (22.3 mJ/cm^2) 4.39 (178.5 mJ/cm^2)	Lim & Harrison, 2016

* More than 5 log CFU reduction were reported

products can also be contaminated with foodborne pathogens due to unhygienic conditions during cutting, slicing, packaging, or storage (Kim et al., 2017). Slicing or kitchen operations on the fruit and vegetable cause spreading of bacteria over cut surfaces. Considering these products are often consumed raw, microbial safety has become an important priority for public health authorities due to recent foodborne outbreaks (Garvey & Rowan, 2019). Washing with aqueous sanitizers is a key step to remove dirt, tissue fluids, and most importantly microorganisms from cut surfaces (Feliziani et al., 2016; Sahoo et al., 2021). Public awareness is increasing for chemically free foods to prevent unintentional consumption of these chemical contaminations. PL has been applied on fresh-cut commodities to avoid the development of harmful effects on the product (Table 8.2). The inactivation of microorganisms seems more difficult on these products because especially mechanical bruises can cause internalization of microbes though the food matrix (Table 8.2).

As far as fruit juices are concerned, heat pasteurization has been the standard method for obtaining microbiologically safe liquid food products; however, conventional thermal processes can cause losses in heat-sensitive ascorbic acid and thiamin and deleteriously affect the sensory and rheological properties of fruit and vegetable juices (de Souza et al., 2020; Fenoglio et al., 2020; Xiang et al., 2020). In the last two decades, increasing consumers' demand for fresh-like foods has led to considerable research to develop milder methods for food preservation. Several studies have shown that the sole PL treatment was effective at inactivating microorganisms in liquid foods, especially those with higher transparency characteristics (Keyser et al., 2008; Palgan et al., 2011). When PL was applied as a sole treatment, there are only a few examples where the reduction of microbial contamination reaches the FDA requirement of 5 log and above (Table 8.3). In the case of apple juice and plum juice, 7 log of *Pseudomonas aeruginosa* reduction after the PL treatment with 12.17–24.35 J/cm^2 was achieved (Hwang et al., 2015). Certain strains of yeasts are frequently involved in the spoilage of beverages, e.g., wine, cider, and fruit juices. Additionally, spoilage by yeast results in the development of turbidity, flocculation, and clumping in juice if thermal processing is insufficient (da Cruz Almeida et al., 2018). The application of UV-LEDs emitting at 275 nm in the inactivation of *Zygosaccharomyces rouxii* in apple juice has recently been tested and 4.86–5.46 log reduction was recorded (Xiang et al., 2020). The *Staphylococcus aureus* strain was the most sensitive to PL treatment on lettuce compared to the other three strains at the tested fluence; approximately 6.56 ± 0.78 log reduction was attained with PL treatment of 16.8 J/cm^2 (Tao et al., 2019).

8.3.2 Combined Hurdle Applications for Fruits and Vegetables

The major limitations of PL are the naturally poor penetrating power of light, the requirement for the product to be transparent, and to have a smooth surface in the case of solid fruits and vegetables, as well as the possible excessive increase in the sample temperature. The majority of up-to-date studies demonstrated that when PL is applied alone, it does not achieve the FDA requirement of 5-log microbial reduction of microorganisms or spores (Table 8.2). Some microorganisms have a mechanism called "photoreactivation" in order to repair the UV-induced DNA damage and might regain viability after UV light treatment and thus raise food safety concerns, especially in the case of coliforms (*Shigella dysenteriae*, *E. coli*, *Salmonella typhimurium*) (Choi et al., 2015; Hu et al., 2012). Furthermore, Kramer and Muranyi (2014) observed a considerable proportion of *E. coli* and *L. innocua* cells entered the "viable but non-culturable" state after PL treatment. Sub-lethal injured microbial cannot be cultured on routine microbiological media but may still retain their capacity to cause infections and risk to consumers (Kramer & Muranyi,

TABLE 8.3   COMBINED APPLICATIONS OF LIGHT AND OTHER FOOD PROCESSING APPLICATIONS

Food Products	Light System	Combined Treatment	Target Microorganism	Reduction (log₁₀ CFU/g, CFU/mL, CFU/mL, spores/g or spores/mL)	Reference
Apple juice	Pulsed Light	Ultrasound	*Saccharomyces cerevisiae*	6.4 and 5.8	Ferrario & Guerrero, 2017*
Apple juice	High Intensity Pulsed Light	Pulsed Electric Field	*Escherichia coli*	5 (minimum)	Caminiti et al., 2011*
Apple juice	Pulsed Light	Ultrasound	*Saccharomyces cerevisiae*	6.4 and 5.8 (commercial and natural apple juices)	Ferrario et al., 2015*
			Alicyclobacillus acidoterrestris spores	3.0 and 2.0 (commercial and natural apple juices)	
Apple juice	Pulsed Light	Thermosonication	*Escherichia coli*	6	Muñoz et al., 2012*
Apple juice	UVC	Ultra-High-Pressure Homogenization	*Alicyclobacillus acidoterrestris* spores	5–6	Sauceda-Gálvez et al., 2020*
Apple juice	Pulsed Light (Continuous Flow-Through)	Ultrasound	*Escherichia coli*	5.9	Ferrario & Guerrero, 2016*
			Salmonella enteritidis	6.3	
			Saccharomyces cerevisiae	3.7	
			Aerobic mesophilic	2.2	
			Mold and yeast counts	2.0	
Apricots, plums, and cauliflowers	LEDs	Hypericin (Food additive, aroma)	*Bacillus cereus*	4.4	Aponiene et al., 2015
			Mesophilic bacteria	0.6–0.72	
Avocados, seedless watermelons, and button mushrooms	Pulsed Light	Malic acid	*Listeria innocua*	0.91–1.1	Ramos-Villarroel et al., 2015
			Escherichia coli	1.92–2.97	

(Continued)

TABLE 8.3 (CONTINUED)   COMBINED APPLICATIONS OF LIGHT AND OTHER FOOD PROCESSING APPLICATIONS

Food Products	Light System	Combined Treatment	Target Microorganism	Reduction (log$_{10}$ CFU/g, CFU/mL, spores/g or spores/mL)	Reference
Black peppercorns	UVC	Cold Plasma	Salmonella spp.	3.7	Bang et al., 2021
Black peppercorns	UVC	Cold Plasma	Mesophilic aerobic bacteria	3.4	Bang et al., 2020
			Bacillus tequilensis spores	1.7	
Blueberries	Pulsed Light	Water-assisted	Escherichia coli O157:H7 strains	3.0 and >5.8 (calyx and skin)	Huang & Chen, 2014*
			Salmonella enterica strains	3.6 and >5.9 (calyx and skin)	
Blueberries	Pulsed Light	Water-assisted	Salmonella enterica serotypes	4.4 and 0.8 (spot and dip ino.)	Cao et al., 2017
Blueberries	UVC	Water-assisted	Salmonella strains	5.45 (spot ino.)	Guo et al., 2019*
Blueberries	Pulsed Light and UV	Water-assisted (H$_2$O$_2$/Chlorine)	Salmonella strains	4.5 to 5.7 (WPL and WUV treatments, spot ino.)	Huang & Chen, 2019a*
				1.8 to 2.3 (dip ino.)	
Tomatoes				4.4 to 5.4 (WPL and WUV treatments, spot ino.)	
				1.9 to 2.5 (dip ino.)	
Lettuce shreds				1.9 to 3.1 (WPL and WUV treatments, spot ino.)	
				1.9 to 2.6 (dip ino.)	
Button mushrooms	UVC	H$_2$O$_2$	Escherichia coli H157:O7	0.87	Guan et al., 2013
			Total aerobic count	0.2–1.4	
Coriander	UV-LED	Slightly Acidic Electrolyzed Water (SAEW)	Salmonella Typhimurium	2.72	Jiang et al., 2020
			Escherichia coli O157:H7	2.42	

(Continued)

TABLE 8.3 (CONTINUED)   COMBINED APPLICATIONS OF LIGHT AND OTHER FOOD PROCESSING APPLICATIONS

Food Products	Light System	Combined Treatment	Target Microorganism	Reduction (log₁₀ CFU/g, CFU/mL, spores/g or spores/mL)	Reference
Date fruits	UVC	Ozone (O₃) and Electrolyzed Water (EW)	Mesophilic bacteria, coliforms	1.05 and 0.82 (O₃ and EW)	Jemni et al., 2014
			Yeasts and molds	1.63 (O₃)	
Dongchimi (Korean watery radish kimchi)	UVC	High Hydrostatic Pressure (400 MPa)	Lactic acid bacteria	4.09	Shahbaz et al., 2017*
		High Hydrostatic Pressure (500 MPa)		6.18	
Fresh-cut apple	Pulsed Light	Gellan Gum	Total mesophilic bacteria	~2	Moreíra et al., 2015
			Psychrophilic bacteria	~2	
			Yeast and molds	~1	
Fresh-cut apple	UVC	Pretreatment [Hot water blanching, dipping into a solution (ascorbic acid and calcium chloride)]	Listeria innocua	0.24	Gómez et al., 2010
			Escherichia coli	~0.4	
			Saccharomyces cerevisiae	0.6	
Fresh-cut broccoli	UVC	Water-assisted	Listeria innocua	1.7 and 2.4 (0.3 and 0.5 kJ/m²)	Collazo et al., 2019
Fresh-cut cantaloupe	Repetitive Pulsed Light	Edible Coating (Pectin, alginate, and gellan gum)	Total aerobic mesophilic count and yeast and mold populations	~3-4	Koh et al., 2017
Fresh-cut cauliflower	UVC	Natural antimicrobial (M and A formulation)	Listeria monocytogenes	1.5 (with antimicrobials)	Tawema et al., 2016
			Escherichia coli O157: H7	2.0 (M formulation)	
			Total yeasts/molds	2.0 (M formulation)	

(Continued)

TABLE 8.3 (CONTINUED)   COMBINED APPLICATIONS OF LIGHT AND OTHER FOOD PROCESSING APPLICATIONS

Food Products	Light System	Combined Treatment	Target Microorganism	Reduction ($\log_{10}$ CFU/g, CFU/mL, spores/g or spores/mL)	Reference
Fresh-cut mango	Pulsed Light (PL)	Alginate Coating (ALC) and Malic Acid Dipping (MA)	Listeria innocua	4.5 and 3.91 (MA-PL and PL-ALC-MA)	Salinas-Roca et al., 2016
			Molds and yeasts	2.07 and 2.09 (PL or PL-ALC-MA)	
			Psychrophilic bacteria counts	1.37 and 1.09 (PL or PL-ALC-MA)	
Fruit juices (clear pear)	UVC	Mild Heat	Escherichia coli, Saccharomyces cerevisiae, Lactobacillus plantarum	5.5–6.7 (general bacteria)	Fenoglio et al., 2020*
Turbid orange-tangerine				5.2–5.6 (general bacteria)	
Orange-banana-mango-kiwi strawberry juices				6.3–6.6 (general bacteria)	
Grape juice (Unripe table)	Pulsed Light	Mild Heat	Saccharomyces cerevisiae	5.10–5.06	Kaya et al., 2020*
Grapes	LEDs	Curcumin (Food additive, aroma)	Escherichia coli	6.3 (approximately)	Aurum and Nguyen, 2019*
Green beans	Pulsed Light	Essential Oil and Chitosan-Based Coating	Listeria innocua	2	Donsì et al., 2015
Green onion	Pulsed Light	Sodium Dodecyl Sulfate (SDS), Chlorine, Citric Acid, Thymol, and Hydrogen Peroxide	Escherichia coli O157:H7	4	Xu et al., 2013
Green onion	Pulsed Light	SDS and Tween 80 surfactant	Salmonella strain	2.6–2.9 and 2.4–2.7 (PL-SDS and PL-Tween 80) respectively	Xu et al., 2015
Green onion	Pulsed UV Light	Thymol and H_2O_2, SDS, Citric acid, Acetic acid in combinations	Salmonella enterica Typhimurium	>4 (spot inoculation)	Xu & Wu, 2014

(Continued)

TABLE 8.3 (CONTINUED) COMBINED APPLICATIONS OF LIGHT AND OTHER FOOD PROCESSING APPLICATIONS

Food Products	Light System	Combined Treatment	Target Microorganism	Reduction ($\log_{10}$ CFU/g, CFU/mL, spores/g or spores/mL)	Reference
Iceberg Lettuce	Pulsed Light	Ultrasound and Chlorine	Salmonella enterica	<2 (all treatments)	Huang & Chen, 2018
Orange juice	UVC	Nisin (Food additive, preservative)	Alicyclobacillus acidoterrestris	1.0	Ferreira et al., 2020
Peach nectar	UVC	Potassium Sorbate (KS) or Sodium Benzoate (NB)	Aspergillus flavus and Aspergillus niger	>3.5 (UVC-NB), ~0.5 (UVC-KS)	Flores-Cervantes et al., 2013
Raspberries Blueberries	Pulsed Light	H_2O_2 water-assisted	Salmonella enterica strains	4.0 >5.6	Huang et al., 2015*
Red pepper flakes	Intense Pulsed Light	Cold Plasma	Aspergillus flavus spores	1.3±0.1	Lee et al., 2020b
			Bacillus pumilus spores	2.3±0.2	
			Escherichia coli O157:H7	>3.8	
Strawberries and raspberries	Pulsed Light	Water-assisted (Chlorine, H_2O_2, and SDS)	Escherichia coli O157:H7	3.3 and 5.3 (strawberry and raspberry)	Huang & Chen, 2015*
			Salmonella enterica serotypes	2.8- and 4.9	
			Murine norovirus (MNV-1)	1.8- and 3.6	
Strawberry	LEDs	Chlorophyllin (Food additive, color)	Listeria monocytogenes	98% decontamination	Luksiene & Paskeviciute, 2011
			Yeasts/microfungi	86%	
			Mesophiles	97%	
Tomato	Pulsed Light	LAPEN wash (short chain organic acids, EDTA and nisin)	Salmonella enterica	>5 log	Leng et al., 2020*

* More than 5 log CFU reduction were reported.

2014; Rowan, 2011). The PL alone is only effective on the surface and may require excessively long treatment times for satisfactory reduction (Søltoft-Jensen & Hansen, 2005).

To combat the ill-effects of a particular food processing application at a higher level, the combination of two interventions with a multitargeted approach (hurdle application) with various antimicrobial actions can be more effective. In each different treatment, the target can be one of the diverse cell systems, e.g., cell membrane, enzyme systems, DNA, pH, Eh, aw, within the microbial cell, and can contribute to disturbing the microbial homeostasis. Even further, the use of treatments with the same target may reduce the chances of reversible damage to the cells (Sahoo et al., 2021). Hurdle applications, in general, accomplish better microbial inactivation while minimizing the negative impact on nutritional and organoleptic quality. The combination of PL with different technologies (hurdle) has been investigated by many researchers. Table 8.3 summarizes some recent studies published on the effect of PL treatments combined together with other food processes. To date, among the PL combined treatments, the noteworthy application is low or high temperature, acidity, natural or chemical preservatives, and other non-thermal processes such as pulsed electric field, high hydrostatic pressure, sonication, etc. (Table 8.3).

Based on the hurdle concept, complementary non-thermal technologies (e.g., ultrasound, high hydrostatic pressure, pulsed electric fields, the addition of antimicrobials) are selected depending on the food properties. Combined application with different modes of action can achieve higher microbial inactivation levels with an additive or a synergistic effect (Ross et al., 2003). The results in Table 8.3 show that PL hurdle application can be an alternative preservation technology to thermal pasteurization for bacterial control in fruit juices. Apple juice contaminated with *E. coli* was exposed to a pulsed electric field (24 kV/cm or 34 kV/cm) followed by a continuous PL at dosages corresponding to 65.4 and 51.5 J/ml (Caminiti et al., 2011). The synergistic inhibition effect on *E. coli* was above the FDA recommendation of 5-log and juice did not show any obvious change in chemical, physical, and sensory attributes (Caminiti et al., 2011). Later, the same group reported effects of selected physicochemical and sensory attributes a blend of orange and carrot juice processed sequentially by pulsed electric fields (24 kV/cm, 18 Hz, 93s), UVC (10.6 J/cm^2), or high-intensity PL (3.3 J/cm^2) with manothermosonication (400 kPa, 35°C, 1000 W, 20 kHz) (Caminiti et al., 2012). No significant change in total phenolics and antioxidant activity of the orange and carrot juice was observed regardless of the treatments applied. The combined treatment ultrasound (10-min and 30-min ultrasound) followed by PL showed an additive effect for a great reduction of *Alicyclobacillus terrestris* and *Saccharomyces cerevisiae* counts in apple juice 5.9 and 6.4 log CFU/g, respectively, and this delayed spoilage for seven days (Ferrario & Guerrero, 2016, 2017). Kwaw et al. (2018) observed that PL treatment (360 µs, 3 Hz, 1.213 J/cm^2, 4 sec) before sonication (28 kHz, 60 W, 15 min) improved significantly the phenolic concentration, flavonoid concentration, anthocyanin concentration in lactic acid fermented mulberry juice. Shahbaz et al. (2017) reported the use of UVC light followed by high hydrostatic pressure (400–500 MPa) at ambient temperature for the control of lactic acid bacteria growth in fermented watery radish kimchi. The UV treatment before high-pressure application inhibited bacterial growth over 6 log via a synergistic effect while the characteristic product quality was preserved (Shahbaz et al., 2017). PL combined with plasma treatment in red pepper powder reduced the number of indigenous mesophilic aerobic bacteria from 4.8 log CFU/g to 1.9 log CFU/g and decreased the number of indigenous yeast and molds by 2.5 log CFU/g, which is below the maximum limits established in spices (Lee et al., 2020a). PL and plasma treatment did not affect the color of red pepper powder though it affected flavor attributes (Lee et al., 2020a).

Organic acids are strongly antimicrobial against food contaminating microorganisms and are used as an antimicrobial treatment on food surfaces. Their antimicrobial activity was attributed to the reduction of the medium pH and intracellular pH, chelation of metal ions, and disruption of membrane permeability (Eswaranandam et al., 2004). Peeling and cutting operations make fresh-cut fruits more susceptible to microbial attack. When PL treatment (180–1100 nm, 12 J/cm^2) and malic acid (2% w/v) were individually applied to fresh-cut watermelon, inactivation of *L. innocua* and *E. coli* changed from 0.91 to 2.97 log CFU/g (Ramos-Villarroel et al., 2015) whereas the treatments were applied sequentially, the observed reduction was more than 5-log, which is the FDA requirement (Ramos-Villarroel et al., 2015). The use of anti-browning malic acid dipping solutions along with PL minimized the development of browning after PL treatment on avocadoes and mushroom slices (Ramos-Villarroel et al., 2012b, 2015). TEM observations of both bacterial cells revealed that the combined application of PL and malic acid caused higher cell damage with agglutination of cytoplasmic content and the disruption of cell membrane in both microorganisms (Ramos-Villarroel et al., 2015). PL used in combination with nisin may not achieve higher inactivation levels of bacteria on the surface compared with PL treatment alone probably due to UV absorption of antimicrobials when they are applied before PL treatment (Proulx et al., 2017). Treatment of PL-LAPEN (short-chain organic acids, EDTA, and nisin) demonstrated strong synergistic inactivation, and greater than 5 log CFU/g *Salmonella* reduction was observed (Leng et al., 2020). GRAS microbial mixture not only reduced natural microbiota of tomato and artificially contaminates tomato but also preserved the firmness and visual appearance quality of tomato with no regrowth possibility of the pathogen during cold storage at 10°C for 3 weeks (Leng et al., 2020). As anticipated, the application of nisin alone did not significantly affect the gram-negative bacteria included in this study. The addition of sodium benzoate before UVC treatment allows *Aspergilli* reductions in spore viability (Flores-Cervantes et al., 2013). Since potassium sorbate absorbs light at ultraviolet wavelength, in a combined treatment it should be added after UVC treatment (Flores-Cervantes et al., 2013). The 80–240 µW/cm^2 UV-LED irradiation for 30 min combined with slightly acidic electrolyzed water (ACC of 60 mg/L) 5-min-washing led to a reduction in *Salmonella* (2.42 log CFU/g) and *E. coli* 2.72 log CFU/g concentrations and delayed deterioration while the quality score of the control group declined significantly after two- days of storage (Jiang et al., 2020).

Coupling PL treatment with antimicrobials offered a more effective solution for reducing the amount of antimicrobials applied to fresh produce. In particular, the combination of PL with natural antimicrobials represents a novel approach for fresh produce and fresh-cut fruits and vegetables. Donsi et al. (2015) evaluated the effects of a modified chitosan-based bioactive coating essential oil nanoemulsion as antimicrobial agent with PL against *L. innocua* inoculated on green beans. The combination of the chitosan-coating application with PL did not show any synergistic or additive antimicrobial effect on *L. innocua* during storage and had a harmful impact on the color properties of fresh green beans (Donsi et al., 2015). Samples were treated with PL treatment and bioactive coating (Donsi et al., 2015). Koh et al. (2017) reported that the combination of pectin, alginate, and gellan coatings with repeated PL treatment (0.9 J/cm^2 every 48 h up to 28 d of storage at 4°C) was effective to reduce fluid loss and retain the firmness of fresh-cut cantaloupes compared to the control. The combination of alginate and repeated PL treatment was the most effective treatment for maintaining microbiological quality, enhanced firmness, and extended the shelf life of fresh-cut cantaloupes by up to 28 days (Koh et al., 2017). PL (20 pulses at fluence of 0.4 J/cm^2 /pulse), alginate coating (2%), and malic acid dipping (2%) treatments on quality and safety aspects of fresh-cut mangoes inoculated with *L. innocua* were evaluated and the tested combined treatments resulted in a maximum of 4.5 log reduction (Salinas-Roca et al., 2016). The firmness of alginate-coated slices progressively decreased over storage. Color parameters and total soluble solids content decreased in all treated mango slices throughout 14 days, while pH was kept similar to that of fresh tissue (Salinas-Roca et al., 2016). 1-methylcyclopropene (SmartFresh®) (600 nL/L) which binds ethylene and delays the fruit ripening (Williamson et al., 2018). PL (9 J/cm²) combined with 1-methylcyclopropene limited ethylene biosynthesis promoted an additional 12 days in the shelf life of melons (Sousa et al., 2019).

Some investigators assessed the influence of PL application with quality-stabilizing dips. Water-assisted PL treatment showed higher efficacy than dry PL treatment for both spot- and dip-inoculated samples. For *Salmonella* inactivation, dry PL treatment achieved 0.9 and 0.6 log reduction of *Salmonella* for spot and dip inoculation on blueberries; while the water-assisted PL treatment reduced *Salmonella* by 4.4 log and 0.8 log for spot and dip inoculation, respectively (Cao et al., 2017). The application of PL to enhance the safety and shelf life has been studied and two major challenges are substantial heating of fruits and samples should be positioned directly toward fruit in all directions (Huang & Chen, 2014; Ramos-Villarroel et al., 2011, 2012b). Reduction of *Salmonella enterica* was achieved in inoculated blueberries (4.2 $\log_{10}$ CFU/g) and iceberg lettuce (1.5 log CFU/g), using water-assisted UV (34.8 kJ/m²) (Guo et al., 2017). After water-assisted 60-s PL treatment, the populations of *E. coli* O157:H7 inoculated on calyx and skin of blueberries were reduced by 3.0 and >5.8 log CFU/g, respectively (Huang & Chen, 2014). Water-assisted PL treatment was highly effective; thus, no viable bacterial cells were recovered from the water used in the wet PL treatments (Huang & Chen, 2014). PL treatment of blueberries and tomato samples showed more than 5 log reduction of *Salmonella* when water-assisted washing was used (Guo et al., 2019; Huang & Chen, 2019a). Low-dose PL immersion-assisted in water allows inactivation and inhibition of native and pathogenic microbiota; nevertheless, the water-assisted PL system is the proper application to be used for fresh-cut or frozen fruits and vegetables; as such, cherries and berries are not washed.

Instead of prolonged PL treatment, a combination of short PL exposure with active sanitizer treatment provides the stipulated inactivation with fewer adverse quality effects. H_2O_2 produces reactive hydroxyl radicals, which demonstrate strong antimicrobial activity (Rosenfeldt et al., 2006). The combination of water-assisted PL treatment with 1% H_2O_2 for 60 s showed significantly higher efficacy by reducing *E. coli* O157:H7 on strawberries and raspberries by 5.3-log units (Huang & Chen, 2015). Under the same treatment conditions, *Salmonella* on strawberries and raspberries was inactivated by 2.8- and 4.9-log units (Huang & Chen, 2015). Meanwhile, no significant increase in the efficacy of treatment was observed when H_2O_2 was used with PL against murine norovirus on both berries (Huang & Chen, 2015). Huang et al. (2015) demonstrated that a combined treatment of 1% H_2O_2 and PL showed a better *Salmonella* inactivation effect than PL alone on raspberries and blueberries. Sodium dodecyl sulfate (SDS) is an FDA-approved food additive which behaves as an anionic surfactant that can be used to release bound pathogens from the plant surface (Javaid et al., 2021). For spot-inoculated green onions, thymol individually and sanitizers formulated as H_2O_2+SDS, citric acid+SDS, thymol+citric acid achieved higher than 5 log CFU/g reduction of *Salmonella* (Xu & Wu, 2014). The energy dose received by green onion was measured as the fluence of 5.0–56.1 J/cm² but severe quality degradation softening tissue and burned appearance was observed even after 30-sec treatments (Xu & Wu, 2014).

Sanitizers, such as hypochlorite, chlorine dioxide, and peroxyacetic acid are used in the washing process to remove soil, plant debris, and microorganisms from most fresh produce and prevent cross-contamination through wash water (Olaimat & Holley, 2012). Chlorine is the most commonly used and it presents the advantages of relatively low cost and ease of use. Some studies have demonstrated that washing with chlorine cannot eliminate pathogens on fresh produce and the decontamination efficacy of chlorine washing is highly affected from organic load and organic matter can diminish its antimicrobial efficacy (Luo, 2007). The chlorine treatment efficacy is low – even a concentration as high as 320 ppm chlorine for 2 min is unable to inactivate *Salmonella* entirely (Zhuang et al., 1995). Chemical sanitizers currently applied for food-contact surfaces are insufficient to control the emergence of outbreaks as pathogens are becoming persistent to these treatments. Additionally, some bacteria form biofilms which were shown to be more resistant to various disinfectants, including chlorine, than planktonic cells (Sanchez-Vizuete et al., 2015). Consumers have fewer reservations about chemical treatments and due to carcinogenic and mutagenic chlorine products, some countries (e.g., Germany, Holland, Switzerland) have reduced chlorine usage for vegetable disinfection (Feliziani et al., 2016; Sahoo et al., 2021). To overcome the limitations of chlorine-based disinfection, the food industry is currently investigating non-thermal techniques. Huang and Chen (2019b) reported more than 5 log inactivation of *Salmonella* after the combination of 1 min PL treatment with 2 min wash at 30°C with either chlorine (10 ppm) or H_2O_2 (1%) in water. The combined PL and chlorine treatment were, in general, more effective in inactivating *Salmonella* on iceberg than sole chlorine wash regardless of the water quality, inoculation method, and sample size (Huang & Chen 2018); nevertheless, the authors observed that the enhanced inactivation effect of PL-chlorine on *Salmonella* contaminated iceberg over chlorine wash was more pronounced for the small-scale (Huang & Chen, 2018). Later, Huang and Chen (2019b) evaluated the performance of decontamination based on H_2O_2 (1%) and chlorine (10-ppm free chlorine) washing along with PL exposure on *Salmonella*-contaminated grape tomatoes. For dip-inoculation, the combined treatments of PL-chlorine and PL-H_2O_2 in the small-scale treatment had *Salmonella* counts in wash water below the detection limit of 2 CFU/mL (Huang & Chen, 2019b). Huang et al. (2015) developed and evaluated a small scaled-up water-assisted PL system, in which berries were washed (with or without H_2O_2) in a flume washer while treated by PL. Compared with chlorine washing, the combination of water-assisted PL and 1% H_2O_2 was the most effective treatment which reduced *Salmonella* on raspberries and blueberries by 4.0 and 5.6 log CFU/g, respectively. The authors concluded that this combined application is an option for the processing of small berries intended for fresh-cut and frozen berry products (Huang et al., 2015). During pre-harvest, pathogens in animal manure or irrigation water may contaminate the plant's surface, transfer to other parts of the plant or even internalize within the tissue (Solomon et al., 2002). Processing of fresh fruit and vegetables, especially soil harvested green onions, potatoes, carrots, etc., produces large amounts of contaminated wastewater that contain organic and inorganic environmental pollutants. The application of PL to decontaminate wastewater derived from fresh produce processing has the potential to decrease the water footprint of fresh-cut vegetables by minimizing the overall requirement for water in industrial plants. Additionally, it would decrease the residuals of toxic chemicals in fresh-cut vegetables by avoiding the use of sanitizers, such as chlorine (Manzocco et al., 2015).

Thermal processing disrupts the integrity of cell walls and internal membranes, thereby facilitating the release of cellular compounds and increasing enzymatic bioaccessibility. Mild thermal processing combined with light processing has the potential to play a pivotal role in minimizing temperature-related damages. Changes in physicochemical characteristics of fresh-cut produce and juices were evaluated by the optimized the decontamination process using thresholds. Fenoglio et al. (2020) explored the feasibility of moderate heat (50°C) combined with PL to reduce *S. cerevisiae* in clear and turbid juices. An additive inactivation effect was observed ranging from 5.2–6.7 log in clear pear juice, while 4.6–4.9 log reductions of *S. cerevisiae* were observed in turbid orange-tangerine and orange-banana-mango-kiwi-strawberry juice (Fenoglio et al., 2020). Similarly, Garcia Carrillo et al. (2018) also achieved 5 log reduction of *S. cerevisiae* cells with increased membrane permeability in carrot-orange juice treated by UV-C (1060 mJ/cm², 10 or 15 min) accompanied by heat (50°C). Inactivation levels reported so far were not high enough to ensure the safety of the treated juices, however, these results represent a starting point for exploring the usefulness of PL treatment in the control of microorganisms in juices. Though some scientists have optimized thermal processing conditions within a hurdle to minimize the negative effects, the impact remains a great challenge in fruit juice processing.

8.4 OTHER APPLICATIONS WITH PL

8.4.1 Drying of Fresh Produce

PL treatment is currently being investigated as drying pretreatment. Limited publications on the application of PL in convective drying of fruits and vegetables are available. The presence of protective layers in fruits and vegetables obstructs moisture removal leading to longer drying time and quality degradation. This has encouraged researchers to investigate the use of pre-treatments for improved permeability and accelerated drying rates. The use of light pretreat-ments ahead of a drying alternative to conventional pretreatments has proven to be beneficial for improving the drying of fruits and vegetables in some cases. PL can modify the surface of materials by pore formation, increased evaporation, moisture reduction, and surface tempera-ture elevations, thereby accelerating mass transfer rate, shortening drying time, and enhancing quality retention in fruits and vegetables. Phimphilai et al. (2012) observed that UV-C assisted hot-air drying on longan fruits, the SEM microscope results of their research further revealed that samples dried using UV-C showed deeper shell layers, which caused a higher rate of mois-ture evaporation as compared to hot air drying. UV-C processing results in increased pore for-mation, thereby increasing the rate of moisture transfer (Onwude et al., 2017). The energy of PL is absorbed by the fruits and converted into internal energy, which results in partial evaporation of the water of the fruit (Braga et al., 2019). PL treatment is being investigated to modify the food surface or structure to withhold better exposure to heating on the road to shorten the air-drying time of fruits (Braga et al., 2019; Fernandes et al., 2011). More specifically, as a secondary effect on surface disinfection, the PL energy is absorbed by the fruit and converted into internal energy, which results in partial evaporation of the water of the fruit (Braga et al., 2019). Braga et al. (2019) applied PL on mangoes as a pretreatment before convective air-drying, PL pretreat-ment reduced the initial moisture content by 4.3 to 15.9% in the samples but did not affect the kinetics of water diffusivity during the subsequent drying process. The significant photo thermal and drying effects during LED treatment could be adequately controlled by a continuous sup-ply of low-temperature or mild hot air fluidization to increase the surface area of exposure and replace the long hot air-drying step in food processing (Subedi & Roopesh, 2020). In general, the PL pretreatment did not reduce the air-drying time and Braga et al. (2019) concluded pulsed UV light pretreatment would not be a proper technology to reduce the air-drying time if the primary goal is to reduce the drying duration.

8.4.2 Plant Material Extraction

The industry is keen to use fruit and vegetable waste to mine high-value ingredients that have the potential for pharmaceutical or nutraceutical functional product development. Steroidal alkaloids are considered possible precursors to produce hormones and antibiotics (Hossain et al., 2015). Potato peel is a good source of these compounds and PL treatment of potato peels at 7.86 J/cm^2 led to higher levels of alpha-solanine and alpha-chaconine was 38.78% and 33.13%

respectively (Hossain et al., 2015). PL is a source of abiotic stress for living plant cells and is known to improve the accumulation of glycoalkaloids (Dale et al., 1993), and pretreatment of plant sources for long repeated durations before extraction could be particularly useful for economically improving the content for the later extraction. The action of PL against cells also involves agglutination of cytoplasmic content leading to disruption of cell membranes and increasing the extraction of bioactive compounds with cell rupture (Kim et al., 2019; Ramos-Villarroel et al., 2013). Kim et al. (2019) developed a hurdle application for maximizing extraction efficiencies of quercetin, a bioactive flavonoid from onion waste skin by physical pretreatment of intense PL (1200 V for 60 s) combined with subcritical water extraction (145°C for 15 min). Instead of ethanol or other solvents, they used water, which is inexpensive and requires less energy (Kim et al., 2019). The maximum concentration of quercetin extracted (22.16 ± 1.61 mg/g onion skin) was obtained for PL at 1200 V for 60 sec with following subcritical water extraction at 145°C for 10 min using pilot-scale treatment. These results proved that there is also an economic potential with developing a PL hurdle technology by maximized extraction efficiencies of plant metabolites from horticulture by-products.

8.5 QUALITY EFFECTS OF POST-LIGHT TREATMENT ON FRUIT AND VEGETABLE PRODUCTS

Nearly all studies reported no or an insignificant difference in the pH value, electrical conductivity, titratable acidity, total soluble solids, and reducing sugar of juice samples following PL irradiation at 200–1200 mJ/cm^2 (Noci et al., 2008; Palgan et al., 2011; Xiang et al., 2020).

The influence of light on quality characteristics of fresh and fresh-cut fruits and vegetables is controversial, as it has been described as constructive or in some cases harmful. PL treatments considerably increased the total lycopene, phenolic content, and antioxidant activities (Koyyalamudi et al., 2011; Liu et al., 2009, 2012; Pataro et al., 2015). PL stimulates higher production of photo protective anthocyanins induced in plant tissues under stressful conditions (Solovchenko & Merzlyak, 2008). The release of lipophilic (carotenoids) and hydrophilic (e.g. polyphenols) bioactive compounds from raw fruit and vegetables are restricted by the cell walls (Schweiggert & Carle, 2017). The decrease in the antioxidant activity of UVC-LEDs irradiation of juices may be due to the degradation of phenolic compounds, which are major contributors to the total antioxidant capacity of apple juice (Xiang et al., 2020). Moreover, UVC-LED irradiation has been reported to cause the degradation of ascorbic acid, a water-soluble antioxidant in fruit and vegetable juices (Zhu et al., 2014). Processing may affect the number of phenolic compounds in the microstructure of the food matrix, consequently affecting the bioaccessibility of phenolic compounds. Scarce studies were conducted to evaluate the nonthermal processing impact on bioaccessible phenolic compounds.

8.5.1 Firmness and Texture Attributes

Firmness is a critical quality indicator in the evaluation of fresh produce for processing, which further affects the partiality and satisfaction of consumers (Ramos et al., 2013). Pombo et al. (2009) reported that UV-C radiation delayed strawberry fruit softening due to a decrease in the transcription of genes encoding enzymes involved in wall degradation, during the first hours after treatment. PL did not have a damaging effect on the mechanical properties of strawberries throughout cold storage and delayed softening (Duarte-Molina et al., 2016; Romero Bernal et al.,

2019a). There were no significant differences in maximal rupture force, mechanical work, and deformability modulus values between PL treated and untreated fruits immediately after treatments (Duarte-Molina et al., 2016). Berries are extremely fragile, highly prone to fungal contamination, and require gentle handling seasonal fruits, usually last from days to weeks. Normally, no post-harvest processing technologies (e.g., washing) are applicable (Barkai-Golan, 2001). PL is a promising tool capable of inducing the synthesis of endogenous polyamines, which could be related to a limitation of accessibility to the cell wall polysaccharides of the hydrolytic enzymes whose activity promotes fruit softening (Charles et al., 2013). Sousa et al. (2019) reported an increase in endogenous polyamines in melon increased postharvest resistance and improved textural quality parameters for better shelf life (Suosa et al., 2019). In the case of fresh-cut produce, the data indicated that there were no significant differences in the texture parameters between fresh-cut cantaloupes exposed to PL (Koh et al., 2016).

8.5.2 Color

Color is an important quality attribute that influences consumer acceptance. The effect of PL irradiation on the color of fruit and vegetables was evaluated in nearly all studies. Generally, no significant differences in color parameters were observed between untreated and PL-treated fresh-cut fruits and vegetables during storage. In some cases, treatment of fresh produce with high-intensity PL improves color with an increasing concentration of anthocyanins and total phenolics (Rodov et al., 2012). Rodov et al. (2012) reported that a PL treatment was enough to develop deepening in color in poorly colored "Brown Turkey" figs. Brief postharvest exposure to PL during storage compensated for the insufficient color development in these fruits (Rodov et al., 2012). The color changes are attributed to the breakage of plant cellular membranes. Changes in cell integrity lead to loss of functional plant cell compartmentalization and consequently result in contact between enzymes and substrates in plant tissue (Gómez et al., 2012b). Similarly, PL processing may increase bio accessibility by destroying the integrity of the cell walls and internal membranes, thereby facilitating the release of the carotenoids (Cervantes-Paz et al., 2016; Thakkar et al., 2009). In some cases, quality defects during storage were reported as yellowing of the stems or loss of green pigmentation, which is considered a major concern for shelf life (Aguilar et al., 2018; Diesler et al., 2019; Xiang et al., 2020).

8.5.3 Dietary/Nutritional Properties

Consumer demand for high-quality preserved foods rich in natural bioactive compounds such as vitamins, phenolic compounds, pigments, and dietary fibers, etc. is drastically increasing. Studies have reported that the nutritional changes induced by intense light pulses are highly different and depend on plant species, the type of product, and PL dose applied. In some studies, UV light was found to contribute toward the enhancement of the nutritional quality of food products, enhancing vitamin content, total carotenoid content, antioxidant capacity, and other phytochemical properties (Cantos et al., 2000; Ganan et al., 2013; Gonzalez-Aguilar et al., 2001; Koutchma, 2008; Tikekar et al., 2011). In other studies, PL technologies showed contradictory results regarding their impact on the nutritional compounds in fruits and vegetables (Aguilo-Aguayo et al., 2017).

Oms-Oliu et al. (2010) observed that a high energy dose of PL treatment caused an initially sharp and then gently continuous decreasing trend in fresh-cut mushrooms during storage. Akgun and Unluturk (2017) observed a complete diminishing of vitamin C after UV

LED treatment in apple juice. The results of Zhang et al. (2015) showed that the red and blue LED-based PL treatment led to higher vitamin C content in citrus varieties. Braga et al. (2019) reported a preparatory application of pulsed UV light (3.6 and 10.8 J/cm^2) increased the concentrations of vitamins C, as well as B1, B3, and B5 (10%–40%) in the air-dried mangoes. Numerous explanations on the increasing level of vitamin C levels after light irradiation were reported. The increase in the pool size of vitamin C can be the indirect result of increasing soluble carbohydrates (Toledo et al., 2003). Later, Leivar et al. (2009) and Zhang et al. (2015) studied the gene expression levels of vitamin C, and the up-regulation of vitamin C biosynthetic genes contributed to increasing. Freshly harvested button mushrooms exposed to pulsed UV light 2.5–9 pulses (1.150 J/cm^2) during storage showed improved vitamin D2 and bioactive ergosterol-derived products including lumisterol$_2$ and tachysterol$_2$ (Koyyalamudi et al., 2011; Kalaras et al., 2012). Similarly, PL tended to result in higher retention of vitamin B1, possibly because the high energy of the PL treatment might have broken the bond of the phosphorylated vitamin, changing its status to free and bioavailable form (Braga et al., 2019). Whereas vitamin B6 decreased by 40 to 50% in pretreated mangoes by exposure to pulsed UV light dosages higher than 10.8 J/cm^2 (Braga et al., 2019). Undoubtedly, further experimental investigations in relation to the fruit cultivar, maturity stage, mode of light, energy delivery, and post-treatment storage conditions are needed to make any advanced conclusion.

8.5.4 Endogenous Enzymes

The reduction of enzymatic activity is one of the main goals for fruit and vegetable processing. Generally, fresh-cut fruit or vegetable are prepared by being washed, cut, or sliced, packaged, and later stored at low temperature. The wounding of the plant tissue brings together enzymes and substrates that are normally compartmentalized within the cell and result in enzymatic deterioration on the surface (Aguilo-Aguayo et al., 2014; Gómez et al., 2012a). To date, PL has been demonstrated to reduce the catalytic activity of only few enzymes. If proven effective, PL can offer a safer and eco-friendly alternative approach to control enzymatic changes in fresh produce compared with traditional additives, such as sulfites, and thermal treatments that have possible adverse effects on fresh products, which involve high energetic costs and environmental impact.

Pectin plays a crucial part in the texture. It is primarily present in the primary cell wall and middle lamella, and is involved in cell wall adhesion (Liu et al., 2020). Pectin methyl esterase is an enzyme that is implicated in the hydrolysis of the ester linkage in pectin. Inactivation of the pectin methyl esterase enzyme is necessary to preserve cloudy juices since cloudiness causes jellification of concentrated juice and an unacceptable appearance for consumers (Liu et al., 2020). PL doses (6 kJ/m^2) have shown to be suitable to preserve the firmness of whole mangoes, by decreasing the pectin methyl esterase activity (Lopes et al., 2016). The hurdle application of PL along with a pulsed electric field caused marginal inactivation (86%) of the enzyme under selected conditions, while a single application of either PL or UV did not result in any reduction (Caminiti et al., 2012). In a more recent study, Pellicer et al. (2020) reported PL at a fluence of 128 J/cm^2 decreased 90% of pectin methyl esterase, and the PL treatment affected the tertiary structure rather than the secondary structure. PL disrupts the tertiary structure of the enzyme; therefore, the exposure of hydrophobic residues led to intermolecular hydrophobic interactions causing that leads to the formation of aggregates (Pellicer et al., 2020). According to the structural analysis results, deactivation resulted in site-sensitive protein fragmentation and changes to the tertiary structure (Janve et al., 2014; Jeon et al, 2019; Pellicer et al., 2019).

Peroxidase is an enzyme that needs to be inhibited to prevent its deleterious effects on the quality of fruits and vegetables. The feasibility of using PL (128 J/cm²) to inactivate peroxidase was tested and PL reduced the activity by more than 95% (Pellicer & Gomez-Lopez, 2017). The inactivation by PL was due to induced significant loss of secondary structure in the α-helix structure (Pellicer & Gómez-Lopez, 2017; Wang et al., 2017). The peptide bonds in enzymes absorb UV at 180–230 nm, while the aromatic residues have peak absorbance between 258 and 280 nm and disulfide bonds absorb near 260 nm (Schmid, 2001). Recent studies have reported that PL wavelength can cause enzyme deactivation through changes in protein structure (Pellicer et al., 2020). Pellicer and Gomez Lopez (2017) studied the structural changes in the horseradish peroxidase enzyme structure and reported far-UV circular dichroism showed changes in the secondary structure, with loss of helical structure and increase in β-sheet fraction.

Polyphenol oxidase enzymes are responsible for enzymatic browning, which is the major source of quality and economic loss in fresh fruit and juice products. Slicing or kitchen operations on the fruit and vegetable combine substrates and enzymes to initiate enzymatic browning reactions (Brennan & Gormley, 1998). Browning represents a problem for fresh-cut fruit and vegetables, especially white-fleshed fruit such as apples and pears. To prevent enzymatic browning, traditionally thermal processing and antioxidant addition are applied to food. Manzocco et al. (2009) studied the effect of UVC (15 W lamps with maximum emission at 253.7 nm) and visible light (430–560 nm) treatments on polyphenol oxidase in apples at 28°C. They found that the UV-C was more effective than visible light in the inactivation of polyphenol oxidase in an aqueous solution showing inhibitions of 100% after 60 min exposure to 13.8 W/m² irradiances. Irradiation for 120 min with a high-pressure mercury lamp of 400 W emitting UV-visible light between 250 and 740 nm effectively inactivated the enzyme in apple and pear juice (Falguera et al., 2011, 2014). Lante et al. (2016) reported that UV-A treatment was successful in controlling the enzymatic browning on fresh-cut apples and pears. After determining the optimal operational conditions of a UV LED illuminator (2.43 10–3 W/m² irradiance, 0.5 cm distance from the sample), the anti-browning effect of UV-A light (390 nm) treatment at 25°C was assessed on fresh-cut apples and pears (Lante et al., 2016). Under these treatment conditions, the color change of fresh-cut apples decreased by approximately 60% after 60 min exposure (Lante et al., 2016). Manzocco et al. (2013) observed that the loss of polyphenol oxidase activity was mainly related to structural modifications, including protein backbone cleavage, unfolding, and aggregation.

The UV spectrum is well known to accelerate lipid oxidation (Rajkovic et al., 2017). Photochemical damage is observed in double bonds of unsaturated fatty acids, where high energy absorbed can cause the production of free radicals, which in turn leads to the autoxidation of the lipids (Ahmed et al., 2016). This oxidation process results in rancidity and subsequent production of undesirable flavors (Aguiló-Aguayo et al., 2014). Few examples of fruit and vegetables in which the oxidation of fatty acids can be a concern were reported. The lipolytic fraction of fresh-cut avocados subjected to PL treatments exhibited minimal peroxide formation for 15 days and the treatments did not cause increased rancidity in the avocados (Aguiló-Aguayo et al., 2014). Structural modification of the protein backbone and resulting inactivation effect in the *Chromobacterium viscosum* lipase because of PL effect was reported (Jeon et al., 2019). Here, when the conditions were reversed, lipase structure would be inverted (Jeon et al., 2019).

The studies that report the inhibition of enzymatic activity in fruits and vegetables by using PL are scarce. Operating conditions that influence the inhibitory effect of PL on enzymes are not still well understood. Hence, to establish a commercial application, enzyme inactivation parameters should be further validated.

8.5.5 Photosynthetic Activity During Storage

Several studies have reported that the short durations (minutes and hours) of post-harvest PL exposure at various wavelengths of light at low (hormetic) doses (1 J/cm^2) causes stress in plant tissue (Ribeiro et al., 2012). This stress stimulates the biosynthesis of defensive secondary metabolites with antimicrobial activity, thereby preparing the plant tissue against succeeding microbial attacks (Aguiló-Aguayo et al., 2017; Gamage et al., 2014; Pataro et al., 2015; Ribeiro et al., 2012). These biochemical changes result in increased/decreased rate of respiration, delayed ripening or maturation, improved nutritional content, delayed chlorophyll degradation, and improved resistance to disease through the enhancement of the "natural defense mechanism"; all of which can contribute to reduction of postharvest deterioration during storage and prolonging the quality (Aguiló-Aguayo et al., 2017; Bett-Garber et al., 2011; Ramos-Villarroel et al., 2012a; Ribeiro et al., 2012). Low-intensity UV induces disease resistance to *B. cinerea* on tomato fruit (Charles et al., 2008). In accordance with the results shown by Charles et al. (2008), some studies demonstrated that not only low-intensity UV-C light (254 nm) but also high-intensity pulsed polychromatic light induces hormetic benefits to various stored fresh produce (Pataro et al., 2015; Scott et al., 2017, 2018). Scott et al. (2017, 2018) compared the hormesis effect of high-intensity PL (505 J, 360 µs at 3.2 pulses/s; 240–1050 nm) and low-intensity UV (3.7 kJ/m^2, 20 W/m^2) on both mature green and ripe tomatoes. When high-intensity pulsed polychromatic light and low-intensity UV-C light treatment was compared, similar transcriptional changes were detected for the genes in ripening and defense (Scott et al., 2017, 2018). They concluded that both light sources can provide abiotic stressors and broad range of resistance against pathogens (Scott et al., 2018).

Anthocyanins and other water-soluble pigments are predominantly concentrated in the skin and determine the color of fruits; fruit color is a major trait of commercial quality in fruits (Patras et al., 2009). Skin coloration is related to fruit maturation and is highly cultivar specific. After harvest, light affects metabolism and contributes to the biosynthesis of pigments and evolution of the color of several fruits and vegetables. Recently, the use of PL has emerged as a potential method to improve or maintain the color tissues of fruit (Lopes et al., 2016; Sousa et al., 2019). The repeated PL-treated melon samples retained the total phenolic content well until the end of the 28-day storage. Abiotic stress imposed by repeated PL treatment stimulated the formation of polyphenolic compounds such as flavonoids and phenylpropanoids (Ramesh et al., 2012). Immature green tomatoes were exposed to different doses of PL (1–8 J/cm^2) and the effects of light treatments on the physicochemical properties and antioxidant compounds of tomato fruits were evaluated during storage of up to 21 days (Pataro et al., 2015). The lycopene concentration in all tomato cultivars increased between 30% and 60% when they were exposed to LED light compared with dark-exposed fruit. One hour of LED light exposure per day for 16 days during postharvest storage increased lycopene concentrations between 30% and 45%, depending on the cultivar and commercial and organoleptic parameters (Nájera et al., 2018). The poor coloration of fruits (e.g., figs) and vegetables due to insufficient sunlight color development can be weak. Brief postharvest exposure to PL is capable of stimulating anthocyanin accumulation during storage and consequently developing a better marketable coloration on the fruit surface (Rodov et al., 2012).

One of the main goals of postharvest technology – to reduce the decline in leaf quality in green vegetables – is to delay senescence symptoms (Page et al., 2001). Periodic PL treatments in intervals have the potential utility of preserving both the appearance and nutritional quality of some green vegetables and lengthened shelf life even at room temperature (Garrido et al., 2016). For instance, continued low-intensity light exposition-retarded tissue browning and maintained soluble sugars and ascorbic acid in fresh-cut romaine lettuces, delayed senescence and yellowing delayed postharvest senescence of spinach and broccoli, and improved nutritional qualities of spinach, and increased the content of monosaccharides and starch in leaves of Chinese kale (Büchert et al., 2011; Favre et al., 2018; Gergoff Grozeff et al., 2013; Lester et al., 2010; Noichinda et al., 2007; Zhan et al., 2012, 2013a). The main visible symptoms of postharvest senescence of the *Brassicaceae* family are due to changes in chlorophyll degradation and changing color. Chlorophyll degradation in fresh-cut green products is due to photo-oxidative stress of high light intensity and the bursting of intracellular membranes, releasing chlorophylls to be bonded to the remaining active chlorophyllase (Aguiló-Aguayo, et al., 2014; Artés et al., 2002). Light treatment has been reported as an effective way to inhibit senescence in pak choi, lettuce, Brussels sprouts, broccoli, kale, and spinach under storage conditions (Hasperué et al., 2016b; Qian et al., 2016; Samuoliené et al., 2012; Song et al., 2020; Toledo et al., 2003). Bárcena et al. (2019) investigated whether low-intensity PL daily treatment can be used to delay postharvest senescence of this plant stored at room temperature. Leaves had 40% higher chlorophyll and protein and 20% higher antioxidant capacity and soluble sugar content than control samples under the white or red light used (Bárcena et al., 2019). Büchert et al. (2011) showed that storage under continuous low-intensity light (12 mol/m² sec) was an efficient treatment to delay postharvest senescence in broccoli. Results in a study by Costa et al. (2013) show that treatment of basil leaves for 2-hour low-intensity white light and red light pulses (30–37 mol/m²s) every day delayed post-harvest senescence of basil leaves, in terms of chlorophyll and protein retention, and decreased ammonium accumulation. Chlorophyll content of basil leaves exposed or unexposed to light decreased after 3 days of storage and then remained constant over time while chlorophyll b content showed significant degradation in control samples, but light pulses decreased this degradation (Costa et al., 2013). In a recent study, Song et al. (2020) reported that daily irradiation of red light (35 Mm/m² sec) for 8 hours inhibited the expression of senescence-associated genes while promoting the expression of putative vitamin C biosynthetic genes. Some report increasing amount of monomer sugar levels and reduction of ascorbic acid degradation especially when green leave vegetables were exposed to light during storage (Costa et al., 2013; Zhan et al., 2013a).

Lately, a few studies have tested the use of light treatments lower than the photosynthesis light compensation point delay/reduce postharvest senescence during storage. Studies that investigated the potential applications of PL as postharvest treatment demonstrated that repetitive short treatments of PL during storage and distribution could significantly promote shelf life and quality parameters of the fruit and vegetable in the cold chain. Treatment conditions such as light intensity and exposure durations directly influence the effectiveness. Even further, different plant species may respond differentially to distinct light spectra or intensities. Further studies are needed to evaluate the nutritional quality and the phytochemical properties of horticulture products for common PL applications during storage.

8.5.6 The Detoxification Mycotoxin and Agrochemical Contamination

The contamination due to filamentous fungi (*Aspergillus, Penicillium, Fusarium, Alternaria,* etc.) can result in the production of mycotoxins with potent toxigenic, carcinogenic, mutagenic, and teratogenic properties (Makinde et al., 2020). Many countries strictly control maximum allowable levels in contaminated products. Although various strategies such as (thermal, physical, chemical, and biological processes) have been researched to help suppress fungal growth and resulting formation of mycotoxins (Basaran & Ozcan, 2009), they have shown undesirable detrimental quality side effects or inefficient reduction levels. Food processors urgently need innovative low-cost strategies to reduce mycotoxins contaminations. PL can be a treatment for mycotoxin-contaminated food products. PL has been verified to be an effective decontamination technique capable of destroying mycotoxins on food surface through the photochemical and potentially photo thermal effects (Pankaj et al., 2018). In the earliest study in our records, Moreau et al. (2013) reported that eight flashes of PL degraded deoxynivalenol by 72%, zearalenone by 85%, aflatoxin B1 by 93%, and ochratoxin by 98% within buffer solutions. PL revealed its reliability as a treatment for degrading patulin from about 50–90% in apple products (Funes et al., 2013). Similarly, Jubeen et al. (2012) examined the effect of UV-C light on aflatoxin stability and found that aflatoxin G1, aflatoxin G2, aflatoxin B1, and aflatoxin B2 in walnut were reduced in a range of 68.96–100% after a 15-min exposure. PL treatment at fluence of 1 J/cm^2 achieved reductions of 41% for zearalenone, 14% for deoxynivalenol, 30% for aflatoxin B1, and 88% for ochratoxin (Moreau et al., 2013). Later, Abuagela et al. (2019) assessed the potential benefits of citric acid followed by PL treatment in peanuts. The results indicated that the total number of aflatoxins (B1, B2, and total) was reduced by 98%, slightly higher than a sole PL application (Abuagela et al., 2019). The degradation rate of mycotoxins during PL treatment depends on various factors, such as the fluence of PL, exposure time, mycotoxin structure, and product features. Recently, Deng et al. (2020) summarized the effect of various non-thermal processing application for the detoxification of mycotoxins. Among all, the deoxynivalenol was found the most resistant to PL due to probably its compact structure (Deng et al., 2020). The susceptibility to PL degradation is most likely due to the structural differences in these compounds. Detoxification of aflatoxin B1 during PL treatment might be due to the disruption of terminal furan double bond and the opening of lactone ring (Abuagela et al., 2018). Possibly, the photodegradation pathway of aflatoxin B1 demonstrated that C8–C9 and C1–C14 bonds within the aflatoxin B1 toxin absorb UV radiation maximum at 362 nm as a result and PL treatment can create a fragmentation (Abuagela et al., 2018; Pankaj et al., 2018). The C8-C9 double bond in the terminal furan ring in aflatoxin B1 potentially makes it more vulnerable to photodegradation as compared with other aflatoxins (Wang et al., 2016). Wang et al. (2016) evaluated the toxicity and mutagenic activity of the degradation by-products from aflatoxin B1 and aflatoxin B2 after PL treatment (Wang et al., 2016). In addition, the effects of PL on the potential formation of chemical compounds that may present a health threat should be evaluated to determine if there are any toxicological or chemical safety concerns. Deng et al. (2020) also concluded that more extensive research is required to determine potential the toxicity of mycotoxin residues after PL degradation.

In a study, Baranda et al. (2014) tested PL for the degradation of several herbicides present in water: simazine, atrazine, phosmet, azinphos-methyl, methyl-parathion, pirimiphos-methyl, and chlorpyrifos-ethyl. PL at about 4.65 J/cm^2 induced 50% reduction of simazine, atrazine, phosmet, azinphos-ethyl, and pirimiphos-ethyl and only a higher dose of PL was applied (9.81 J/cm^2) induced 50% reduction for chlorpyrifos-ethyl. According to these results, PL technology presents a potential for the removal of pesticide residues in water and possibly on the surface of fresh produce (Baranda et al., 2014).

8.5.7 Microbial Inactivation on Inorganic Contact Surfaces and Food Packaging

PL can be applied to reduce microbial load in foods, water used for cleaning, food packaging materials, as well as the surfaces in the processing plant environments, devices used, and the air involved in food processes. A great variance in the reports regarding the effect of PL treatment on fresh produce-related food facility surface materials was observed. Kim et al. (2002) (90 mJ/

cm^2) and Haughton et al. (2011) (0.192 J/cm^2) also reported more than 4 logs reduced populations of the pathogens L. *monocytogenes, S. typhimurium, Salmonella enteritidis and E. coli* O157:H7 on various food contact surfaces. In another study, four food contact surfaces (stainless steel, high-density polyethylene, polyvinyl chloride, and waxed cardboard) were chosen to show the effect of UV on the commonly used contact surfaces in the fresh food industry (Lim & Harrison, 2016). PL exposure for 5-second (3.3 mJ/cm^2) reduced microbial populations maximum by 2.93 log CFU/coupon while increasing the duration to 30 seconds (19.7 mJ/cm^2) resulted in 4.32 log CFU/coupon reduction (Lim & Harrison, 2016).

Biofilms are a complex population of microorganisms on food products and abiotic surfaces. The extracellular polymeric substance matrix is crucial for attachment on the inorganic surface and provides mechanical integrity while protecting the microbial community from antimicrobial agents (Jayathilake et al., 2017). Natural microflora is a part of the phylosphere of horticulture products, where they can be forming micro-colonies and biofilms on the surface, and even internalized, which would decrease PL efficacy (Izquier & Gomez-Lopez, 2011). Also, the formation of foodborne pathogen biofilms decreases the efficacy of commonly used sanitizers (Critzer & Doyle, 2010). The antimicrobial efficacy of pulsed UV light treatment on nascent biofilms formed by E. *coli* O157:H7 and L. *monocytogenes* on the fresh produce lettuce leaves has been reported (Montgomery & Banerjee, 2015).

8.5.8 Microbial Inactivation During Indirect/In-Package Treatment

At present, PL is applied for the decontamination of bottles, cups, foils, and flexible food packages at an industrial scale that can be processed in continuous flows at 7000–90,000 cups and up to 90,000 bottles per hour (Koutchma, 2019; Mahendran et al., 2019). Lately, efforts are concentrated on a new approach of PL treatment ready-to-eat products within the package to prevent post-treatment contamination. Fresh-cut vegetables and fruits are typically marketed in flexible packages with a high headspace volume, and the PL can be used for in-package pasteurization provided that the packaging material has high PL transmissivity. In this sense, PL treatment is an outstanding solution for in-package non-contact set-up if proper packaging material is available. Studies dealing with fruits and vegetables in package PL treatment are very rare. Ramos-Villarroel et al. (2012a, 2012b, 2015) evaluated the effect of PL treatment on a number of fresh products in polypropylene film packaging. The application of PL (12 J/cm^2) to fresh-cut mushrooms through a polypropylene film caused a reduction of E. *coli* and L. *innocua* by 3.03 and 2.66 log CFU/g and 3.01 and 2.79 log CFU/g with fresh-cut watermelons, respectively (Ramos-Villarroel et al., 2012a; 2012b). Slightly reduced PL (10.68 J/cm^2) demonstrated a lower reduction in the populations of E. *coli* and L. *innocua* from fresh-cut avocado by 2.47 and 1.35 log CFU/g (Ramos-Villarroel et al., 2015). L. *innocua* is less sensitive to intense light pulses than E. *coli* (Ramos-Villarroel et al., 2012a). Lettuce microbial load range from 5 to 9 log units, depending on cultivation, harvesting, and handling procedures and cutting for ready-to-eat salad processes increase in microbial load (Barth et al., 2009). The efficacy of PL treatment on E. *coli* O157:H7 decontamination in romaine lettuce packaged in polyethylene film of three thicknesses demonstrated that efficacy on the inactivation was dependent on the packaging film thickness (Mukhopadhyay et al., 2021). Both direct and in-package PL treatment exhibited similar sigmoidal inactivation; nevertheless, indirect PL treatments for packaged lettuce provided slightly lower (2.52–2.18 log CFU/g) compared with direct PL treatment (2.68 log CFU/g) (Mukhopadhyay et al., 2021). Glass and polymeric plastic materials can be considered appropriate for in-package PL application (Eie, 2009). In several studies enhanced transpiration led to greater moisture condensation inside the packaging film if the product was exposed to light within a package (Glowacz et al., 2015; Xiao et al., 2014). The reduction in inactivation efficacy was due to the reduction in UV light transmission ability of PL through these plastic films. Alternative light transparent or selective light-permeable packaging material needs to be developed for commercial in-package PL treatments. The transmittance of a certain polymeric material is strongly influenced by UV protective agents integrated on the surface of the packaging material to protect either the packaging material itself or the packaged commodity (Pospíšil & Nešpůrek, 2008). Furthermore, the absorbed PL radiation induces photochemical reactions and changes in polymers; therefore, packaging materials for PL should be selected as materials that withstand intense PL treatment against changes in strength, extensibility, permeation, migration as well as discoloration (Heinrich et al., 2015). Further studies are expected to concentrate not only on antimicrobial efficiency but also on PL and its interactions with packaging materials should more intensely address chemical migration to food and its safety issues (Heinrich et al., 2015).

8.6 MATHEMATICAL MODELINGS

Mathematical models for PL processing can provide increased understanding and predictive capabilities that can be applied to identify potential treatment non-uniformities. Even further, models increase efficiency in the prediction of product stability during storage, determine/modify optimum process conditions, and development of new equipment proper for each vegetable or fruit on *crop* by *crop* basis. A modeling approach that will minimize experimental trials, providing adequate information on the process parameters, can be applied to various horticultural products. Predicting levels and spatial distribution of *E. coli* or *L. innocua* inactivation in PL treatment of liquids with different composition and optical properties were evaluated for model development to validate the calculations with experimental data (Hsu & Moraru, 2011). The predictions of microbial inactivation, especially for the opaque and turbid substrates were more predictive (Hsu & Moraru, 2011). Ferrario et al. (2013) evaluated the inoculum size and treatment time with the reductions achieved after PL exposure and revealed that lower initial inoculum sizes were associated with higher inactivation rates. PL inactivation of *P. aeruginosa* for different pulse repetition rates (2–15 Hz), widths (0.15–1.5 ms), and pulse repetition rate fixed at 10 Hz were compared using the double Weibull model (Yi et al., 2017). Preetha et al. (2021) compared the Weibull, Biphasic, and log-linear plus tail mathematical models to predict the survival curves of *E. coli* in tender coconut water, orange, and pineapple juice. Inactivation kinetics of *E. coli* for the juices used in this study were best fitted by the Weibull model (Preetha et al., 2021). Several researchers are currently developing mathematical models to predict the microbial load and quality changes during sole or PL-combined hurdle treatments. To the best of our knowledge, no models have been established to describe the relationship between microbial inactivation and characteristics such as the color, composition, surface structure of the fruits and vegetables, and the penetration of light. Including variability in predictions can help food manufacturers to more accurately estimate the product shelf life, and comply with food safety criteria (Buzrul et al., 2005).

8.7 COMMERCIALIZATION ASPECTS

Intense and short pulses of UV light applications by employing inert-gas flash lamps were first introduced for microbial inactivation in the late 1970s in Japan (Oms-Oliu et al., 2010). After initial launch of the PL process as Foodco® (Maurice, 1994), it was further industrialized by PurePulse Technologies under the trade name PureBright® (Dunn et al., 1995). PL treatment of foods has been approved by the US FDA (1996) under the code 21CFR179.41 "for production, processing, and handling of foods" to decontaminate foodstuffs under certain recommended conditions (wavelengths between 200 and 1000 nm, with a pulse width not exceeding 2ms, the total cumulative treatment not exceeding 12.0 J/cm^2) (FDA, 1996). Since the FDA's approval, there is ever-increasing scientific research on PL treatment of food, food contact surfaces, and processing environments. UV-treated food products for the control of pathogenic microorganisms mostly in the fruit juice and seafood have already been approved in the USA, Great Britain, India, Israel, Canada, Australia, New Zealand, and China (Pisanello & Caruso, 2018). In Europe, although EFSA has a good opinion of the technology, the legal status of UV light and pulsed light has a different approach, since the legislation is not technology-oriented, but food- and food ingredient-oriented according to the novel food regulation (Arroyo et al., 2017; Lavilla & Gayan, 2018).

There is no large-scale or common application of PL in the food industry globally yet. The forerunner commercial company producing PL equipment for disinfection is Purepulse Technologies Inc. (Xenon Corp., USA), which has commercialized the PureBright system. Nowadays, there are several companies worldwide that design, construct, and commission PL processing units, and information regarding devices for industrial applications can be found at the websites of these companies (Table 8.4, Figures 8.1 and 8.2). Nevertheless, shortcomings still impede the widespread commercialization of PL technology. Major restrictions are only effective on the surface; low penetration in opaque liquids and solid surfaces, lack of accurate dosing of PL and each wavelength, stress and respiration rate caused fresh produce after PL treatment, absorption of effective wavelengths by certain food components, changes in organoleptic properties, production of off-flavors, change in product color, microbial resistance to light treatment, and change in effectiveness with decontamination levels. To solve individual limitations, the industry must implement different strategies by combining sustainable techniques if possible in conjunction with proper selection of their parameters.

Two major technical drawbacks of PL inactivation of pathogens on fresh produce include sample heating and shadow effect if the instrument is not properly designed. In order for PL systems to become feasible for fresh produce, the issue of sample heating and homogeneous treatment must be addressed. Heat can originate from the absorption of infrared part of PL on the surface and some studies reported a significant temperature increase in fresh produce such as blueberries, avocado, lettuce, etc. (Huang & Chen, 2014; Ramos-Villarroel et al., 2011). As an example, the temperature of PL-treated lettuce samples at fluence of 16.8 J/cm^2 PL treatment was 14.7±1.5°C (Tao et al., 2019). Commercial PL treatment in a continuous flow system may increase temperature depending on the in-stay time and the distance of the liquid product from the light source (Pataro et al., 2011). Increasing the fluence delivered to the sample from 1.8 to 5.5 J/cm^2, the temperature increases of the juice increased from 18 to 24°C inside the chamber (Pataro et al., 2011). Uncontrolled heating can significantly alter the sensorial quality, textural properties, and cause nutrient deterioration. A cooling system should be incorporated into PL equipment and temperature must be monitored carefully during processing to prevent any impact on freshness. LED systems are also alternative PL options. Essentially, they are semiconductor devices able to exhibit cool photon emitting surfaces with lower heat outputs and prevent overheating (Mitchell et al., 2012). The other major factor limiting the PL commercialization is the shadowing effect: an area that will remain not treated equally by the light applied (Gomez-Lopez et al., 2007). The complexity of food matrices might make it more challenging for the highly

TABLE 8.4   COMPANIES THAT PRODUCE PULSED LIGHT & UV LIGHT SYSTEMS FOR FOOD-RELATED INDUSTRIES

Company	Light System	Model	Wavelength	System Features	Microorganism and Reduction	Websites
Claranor, France	Pulsed Light	–	White light spectrum	Energy of the lamp/flash duration = 300J/0.3 ms = 1MW	3–5 log CFU reduction mold and bacteria, 4 log CFU reduction on the reference germ of dairy products *Aspergillus brasiliensis*	https://www.claranor.com/en/
MicroTek, United Kingdom	High-Intensity UVC Light	MP (0–10) Fluid Disinfection	254 nm	Low-Pressure Mercury Argon lamps	N.A.	http://www.microtekprocesses.com/
Xenon, United States of America	Pulsed Light	Xenon X-1100, Xenon Z-2000, Xenon Z-1000, Xenon CIXL	From UV to infrared or UVC, UVB and UVA (200–400 nm)	Pulse Energy and Pulse Rate Options; 1 – 505 J/pulse @ 3pps and 2 – 9.4 J/pulse @ 100pps.	3–6 log CFU reduction (general), destruction of *B. subtilis*, reduction of >6 log CFU/ml (for 1 sec. treatment time)	https://xenoncorp.com/
wek-tec e. K., Germany	Intense Pulsed Light	Custom pilot or small production systems for sterilization	Customized output in selected spectral intervals for UV, white and IR light	Customised flash lamps for a task	Goals are set by a customer using the below line or new test data	http://www.wek-tec.de/
		Three bench-top R&D systems XeMaticA: air cooled Basic 1L, 2L and water cooled, one or two lamps system		Standard single pulse energy 150–500 J, custom pulses: 150–1000 J, pulse lengths ca. 0.1–0.4 ms	Up to 6 logs for all common bacteria and spores, 0.5–1 kJ pulses fully destroy UV resistant spores *Bacillus pumilos*	
Xenex, United States of America	Pulsed UV Light (for hospital air)	LightStrike™	200–315 nm	N.A.	99.99% of SARS-CoV-2, 95% of *Clostridium difficile* spores, 99.99% of Methiciline-resistant *Staphylococcus aureus*, 100% Vancomycin-resistant Enterococci deactivated in the environment	https://xenex.com/

(Continued)

TABLE 8.4 (CONTINUED) COMPANIES THAT PRODUCE PULSED LIGHT & UV LIGHT SYSTEMS FOR FOOD-RELATED INDUSTRIES

Company	Light System	Model	Wavelength	System Features	Microorganism and Reduction	Websites
Pulsed Light Power, Netherlands	Pulsed UV-A, B Light	-	240 nm and upwards	Flash energy: 1,850 Joule / flashFlash Frequency: 1 flash / second	*Listeria* and *Salmonella* bacteria reduced by 99.85%	https://www.pulsedlightpower.com/
ZHONGWU, China	Pulsed Light (For Solid Food)	ZWL-G1-E8	N.A.	Lamp module quantity: 4Production capacity (KG/H): 100-1000Lamp life (million pulses) ≥8, 50/60Hz	Reduction of mold and spores up to 6 log	https://www.zwpulsedlight.com/
	Pulsed Light	ZWP-UY Series (Disinfection Robot) (For Air)	Inert-gas flash lamps (200-1100nm)	3-6 Hz, Lamp life (million pulses) >300	Killing viruses, *Clostridium difficile*, Methicilline-resistant *Staphylococcus aureus*, Vancomycin-resistant *Enterococci*, multidrug resistance organisms, etc.) and other environmental bacteria	
Phoxene, France	Pulsed Light	DLP CUP	N.A.	A treatment in 0.1 seconds by flash of light. Maximum processing capacity 2400 pots/hour	3-6 log CFU decontamination in 0.1 seconds. *Aspergillus Niger, Aspergillus brasiliensis, Bacillus Subtilis, E. coli,* group B Streptococcus, *Staphylococcus aureus*> 3-6 log CFU guaranteed in 1 flash. >4 log CFU reduction on *Aspergillus brasiliensis*	https://www.phoxene.com/

(Continued)

TABLE 8.4 (CONTINUED)   COMPANIES THAT PRODUCE PULSED LIGHT & UV LIGHT SYSTEMS FOR FOOD-RELATED INDUSTRIES

Company	Light System	Model	Wavelength	System Features	Microorganism and Reduction	Websites
Montena, China	Pulsed Light	-	N.A.	Max. flashing frequency of 1.4 HzFluency of one light pulse up to 2.2 J/cm²Designed for 24 hours/7 days operation3 lamp cassettesTypical conveyor speed: 0.65 m/sOperating and amortization costs ~ €1ct	>5–6 log CFU *Aspergillus niger, Strain Globigii, Bacillus subtilis var. Niger, Geobacillus stearo thermophilus, Bacillus pumilus, Alicyclobacillus acidoterrestris,* Reduction of mycotoxins, and other Trichothecens, Fumonisins, Zearaleon. Aflatoxin generated by *Aspergillus*	https://www.montena.com/system/home/
Sanodev, France	Pulsed Light	LP.Box	200–1100 nm	Light flashes (between 1 μs and 100 ms by flash)	4 log CFU reduction (or even 5 log CFU depending on the pathogens) in 5 seconds and at a distance of 50 cm from the target. Degradation of phytotoxins, mycotoxins, alkaloids, phytosanitary residues and but also to decontaminate containers or samples.	https://sanodev.com/en/accueil-english/
	Pulsed Light	Agro Clean Light	200–1200 nm	The emission of polychromatic light covering a broad spectrum by light flashes (between 1 μs and 100 ms by flash)	N.A.	

(Continued)

TABLE 8.4 (CONTINUED)   COMPANIES THAT PRODUCE PULSED LIGHT & UV LIGHT SYSTEMS FOR FOOD-RELATED INDUSTRIES

Company	Light System	Model	Wavelength	System Features	Microorganism and Reduction	Websites
STERIXENE, France	Pulsed Light	-	400 nm max	The light exposure limit values are 3 mJ/cm² in the UV and 10 mJ/cm² in the visible.	*Aspergillus brasiliensis, Bacillus atrophaeus, Bacillus cereus, Escherichia coli, Listeria monocytogenes, Salmonella,* etc. of 2–5 log CFU reduction	https://en.sterixene.com/
Lamplic Science, China	Pulsed UV-Light	Lamplic UV pulsed light	N.A.	10–100 microseconds	2 log/cm CFU *Salmonella, Listeria monocytogenes* (in meat). Total aerobic count, lactic acid bacteria, enteric bacteria, and *Pseudomonas* count are similarly reduced by about 1–3 log CFU. 1–3 log CFU decrease for media, packaging, or related sample surfaces, as well as complex surfaces (general)	http://en.szlamplic.com/

All data are taken from each company's website in May–July 2021.
N.A. Not available

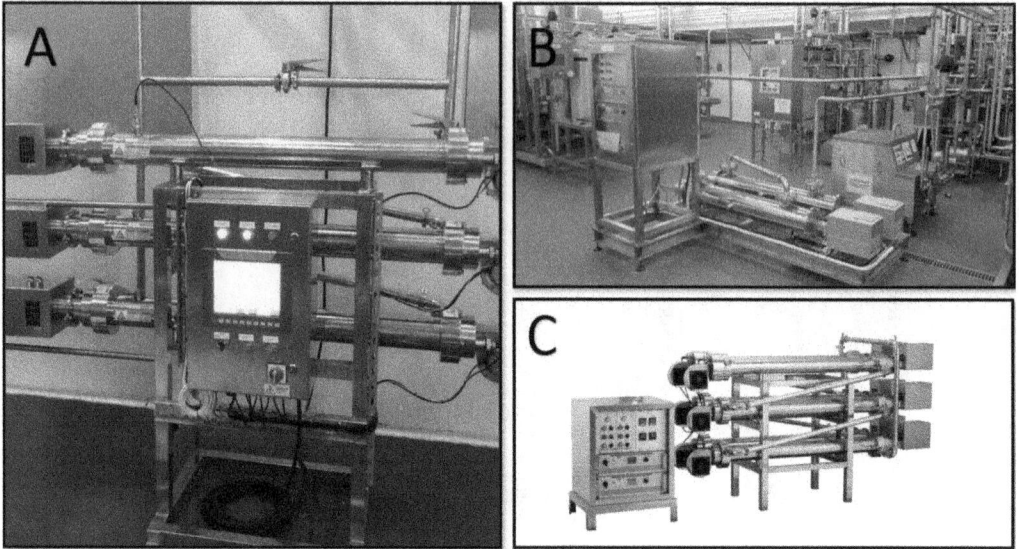

Figure 8.1   A) MP5 Fluid Disinfection chamber on coconut water, B) MP3 chambers on milk, C) Pulsing chamber system (All systems manufactured by MicroTek Processes LTD, UK.) (Photo courtesy of MicroTek Processes Ltd.).

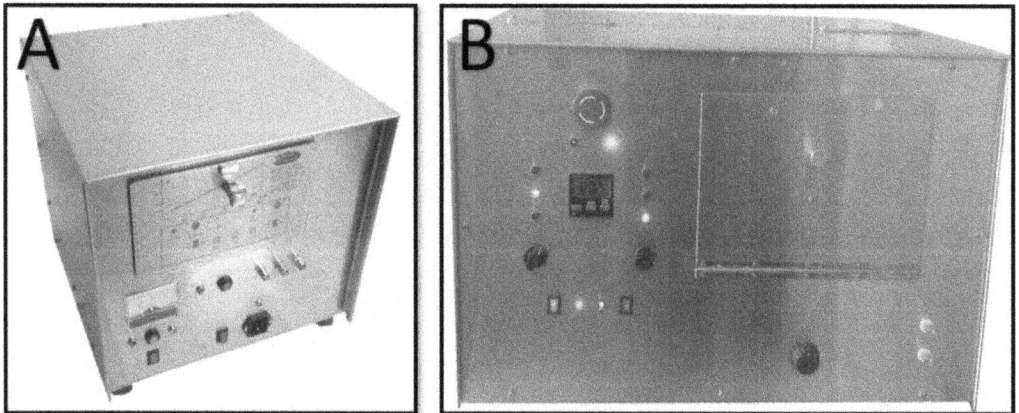

Figure 8.2   A) Intense Pulsed (UV) Light System-Basic XeMaticA-1L, B) Intense Pulsed UV Light System-XeMaticA-RepRate-Front-V3a (All systems manufactured by Wek-Tec, Gottmadingen, GERMANY) (Photo courtesy of Wek-Tec. International UG).

directional, coherent PL beam from reaching its target as microbial cells and spores can harbor in small crevices on the surface. The shadowing effect can be avoided by minimizing the depth of liquid material by using turbulent treatment instead of the static treatment (Xu et al., 2019). Even further, most of the lab level studies were conducted as batch treatment to make light to better contact with microorganisms; liquid foods can be stirred or flow through continuously designed equipment (Artíguez et al., 2011; Hwang et al., 2017; Yi et al., 2016). Various PL operating systems including continuous-flow apparatus (laminar, turbulent flow, spiral continuous flow-through tube reactors, etc.), thin-film devices, static and dynamic mixers were developed and validated for various beverage pasteurization (Bintsis et al., 2000; Guerrero-Beltrán & Barbosa-Cánovas, 2005; Koutchma et al., 2007; Koutchma, 2009). The penetration depth also limits the PL efficacy on solid foods (Elmnasser et al., 2007), so a 360° uniform exposure of PL to the surface must be ensured for efficient decontamination. Huang and Chen (2014) developed a prototype water-assisted PL system, in which blueberries were treated with PL while being immersed in clear tap water agitated by stirring movement. In most studies, the treatment chamber constructed from acrylic material or stainless steel with the same design to provide cyclonic flow was compared in a combined study by Hwang et al. (2017). The stainless steel not only congested all generated PL within the chamber but also prevented scratches to be formed on the surface during cyclonic movement of the dry seeds (Hwang et al., 2017). Changing the material of the treatment chamber to stainless steel resulted in light reflections and more than one log reduction was achieved regardless of the voltage when the PL pulse duration was 3.0 ms (Hwang et al., 2017). Lately, design changes made in the treatment chamber with reflective surfaces (e.g., titanium dioxide) to enhance irradiation efficacy (Garvey & Rowan, 2019). Further innovatively designed and practically applicable engineering resolutions are needed to solve the major drawbacks of overheating and shadowing effect while achieving better homogeneity.

Nearly all studies on PL have been conducted on pilot-scale equipment in which conditions of light exposure as well as mass or volume of food product would obviously be different from those would be performed under commercial conditions. The applicability of process conditions by the industry will be based on the evaluation of primary study results obtained from small-scale equipment. This is surely a challenge to the relevance of the results obtained at lab scale. Testing details such as sample preparation and recovery of microorganisms, etc. are among other factors. For the commercialization of PL technology for food and beverage applications there is a lack of control of operational dose and standardized validation procedures. The standardized operational dose and its validation procedure are necessary (Koutchma, 2019).

Engineers and system designers are in need of specific, sustainable microbial inactivation data in relation to light parameters for designing and scaling-up the technology on a commercial level. Published information on PL technology for different food types reveals significant variance and therefore needs harmonization and standardization at even the development stage. Light classifications based on various wavelengths vary throughout the literature and a consistent, accurate description is necessary to have comparable conclusions in different research activities. Proper reporting of treatment conditions will enable researchers to better compare results from different studies. Two recent reviews by Schottroff et al. (2018) and Rowan (2019) highlighted that published research outcomes for PL treatments for even the same food items are incomparable in most cases since PL operational parameters, food matrixes, and targeted microorganism details in publications are insufficient or lack process harmonization. Rowan (2019) reviewed the status of PL for non-thermal food-surface treatments in order to address the rationale and efficacy of PL disinfection performance on a variety of food. The author listed recommendations for future studies on microbial inactivation by PL for harmonization of research and for industry needs. Among the key parameters of PL treatment, a full description of equipment if a commercial instrument is used and PL process parameters (how fluence (J/cm^2), exposure time is calculated/measured, number of pulses and voltage applied, frequency of pulsing, pulse width, peak power, distance of target area from PL source, etc.) are suggested for comparable results. Treatment should be reported in terms of fluence at the substrate surface or within the substrate, rather than fluence merely at the lamp surface, unfortunately, this is the case in the majority of the publications. Advanced measuring for the light integrated into the food matrix is essential to determine the effectiveness of PL treatment on foods with an irregular surface and to implement a uniform treatment. Proper determination of the fluence received by the treated food item is the most significant factor in characterizing PL treatment; especially the energy received by the sample, which is considerably different from the energy delivered by the light source (Gomez-Lopez et al., 2007). It is difficult to compare the effectiveness of PL without the complex influence of critical factors such as: type of microorganism, inoculum size, food composition, and natural competitive microbiota on the product. Even the type of inoculation (spot or dip) can cause different results in PL reduction of surface contamination (Cao et al., 2017). Factors such as distance from the light source, applied energy intensity that reaches the plant surface need to be standardized for comparable results (Ogihara et al., 2013). Additionally, as most of the microbicidal effect of PL results from the UV wavelengths, knowing the proportion of total fluence contributed by UV wavelengths will allow even more accurate comparison of PL effectiveness (Schottroff et al., 2018).

High initial investment costs, maintenance for PL equipment, and difficulty in standardizing process parameters for each product are of great concern for commercial use (Cacace & Palmieri, 2014; Heinrich et al., 2015; Lavilla & Gayan, 2018). According to a recent survey among food producers, 41.6% of respondents indicated that a high initial investment was the biggest obstacle preventing them from using non-thermal food processing technology (Khouryieh, 2021). The high investment cost depends on the size and amenities; however, some value-added products can justify the cost. With lower initial investment and operating costs, a promising opportunity exists for adopting PL processing in the small or large-scale fruit and vegetable processing industry. For instance, to decontaminate 10,000 cups, the total operational cost is estimated to be €42, whereas the estimated cost to perform the same decontamination with per-acetic acid is around €266 (Mahendran et al., 2019). The comparison of the electrical efficiency and lifetime between pulsed and continuous UV sources is not straightforward because, unlike mercury lamps, PL performance depends on operating conditions (Mahendran et al., 2019; Nair & Dhoble, 2021). Recently developed LEDs based applications are characterized by a low-energy input and low-cost operation (Nair & Dhoble, 2021) and thus appear more approachable by the food industry (Cossu et al., 2021). Scientific and commercial communities are keen to exploit the improved effectiveness of light-based food processors that can be scaled and manufactured in robust, functional, and durable materials at a low cost. Nearly all companies producing PL systems contacted in this study offer service of product-specific, custom-made processing equipment (Table 8.1). Future pricing will depend on demonstrating effective and high-yield processing equipment with meaningful market demand.

Few studies have examined electrical energy use efficiency for food applications. Optimization of system design and process conditions for the best product quality with least energy used and low operational costs, and minimal environmental impacts are currently the main goals and driving force for industrial scaling-up (Lavilla & Gayan, 2018). It is estimated that up to 60% of energy input can be converted to light in PL systems, but this includes wavelengths in the visible and infrared range, which generally do not have antimicrobial effects (Heering, 2004). Innovative nonthermal food-processing technologies can have an enormous effect in reducing carbon and water footprints in developing countries when more economic and practically usable food preservation systems have become available to food producers. PL as a highly efficient eco-friendly, green technology processing application can deliver food products without hazardous microorganisms and enzymes. PL allows the decontamination of food packed and unpacked and contact surfaces. Sun is the main source of free UV light and in the future food processing equipment made of new materials only using sunlight would be extremely valuable for green technologies. Packaging materials that do or do not transmit the wavelengths responsible for the activation of the photo repair mechanism. Apart from the antimicrobial efficacy of UV and PL types of technologies, the differences in their economic impact for each potential combination applications must also be taken into account. A detailed economic comparison needs to be carried out once the technical sustainability has been foresighted. LED-based technology is characterized by a low-energy input and low cost of operation thus more approachable to the food industry (Cossu et al., 2021; Nair & Dhoble, 2021). From an environmental point of view, LED has many prominent advantages including low energy consumption (0.07-2 W), relatively less heat dissipation, and low cost (0.5-500 $/piece) (Cossu et al., 2021; Nair & Dhoble, 2021). Future innovation in PL technology is expected to seek to improve fluence efficiency through the use of UV-LED technology to meet the demands of fresh-cut fruit and vegetable processing. The researchers need to address future commercial needs in comparing wavelength efficacy and operational conditions of UV-LED germicidal spectrum of microorganisms of relevance to fruit and vegetables.

Nevertheless, it is necessary to understand the microbial inactivation route for the product parameters of the commercial utilization of LED technology. Exposure to LED light induces a photothermal effect and conditional on the length of the duration can inactivate microorganisms (Prasad et al., 2020). Some bacteria produce endogenous photosensitizers, which are molecules with cyclic structures of flavin and porphyrins that have the capacity to absorb light (Hessling et al., 2017). Their main absorption peaks correspond to the violet-blue range. The effectiveness of LED technology is based on the absorption of light by endogenous naturally present inside the bacterial cells or exogenously added photosensitizers (Angarano et al., 2020). The physicochemical process of microbial inactivation by LED-based photosensitization is also based on the interaction between LED light and an oxygen-dependent photosensitizer which can be present/added to food for different purposes (e.g., curcumin, porphyrin, anthraquinonic, and xanthene) (de Freitas & Hamblin, 2016). The outcome of the interaction between light and photosensitizers is strictly dependent on oxygen. De Oliveira et al. (2018) reported that the inactivation of approximately 2–4 $\log_{10}$ CFU/g of *E. coli* on grapes without quality loss was achieved when a combination of 1.6 mM curcumin and 465–470 nm LEDs was applied (de Oliveria et al., 2018). The use of antimicrobials as food preservatives in combination with PL can cause changes in the organoleptic properties and unless the photosensitizer is a commonly present food component, photosensitizer-based LED applications are not practically applicable. The major benefit of applying LEDs is great flexibility which allows key parameters to be modified based on specific needs of food application. However, the knowledge on how the light intensity and spectrum provided by LEDs interactively affect fresh produce under controlled environments still remains limited.

The LED technology has also been investigated in agriculture pre-harvest (cultivation of crops, production, delaying of fruit ripening during storage) and postharvest storage (Massa et al., 2008). As a nonthermal postharvest intervention method, LED systems have emerged as a clean and efficient artificial lighting technique for use in horticulture activities (Urrestarazu et al., 2016). These emerging light-based strategies can be used not only to control microbiological activity to extend the shelf life of fresh-cut products but also to control enzymatic browning and thus preserve the organoleptic and alimentary qualities of produce better than conventional treatments. In addition, it could also be considered a pre-treatment to preserve the natural appearance of fruits and vegetables being processed for the manufacturing of snacks. Other commercial applications of LED-based devices include the use of these light systems within storage area and water retreatment for fresh produce water systems. Undoubtfully, the optimization of LED-based treatment wherever the application is available (warehouses, handling step, packaging application, post-processing, and storage) with classified conditions for industrial options are needed in future studies.

Uncontrolled exposure to PL may cause certain adverse effects on horticulture products such as surface discoloration, browning on the surface, cellular membrane damage and linkage, water loss, drying, and textural changes during storage. Further information is needed to select the most effective spectral region to maximize microbial inactivation while minimizing any deterioration in the quality of fresh produce. In the literature, there is a lack of evidence showing the effect of different spectral regions on the quality attributes of fresh produce. The quality of horticulture products is a very broad concept and needs to be described with quantifiable parameters for comparison. Effects of PL on the sensory quality of food are unclear or contradicting. Especially the sensory properties of PL exposed foods may need to be addressed more in detail since available data is often hardly comparable and ambiguous. In the majority of the studies, neither the trained sensory panelists nor the consumer panelists observed significant differences in the sensory properties (such as color hue, color intensity, and whiteness). Photooxidation reactions through production of activated free radicals can damage color. Responses to PL are highly different for each fruit and vegetable species, and eventually it should be evaluated on a *crop-by-crop* basis. Establishing the process for different food products and understanding the mode of action of light to each food component is a key step toward optimization of the technology for specific applications in the food processing industry. Critical parameters depending on the physicochemical properties of each fresh produce with special characteristics should be determined when designing experiments to assess the suitability of PL.

8.8 CONCLUSION AND FUTURE PERSPECTIVES

The relatively low cost of energy, flexibility of the treatment process, the lower environmental impact, and the capacity to be combined in processing lines with various hurdle treatments are some of the main advantages to using PL technology. PL-based technologies have been shown to have strong potential for selected food and food preparation surface, and air and water applications. PL also have potential to decontaminate packaging and processing equipment. In fact, the major expectation in the future is the PL application on fresh-cut fruits and vegetable products within packaging.

Despite extreme research activities, the inactivation performance of PL on food surface-related microorganisms is still below the FDA-recommended 5-log pathogen reduction standards. Also, high power and duration PL may cause adverse physical effects on the surface. The effectiveness of sole PL application against microbial and enzymatic inactivation has not been completely successful and therefore it is more logical that PL can be considered as a complementary technology within a hurdle. The use of PL in such conditions may become relevant especially if or when coupled with other supporting systems concomitant with the ongoing improvements in the efficiency of the systems along with the decline in their instrumental prices. Multi-hurdled approaches aimed at mitigating the drawbacks of singular pretreatment could present a significant breakthrough in novel pre- or actual food treatment. To date, none of these hurdle applications have been adopted by the food industry due to mostly lack of efficiency, to a lesser extent due to economic concerns, difficulty to apply to actual food processing systems, or the adverse effects on product quality. Multidisciplinary studies should be performed to design new hurdle processes using available options with functionally synergistic effects. Obviously, any hurdle application developed must be affordable and safe to carry out, with no adverse effect on quality, and be approved by applicable regulatory agencies.

REFERENCES

Abuagela, M. O., Iqdiam, B. M., Mostafa, H., Gu, L., Smith, M. E., & Sarnoski, P. J. (2018). Assessing pulsed light treatment on the reduction of aflatoxins in peanuts with and without skin. *International Journal of Food Science and Technology, 53*(11), 2567–2575.

Abuagela, M. O., Iqdiam, B. M., Mostafa, H., Marshall, S. M., Yagiz, Y., Marshall, M. R., … Sarnoski, P. (2019). Combined effects of citric acid and pulsed light treatments to degrade B-aflatoxins in peanut. *Food and Bioproducts Processing, 117*, 396–403.

Agüero, M. V., Jagus, R. J., Martín-Belloso, O., & Soliva-Fortuny, R. (2016). Surface decontamination of spinach by intense pulsed light treatments: Impact on quality attributes. *Postharvest Biology and Technology, 121*, 118–125.

Aguilar, K., Garvín, A., & Ibarz, A. (2018). Effect of UV-Vis processing on enzymatic activity and the physicochemical properties of peach juices from different varieties. *Innovative Food Science and Emerging Technologies, 48*, 83–89.

Aguiló-Aguayo, I., Gangopadhyay, N., Lyng, J. G., Brunton, N., & Rai, D. K. (2017). Impact of pulsed light on color, carotenoid, polyacetylene and sugar content of carrot slices. *Innovative Food Science and Emerging Technologies, 42*(March), 49–55.

Aguiló-Aguayo, I., Oms-Oliu, G., Martín-Belloso, O., & Soliva-Fortuny, R. (2014). Impact of pulsed light treatments on quality characteristics and oxidative stability of fresh-cut avocado. *LWT - Food Science and Technology, 59*(1), 320–326.

Aguirre, J. S., García de Fernando, G., Hierro, E., Hospital, X. F., Espinosa, I., & Fernández, M. (2018). Characterization of damage on Listeria innocua surviving to pulsed light: Effect on growth, DNA and proteome. *International Journal of Food Microbiology, 284*(February), 63–72.

Ahmed, M., Pickova, J., Ahmad, T., Liaquat, M., Farid, A., & Jahangir, M. (2016). Oxidation of lipids in foods. *Sarhad Journal of Agriculture, 32*(3), 230–238.

Akgün, M. P., & Ünlütürk, S. (2017). Effects of ultraviolet light emitting diodes (LEDs) on microbial and enzyme inactivation of apple juice. *International Journal of Food Microbiology, 260*, 65–74.

Angarano, V., Akkermans, S., Smet, C., Chieffi, A., & Van Impe, J. F. M. (2020). The potential of violet, blue, green and red light for the inactivation of P. fluorescens as planktonic cells, individual cells on a surface and biofilms. *Food and Bioproducts Processing, 124*, 184–195.

Aponiene, K., Paskeviciute, E., Reklaitis, I., & Luksiene, Z. (2015). Reduction of microbial contamination of fruits and vegetables by hypericin-based photosensitization: Comparison with other emerging antimicrobial treatments. *Journal of Food Engineering, 144*, 29–35.

Arroyo, C., Dorozko, A., Gaston, E., O'Sullivan, M., Whyte, P., & Lyng, J. G. (2017). Light based technologies for microbial inactivation of liquids, bead surfaces and powdered infant formula. *Food Microbiology, 67*, 49–57.

Artés, F., Minguez, M. I., & Hornero, D. (2002). Analysing changes in fruit pigments. In D. MacDougall (Ed.), *Color in food* (pp. 248–282). Amsterdam: Elsevier.

Artíguez, M. L., Lasagabaster, A., & de Marañón, I. M. (2011). Factors affecting microbial inactivation by Pulsed Light in a continuous flow-through unit for liquid products treatment. *Procedia Food Science, 1*, 786–791.

Aune, D., Giovannucci, E., Boffetta, P., Fadnes, L. T., Keum, N., Norat, T., … Tonstad, S. (2017). Fruit and vegetable intake and the risk of cardiovascular disease, total cancer and all-cause mortality—A systematic review and dose-response meta-analysis of prospective studies. *International Journal of Epidemiology, 46*(3), 1029–1056.

Aurum, F. S., & Nguyen, L. T. (2019). Efficacy of photoactivated curcumin to decontaminate food surfaces under blue light emitting diode. *Journal of Food Process Engineering, 42*(3), e12988.

Banach, J. L., van Bokhorst-van de Veen, H., van Overbeek, L. S., van der Zouwen, P. S., van der Fels-Klerx, H. J., & Groot, M. N. N. (2017). The efficacy of chemical sanitizers on the reduction of Salmonella typhimurium and Escherichia coli affected by bacterial cell history and water quality. *Food Control, 81*, 137–146.

Bang, I. H., In, J., & Min, S. C. (2021). Inactivation of Salmonella on black peppercorns using an integrated ultraviolet-C and cold plasma intervention. *Food Control, 119*, 107498.

Bang, I. H., Kim, Y. E., Lee, S. Y., & Min, S. C. (2020). Microbial decontamination of black peppercorns by simultaneous treatment with cold plasma and ultraviolet C. *Innovative Food Science and Emerging Technologies, 63*, 102392.

Baranda, A. B., Fundazuri, O., & Martínez De Marañón, I. (2014). Photodegradation of several triazidic and organophosphorus pesticides in water by pulsed light technology. *Journal of Photochemistry and Photobiology A: Chemistry, 286*, 29–39.

Bárcena, A., Martínez, G., & Costa, L. (2019). Low intensity light treatment improves purple kale (Brassica oleracea var. sabellica) postharvest preservation at room temperature. *Heliyon, 5*(9), e02467.

Barkai-Golan, R. (2001). *Postharvest diseases of fruits and vegetables: Development and control.* Amsterdam: Elsevier.

Barth, M., Hankinson, T. R., Zhuang, H., & Breidt, F. (2009). Microbiological spoilage of fruits and vegetables. In W. H. Sperber & M. P. Doyle (Eds.), *Compendium of the microbiological spoilage of foods and beverages* (pp. 135–183). New York: Springer.

Basaran, P., & Ozcan, M. (2009). Occurrence of aflatoxins in various nuts commercialized in Turkey. *Journal of Food Safety, 29*, 95–105.

Bett-Garber, K. L., Greene, J. L., Lamikanra, O., Ingram, D. A., & Watson, M. A. (2011). Effect of storage temperature variations on sensory quality of fresh-cut cantaloupe melon. *Journal of Food Quality, 34*(1), 19–29.

Bialka, K. L., Demirci, A., & Puri, V. M. (2008). Modeling the inactivation of Escherichia coli O157:H7 and Salmonella enterica on raspberries and strawberries resulting from exposure to ozone or pulsed UV-light. *Journal of Food Engineering, 85*(3), 444–449.

Bintsis, T., Litopoulou-Tzanetaki, E., & Robinson, R. K. (2000). Existing and potential applications of ultraviolet light in the food industry - A critical review. *Journal of the Science of Food and Agriculture, 80*(6), 637–645.

Birmpa, A., Sfika, V., & Vantarakis, A. (2013). Ultraviolet light and ultrasound as non-thermal treatments for the inactivation of microorganisms in fresh ready-to-eat foods. *International Journal of Food Microbiology, 167*(1), 96–102.

Braga, T. R., Silva, E. O., Rodrigues, S., & Fernandes, F. A. N. (2019). Drying of mangoes (Mangifera indica L.) applying pulsed UV light as pretreatment. *Food and Bioproducts Processing, 114*, 95–102.

Brem, R., Guven, M., & Karran, P. (2017). Oxidatively-generated damage to DNA and proteins mediated by photosensitized UVA. *Free Radical Biology and Medicine, 107*, 101–109.

Brennan, M. H., & Gormley, T. R. (1998). Extending the shelf life of fresh sliced mushrooms. *Research report no 2. The national food centre, Dunsinea, Castleknock, Dublin*, 30.

Büchert, A. M., Civello, P. M., & Martínez, G. A. (2011). Effect of hot air, UV-C, white light and modified atmosphere treatments on expression of chlorophyll degrading genes in postharvest broccoli (Brassica oleracea L.) florets. *Scientia Horticulturae, 127*(3), 214–219.

Buzrul, S., Alpas, H., & Bozoglu, F. (2005). Use of Weibull frequency distribution model to describe the inactivation of Alicyclobacillus acidoterrestris by high pressure at different temperatures. *Food Research International, 38*(2), 151–157.

Cacace, D., & Palmieri, L. (2014). High-intensity pulsed light technology. In Da-Wen Sun (Ed.), *Emerging technologies for food processing* (2nd ed.). New York: Elsevier.

Caminiti, I. M., Noci, F., Morgan, D. J., Cronin, D. A., & Lyng, J. G. (2012). The effect of pulsed electric fields, ultraviolet light or high intensity light pulses in combination with manothermosonication on selected physico-chemical and sensory attributes of an orange and carrot juice blend. *Food and Bioproducts Processing, 90*(3), 442–448.

Caminiti, I. M., Palgan, I., Noci, F., Muñoz, A., Whyte, P., Cronin, D. A., … Lyng, J. G. (2011). The effect of pulsed electric fields (PEF) in combination with high intensity light pulses (HILP) on Escherichia coli inactivation and quality attributes in apple juice. *Innovative Food Science and Emerging Technologies, 12*(2), 118–123.

Cantos, E., Garcia-Viguera, C., De Pascual-Teresa, S., & Tomas-Barberan, F. A. (2000). Effect of postharvest ultraviolet irradiation on resveratrol and other phenolics of cv. Napoleon table grapes. *Journal of Agricultural and Food Chemistry, 48*(10), 4606–4612.

Cao, X., Huang, R., & Chen, H. (2017). Evaluation of pulsed light treatments on inactivation of Salmonella on blueberries and its impact on shelf-life and quality attributes. *International Journal of Food Microbiology*, *260*, 17–26.

Cao, X., Huang, R., & Chen, H. (2019). Evaluation of food safety and quality parameters for shelf life extension of pulsed light treated strawberries. *Journal of Food Science*, *84*(6), 1494–1500.

CDC (Centers for Disease Control and Prevention). (2020). Foodborne illness and outbreaks. Retrieved from https://www.cdc.gov/foodsafety/outbreaks/multistate-outbreaks/outbreaks-list.

Cervantes-Paz, B., Ornelas-Paz, J. de J., Pérez-Martínez, J. D., Reyes-Hernández, J., Zamudio-Flores, P. B., Rios-Velasco, C., … Ruiz-Cruz, S. (2016). Effect of pectin concentration and properties on digestive events involved on micellarization of free and esterified carotenoids. *Food Hydrocolloids*, *60*, 580–588.

Charles, M. T., Goulet, A., & Arul, J. (2008). Physiological basis of UV-C induced resistance to Botrytis cinerea in tomato fruit: IV. Biochemical modification of structural barriers. *Postharvest Biology and Technology*, *47*(1), 41–53.

Charles, F., Vidal, V., Olive, F., Filgueiras, H., & Sallanon, H. (2013). Pulsed light treatment as new method to maintain physical and nutritional quality of fresh-cut mangoes. *Innovative Food Science and Emerging Technologies*, *18*, 190–195.

Choi, D. S., Park, S. H., Choi, S. R., Kim, J. S., & Chun, H. H. (2015). The combined effects of ultraviolet-C irradiation and modified atmosphere packaging for inactivating Salmonella enterica serovar Typhimurium and extending the shelf life of cherry tomatoes during cold storage. *Food Packaging and Shelf Life*, *3*, 19–30.

Cliffe-Byrnes, V., & O'Beirne, D. (2005). Effects of chlorine treatment and packaging on the quality and shelf-life of modified atmosphere (MA) packaged coleslaw mix. *Food Control*, *16*(8), 707–716.

Collazo, C., Charles, F., Aguiló-Aguayo, I., Marín-Sáez, J., Lafarga, T., Abadias, M., & Viñas, I. (2019). Decontamination of Listeria innocua from fresh-cut broccoli using UV-C applied in water or peroxyacetic acid, and dry-pulsed light. *Innovative Food Science and Emerging Technologies*, *52*(December 2018), 438–449.

Cossu, M., Ledda, L., & Cossu, A. (2021). Emerging trends in the photodynamic inactivation (PDI) applied to the food decontamination. *Food Research International*, *144*, 110358.

Costa, L., Millan Montano, Y., Carrión, C., Rolny, N., & Guiamet, J. J. (2013). Application of low intensity light pulses to delay postharvest senescence of Ocimum basilicum leaves. *Postharvest Biology and Technology*, *86*, 181–191.

Critzer, F. J., & Doyle, M. P. (2010). Microbial ecology of foodborne pathogens associated with produce. *Current Opinion in Biotechnology*, *21*(2), 125–130.

da Cruz Almeida, E. T., de Medeiros Barbosa, I., Tavares, J. F., Barbosa-Filho, J. M., Magnani, M., & de Souza, E. L. (2018). Inactivation of spoilage yeasts by Mentha spicata L. and M. × Villosa Huds. essential oils in cashew, guava, mango, and pineapple juices. *Frontiers in Microbiology*, *9*, 1111.

Dale, M. F. B., Griffiths, D. W., Bain, H., & Todd, D. (1993). Glycoalkaloid increase in Solarium tuberosum on exposure to light. *Annals of Applied Biology*, *123*(2), 411–418.

De Freitas, L. F., & Hamblin, M. R. (2016). Antimicrobial photoinactivation with functionalized fullerenes. In A. Grumezescu (Ed.), *Nanobiomaterials in antimicrobial therapy: Applications of nanobiomaterials* (pp. 1–27). New York: Elsevier.

de Oliveira, E. F., Tikekar, R., & Nitin, N. (2018). Combination of aerosolized curcumin and UV-A light for the inactivation of bacteria on fresh produce surfaces. *Food Research International*, *114*, 133–139.

de Sousa, A. E. D., de Almeida Lopes, M. M., Moreira, A. D. R., Nunes Macedo, J. J., Moura, C. F. H., de Aragão, F. A. S., … de Oliveira Silva, E. (2019). Induction of postharvest resistance in melon using pulsed light as abiotic stressor. *Scientia Horticulturae*, *246*, 921–927.

de Souza, V. R., Popović, V., Bissonnette, S., Ros, I., Mats, L., Duizer, L., … Koutchma, T. (2020). Quality changes in cold pressed juices after processing by high hydrostatic pressure, ultraviolet-c light and thermal treatment at commercial regimes. *Innovative Food Science and Emerging Technologies*, *64*, 102398.

Deng, L. Z., Tao, Y., Mujumdar, A. S., Pan, Z., Chen, C., Yang, X. H., … Xiao, H. W. (2020). Recent advances in non-thermal decontamination technologies for microorganisms and mycotoxins in low-moisture foods. *Trends in Food Science and Technology*, *106*(October), 104–112.

Diesler, K., Golombek, P., Kromm, L., Scharfenberger-Schmeer, M., Durner, D., Schmarr, H. G., … Fischer, U. (2019). UV-C treatment of grape must: Microbial inactivation, toxicological considerations and influence on chemical and sensory properties of white wine. *Innovative Food Science and Emerging Technologies, 52*, 291–304.

Donsì, F., Marchese, E., Maresca, P., Pataro, G., Vu, K. D., Salmieri, S., … Ferrari, G. (2015). Green beans preservation by combination of a modified chitosan based-coating containing nanoemulsion of mandarin essential oil with high pressure or pulsed light processing. *Postharvest Biology and Technology, 106*, 21–32.

Duarte-Molina, F., Gómez, P. L., Castro, M. A., & Alzamora, S. M. (2016). Storage quality of strawberry fruit treated by pulsed light: Fungal decay, water loss and mechanical properties. *Innovative Food Science and Emerging Technologies, 34*, 267–274.

Dunn, J., Ott, T., & Clark, W. (1995). Pulsed-light treatment of food and packaging. *Food Technology, 49*(9), 95–98. Retrieved from https://www.scopus.com/inward/record.uri?eid=2-s2.0-0029378632&partnerID=40&md5=4340e5096eff9b3174ef74189b4ba7ff.

Eie, T. (2009). Light protection from packaging. In K. L. Yam (Ed.), *The Wiley Encyclopaedia of Packaging Technology* (pp. 655–659). Hoboken, NJ: Wiley.

Elmnasser, N., Guillou, S., Leroi, F., Orange, N., Bakhrouf, A., & Federighi, M. (2007). Pulsed-light system as a novel food decontamination technology: A review. *Canadian Journal of Microbiology, 53*(7), 813–821.

Eswaranandam, S., Hettiarachchy, N. S., & Johnson, M. G. (2004). Antimicrobial activity of citric, lactic, malic, or tartaric acids and nisin-incorporated soy protein film against Listeria monocytogenes, Escherichia coli O157: H7, and Salmonella gaminara. *Journal of Food Science, 69*(3), FMS79–FMS84.

Falguera, V., Garvín, A., Garza, S., Pagán, J., & Ibarz, A. (2014). Effect of UV–Vis photochemical processing on pear juices from six different varieties. *Food and Bioprocess Technology, 7*(1), 84–92.

Falguera, V., Pagán, J., & Ibarz, A. (2011). Effect of UV irradiation on enzymatic activities and physico-chemical properties of apple juices from different varieties. *LWT - Food Science and Technology, 44*(1), 115–119.

Farrell, H., Hayes, J., Laffey, J., & Rowan, N. (2011). Studies on the relationship between pulsed UV light irradiation and the simultaneous occurrence of molecular and cellular damage in clinically-relevant Candida albicans. *Journal of Microbiological Methods, 84*(2), 317–326.

Favre, N., Bárcena, A., Bahima, J. V., Martínez, G., & Costa, L. (2018). Pulses of low intensity light as promising technology to delay postharvest senescence of broccoli. *Postharvest Biology and Technology, 142*, 107–114.

FDA. (1996). Code of federal regulations Title 21, Part 179-Irradiation in the Pro- duction. *Processing and Handling of Food*. Retrieved from https://www.gpo.gov/fdsys/pkg/CFR-2000-title21-vol3/pdf/CFR-2000-title21-vol3-sec179-41.pdf.

Feliziani, E., Lichter, A., Smilanick, J. L., & Ippolito, A. (2016). Disinfecting agents for controlling fruit and vegetable diseases after harvest. *Postharvest Biology and Technology, 122*, 53–69.

Fenoglio, D., Ferrario, M., Schenk, M., & Guerrero, S. (2020). Effect of pilot-scale UV-C light treatment assisted by mild heat on E. coli, L. plantarum and S. cerevisiae inactivation in clear and turbid fruit juices. Storage study of surviving populations. *International Journal of Food Microbiology, 332*, 108767.

Fernandes, F. A. N., Rodrigues, S., Law, C. L., & Mujumdar, A. S. (2011). Drying of exotic tropical fruits: A comprehensive review. *Food and Bioprocess Technology, 4*(2), 163–185.

Ferrario, M., Alzamora, S. M., & Guerrero, S. (2013). Inactivation kinetics of some microorganisms in apple, melon, orange and strawberry juices by high intensity light pulses. *Journal of Food Engineering, 118*(3), 302–311.

Ferrario, M., Alzamora, S. M., & Guerrero, S. (2015). Study of the inactivation of spoilage microorganisms in apple juice by pulsed light and ultrasound. *Food Microbiology, 46*, 635–642.

Ferrario, M., & Guerrero, S. (2016). Effect of a continuous flow-through pulsed light system combined with ultrasound on microbial survivability, color and sensory shelf life of apple juice. *Innovative Food Science and Emerging Technologies, 34*, 214–224.

Ferrario, M., & Guerrero, S. (2017). Impact of a combined processing technology involving ultrasound and pulsed light on structural and physiological changes of Saccharomyces cerevisiae KE 162 in apple juice. *Food Microbiology, 65*, 83–94.

Ferrario, M., Guerrero, S., & Alzamora, S. M. (2014). Study of pulsed light-induced damage on Saccharomyces cerevisiae in apple juice by flow cytometry and transmission electron microscopy. *Food and Bioprocess Technology, 7*(4), 1001–1011.

Ferrario, M. I., & Guerrero, S. N. (2018). Inactivation of Alicyclobacillus acidoterrestris ATCC 49025 spores in apple juice by pulsed light: Influence of initial contamination and required reduction levels. *Revista Argentina de Microbiologia, 50*(1), 3–11.

Ferreira, T. V., Mizuta, A. G., de Menezes, J. L., Dutra, T. V., Bonin, E., Castro, J. C., … de Abreu Filho, B. A. (2020). Effect of ultraviolet treatment (UV–C) combined with nisin on industrialized orange juice in Alicyclobacillus acidoterrestris spores. *LWT, 133*, 109911.

Flores-Cervantes, D. X., Palou, E., & López-Malo, A. (2013). Efficacy of individual and combined UVC light and food antimicrobial treatments to inactivate Aspergillus flavus or A. niger spores in peach nectar. *Innovative Food Science and Emerging Technologies, 20*, 244–252.

Funes, G. J., Gómez, P. L., Resnik, S. L., & Alzamora, S. M. (2013). Application of pulsed light to patulin reduction in McIlvaine buffer and apple products. *Food Control, 30*(2), 405–410.

Gamage, G. R., Heyes, J. A., Palmer, J. S., & Wargent, J. (2014). Antimicrobial effect of UV-C treated fresh-cut broccoli against Listeria monocytogenes. *XXIX international horticultural congress on horticulture: Sustaining lives, livelihoods and landscapes (IHC2014): 1120*, 187–192.

Ganan, M., Hierro, E., Hospital, X. F., Barroso, E., & Fernández, M. (2013). Use of pulsed light to increase the safety of ready-to-eat cured meat products. *Food Control, 32*(2), 512–517.

García Carrillo, M., Ferrario, M., & Guerrero, S. (2018). Effectiveness of UV-C light assisted by mild heat on Saccharomyces cerevisiae KE 162 inactivation in carrot-orange juice blend studied by flow cytometry and transmission electron microscopy. *Food Microbiology, 73*, 1–10.

Garrido, Y., Tudela, J. A., Hernández, J. A., & Gil, M. I. (2016). Modified atmosphere generated during storage under light conditions is the main factor responsible for the quality changes of baby spinach. *Postharvest Biology and Technology, 114*, 45–53.

Garvey, M., & Rowan, N. J. (2019). Pulsed UV as a potential surface sanitizer in food production processes to ensure consumer safety. *Current Opinion in Food Science, 26*, 65–70.

Gergoff Grozeff, G. E., Chaves, A. R., & Bartoli, C. G. (2013). Low irradiance pulses improve postharvest quality of spinach leaves (Spinacia oleraceae L. cv Bison). *Postharvest Biology and Technology, 77*, 35–42.

Ghate, V., Kumar, A., Zhou, W., & Yuk, H.-G. (2016). Irradiance and temperature influence the bactericidal effect of 460-nanometer light-emitting diodes on Salmonella in orange juice. *Journal of Food Protection, 79*(4), 553–560.

Ghate, V., Leong, A. L., Kumar, A., Bang, W. S., Zhou, W., & Yuk, H. G. (2015). Enhancing the antibacterial effect of 461 and 521 nm light emitting diodes on selected foodborne pathogens in trypticase soy broth by acidic and alkaline pH conditions. *Food Microbiology, 48*, 49–57.

Ghate, V., Zelinger, E., Shoyhet, H., & Hayouka, Z. (2019). Inactivation of Listeria monocytogenes on paperboard, a food packaging material, using 410 nm light emitting diodes. *Food Control, 96*, 281–290.

Giese, N., & Darby, J. (2000). Sensitivity of microorganisms to different wavelengths of UV light: Implications on modeling of medium pressure UV systems. *Water Research, 34*(16), 4007–4013.

Glowacz, M., Mogren, L. M., Reade, J. P. H., Cobb, A. H., & Monaghan, J. M. (2015). High-but not low-intensity light leads to oxidative stress and quality loss of cold-stored baby leaf spinach. *Journal of the Science of Food and Agriculture, 95*(9), 1821–1829.

Gómez, P. L., Alzamora, S. M., Castro, M. A., & Salvatori, D. M. (2010). Effect of ultraviolet-C light dose on quality of cut-apple: Microorganism, color and compression behavior. *Journal of Food Engineering, 98*(1), 60–70.

Gómez, P. L., García-Loredo, A., Nieto, A., Salvatori, D. M., Guerrero, S., & Alzamora, S. M. (2012a). Effect of pulsed light combined with an antibrowning pretreatment on quality of fresh cut apple. *Innovative Food Science and Emerging Technologies, 16*, 102–112.

Gómez-López, V. M., Devlieghere, F., Bonduelle, V., & Debevere, J. (2005). Factors affecting the inactivation of micro-organisms by intense light pulses. *Journal of Applied Microbiology, 99*(3), 460–470.

Gómez-López, V. M., Ragaert, P., Debevere, J., & Devlieghere, F. (2007). Pulsed light for food decontamination: A review. *Trends in Food Science and Technology, 18*(9), 464–473.

Gómez, P. L., Salvatori, D. M., García-Loredo, A., & Alzamora, S. M. (2012b). Pulsed light treatment of cut apple: Dose effect on color, structure, and microbiological stability. *Food and Bioprocess Technology*, *5*(6), 2311–2322.

González-Aguilar, G. A., Wang, C. Y., Buta, J. G., & Krizek, D. T. (2001). Use of UV-C irradiation to prevent decay and maintain postharvest quality of ripe "Tommy Atkins" mangoes. *International Journal of Food Science and Technology*, *36*(7), 767–773.

Gouma, M., Gayán, E., Raso, J., Condón, S., & Álvarez, I. (2015). Inactivation of spoilage yeasts in apple juice by UV–C light and in combination with mild heat. *Innovative Food Science and Emerging Technologies*, *32*, 146–155.

Gray, N. F. (2014). Chapter thirty-four ultraviolet disinfection. In S. L. Percival, M. V. Yates, & N. F. Gra (Eds.), *Microbiology of waterborne diseases microbiological aspects and risks* (pp. 617–630). New York: Elsevier.

Guan, W., Fan, X., & Yan, R. (2013). Effect of combination of ultraviolet light and hydrogen peroxide on inactivation of Escherichia coli O157:H7, native microbial loads, and quality of button mushrooms. *Food Control*, *34*(2), 554–559.

Guerrero-Beltrán, J. A., & Barbosa-Cánovas, G. V. (2005). Reduction of Saccharomyces cerevisiae, Escherichia coli and Listeria innocua in apple juice by ultraviolet light. *Journal of Food Process Engineering*, *28*(5), 437–452.

Guo, S., Huang, R., & Chen, H. (2017). Application of water-assisted ultraviolet light in combination of chlorine and hydrogen peroxide to inactivate Salmonella on fresh produce. *International Journal of Food Microbiology*, *257*, 101–109.

Guo, S., Huang, R., & Chen, H. (2019). Evaluating a combined method of UV and washing for sanitizing blueberries, tomatoes, strawberries, baby spinach, and lettuce. *Journal of Food Protection*, *82*(11), 1879–1889.

Gwynne, P. J., & Gallagher, M. P. (2018). Light as a broad-spectrum antimicrobial. *Frontiers in Microbiology*, *9*, 119.

Hanning, I. B., Nutt, J. D., & Ricke, S. C. (2009). Salmonellosis outbreaks in the united states due to fresh produce: Sources and potential intervention measures. *Foodborne Pathogens and Disease*, *6*(6), 635–648.

Hasperué, J. H., Rodoni, L. M., Guardianelli, L. M., Chaves, A. R., & Martínez, G. A. (2016). Use of LED light for Brussels sprouts postharvest conservation. *Scientia Horticulturae*, *213*, 281–286.

Haughton, P. N., Lyng, J. G., Cronin, D. A., Morgan, D. J., Fanning, S., & Whyte, P. (2011). Efficacy of UV light treatment for the microbiological decontamination of chicken, associated packaging, and contact surfaces. *Journal of Food Protection*, *74*(4), 565–572.

Heering, W. (2004). UV sources–basics, properties and applications. *IUVA News*, *6*(4), 7–13.

Heinrich, V., Zunabovic, M., Bergmair, J., Kneifel, W., & Jäger, H. (2015). Post-packaging application of pulsed light for microbial decontamination of solid foods: A review. *Innovative Food Science and Emerging Technologies*, *30*, 145–156.

Hessling, M., Spellerberg, B., & Hoenes, K. (2017). Photoinactivation of bacteria by endogenous photosensitizers and exposure to visible light of different wavelengths–a review on existing data. *FEMS Microbiology Letters*, *364*(2), fnw270.

Hinds, L. M., Charoux, C. M. G., Akhter, M., O'Donnell, C. P., & Tiwari, B. K. (2020). Effectiveness of a novel UV light emitting diode based technology for the microbial inactivation of Bacillus subtilis in model food systems. *Food Control*, *114*, 106910.

Hinds, L. M., O'Donnell, C. P., Akhter, M., & Tiwari, B. K. (2019). Principles and mechanisms of ultraviolet light emitting diode technology for food industry applications. *Innovative Food Science and Emerging Technologies*, *56*(April), 102153.

Hossain, M. B., Aguiló-Aguayo, I., Lyng, J. G., Brunton, N. P., & Rai, D. K. (2015). Effect of pulsed electric field and pulsed light pre-treatment on the extraction of steroidal alkaloids from potato peels. *Innovative Food Science and Emerging Technologies*, *29*, 9–14.

Hsu, L., & Moraru, C. I. (2011). Quantifying and mapping the spatial distribution of fluence inside a pulsed light treatment chamber and various liquid substrates. *Journal of Food Engineering*, *103*(1), 84–91.

Huang, R., & Chen, H. (2018). Evaluation of inactivating Salmonella on iceberg lettuce shreds with washing process in combination with pulsed light, ultrasound and chlorine. *International Journal of Food Microbiology, 285*, 144–151.

Huang, R., & Chen, H. (2019a). Comparison of water-assisted decontamination systems of pulsed light and ultraviolet for salmonella inactivation on blueberry, tomato, and lettuce. *Journal of Food Science, 84*(5), 1145–1150.

Huang, R., & Chen, H. (2019b). Sanitation of tomatoes based on a combined approach of washing process and pulsed light in conjunction with selected disinfectants. *Food Research International, 116*(September 2018), 778–785.

Huang, Y., & Chen, H. (2014). A novel water-assisted pulsed light processing for decontamination of blueberries. *Food Microbiology, 40*, 1–8.

Huang, Y., & Chen, H. (2015). Inactivation of Escherichia coli O157:H7, Salmonella and human Norovirus surrogate on artificially contaminated strawberries and raspberries by water-assisted pulsed light treatment. *Food Research International, 72*, 1–7.

Huang, Y., Sido, R., Huang, R., & Chen, H. (2015). Application of water-assisted pulsed light treatment to decontaminate raspberries and blueberries from Salmonella. *International Journal of Food Microbiology, 208*, 43–50.

Hu, X., Geng, S., Wang, X., & Hu, C. (2012). Inactivation and photorepair of enteric pathogenic microorganisms with ultraviolet irradiation. *Environmental Engineering Science, 29*(6), 549–553.

Hwang, H. J., Cheigh, C. I., & Chung, M. S. (2015). Relationship between optical properties of beverages and microbial inactivation by intense pulsed light. *Innovative Food Science and Emerging Technologies, 31*, 91–96.

Hwang, H. J., Cheigh, C. I., & Chung, M. S. (2017). Construction of a pilot-scale continuous-flow intense pulsed light system and its efficacy in sterilizing sesame seeds. *Innovative Food Science and Emerging Technologies, 39*, 1–6.

Hyun, J. E., & Lee, S. Y. (2020). Blue light-emitting diodes as eco-friendly non-thermal technology in food preservation. *Trends in Food Science and Technology, 105*(August), 284–295.

Ignat, A., Manzocco, L., Maifreni, M., Bartolomeoli, I., & Nicoli, M. C. (2014). Surface decontamination of fresh-cut apple by pulsed light: Effects on structure, color and sensory properties. *Postharvest Biology and Technology, 91*, 122–127.

Iwu, C. D., & Okoh, A. I. (2019). Preharvest transmission routes of fresh produce associated bacterial pathogens with outbreak potentials: A review. *International Journal of Environmental Research and Public Health, 16*(22), 4407.

Izquier, A., & Gómez-López, V. M. (2011). Modeling the pulsed light inactivation of microorganisms naturally occurring on vegetable substrates. *Food Microbiology, 28*(6), 1170–1174.

Janve, B. A., Yang, W., Marshall, M. R., Reyes-De-Corcuera, J. I., & Rababah, T. M. (2014). Nonthermal inactivation of Soy (Glycine max Sp.) lipoxygenase by pulsed ultraviolet light. *Journal of Food Science, 79*(1), C8–C18.

Javaid, A. B., Xiong, H., Xiong, Z., Ullah, I., & Wang, P. (2021). Effects of xanthan gum and sodium dodecyl sulfate on physico-chemical, rheological and microstructure properties of non-fried potato instant noodles. *Food Structure, 28*, 100172.

Jayathilake, P. G., Jana, S., Rushton, S., Swailes, D., Bridgens, B., Curtis, T., & Chen, J. (2017). Extracellular polymeric substance production and aggregated bacteria colonization influence the competition of microbes in biofilms. *Frontiers in Microbiology, 8*, 1865.

Jemni, M., Gómez, P. A., Souza, M., Chaira, N., Ferchichi, A., Otón, M., & Artés, F. (2014). Combined effect of UV-C, ozone and electrolyzed water for keeping overall quality of date palm. *LWT - Food Science and Technology, 59*(2P1), 649–655.

Jeon, M. S., Park, K. M., Yu, H., Park, J. Y., & Chang, P. S. (2019). Effect of intense pulsed light on the deactivation of lipase: Enzyme-deactivation kinetics and tertiary structural changes by fragmentation. *Enzyme and Microbial Technology, 124*, 63–69.

Jiang, Y., Ai, C., Liao, X., Liu, D., & Ding, T. (2020). Effect of slightly acidic electrolyzed water (SAEW) and ultraviolet light illumination pretreatment on microflora inactivation of coriander. *LWT, 132*, 109898.

Jo, W.-K., & Tayade, R. J. (2014). New generation energy-efficient light source for photocatalysis: LEDs for environmental applications. *Ind. Eng. Chem. Res.*, *53*(6), 2073–2084.

Jubeen, F., Bhatti, I. A., Khan, M. Z., Zahoor-Ul, H., & Shahid, M. (2012). Effect of UVC irradiation on aflatoxins in ground nut (Arachis hypogea) and tree nuts (Juglans regia, Prunus duclus and pistachio vera). *Journal of the Chemical Society of Pakistan*, *34*(6), 1366–1374.

Jung, Y., Jang, H., & Matthews, K. R. (2014). Effect of the food production chain from farm practices to vegetable processing on outbreak incidence. *Microbial Biotechnology*, *7*(6), 517–527.

Kalaras, M. D., Beelman, R. B., Holick, M. F., & Elias, R. J. (2012). Generation of potentially bioactive ergosterol-derived products following pulsed ultraviolet light exposure of mushrooms (Agaricus bisporus). *Food Chemistry*, *135*(2), 396–401.

Kaya, Z., & Unluturk, S. (2016). Processing of clear and turbid grape juice by a continuous flow UV system. *Innovative Food Science and Emerging Technologies*, *33*, 282–288. https://doi.org/10.1016/j.ifset.2015.12.006.

Kaya, Z., Unluturk, S., Martin-Belloso, O., & Soliva-Fortuny, R. (2020). Effectiveness of pulsed light treatments assisted by mild heat on Saccharomyces cerevisiae inactivation in verjuice and evaluation of its quality during storage. *Innovative Food Science and Emerging Technologies*, *66*, 102517. https://doi.org/10.1016/J.IFSET.2020.102517.

Keyser, M., Muller, I. A., Cilliers, F. P., Nel, W., & Gouws, P. A. (2008). Ultraviolet radiation as a non-thermal treatment for the inactivation of microorganisms in fruit juice. *Innovative Food Science and Emerging Technologies*, *9*(3), 348–354. https://doi.org/10.1016/J.IFSET.2007.09.002.

Khouryieh, H. A. (2021). Novel and emerging technologies used by the U.S. food processing industry. *Innovative Food Science and Emerging Technologies*, *67*, 102559.

Kim, M. J., Bang, W. S., & Yuk, H. G. (2017). 405 ± 5 nm light emitting diode illumination causes photodynamic inactivation of Salmonella spp. on fresh-cut papaya without deterioration. *Food Microbiology*, *62*, 124–132.

Kim, Y. H., Jeong, S. G., Back, K. H., Park, K. H., Chung, M. S., & Kang, D. H. (2013). Effect of various conditions on inactivation of Escherichia coli O157:H7, Salmonella Typhimurium, and Listeria monocytogenes in fresh-cut lettuce using ultraviolet radiation. *International Journal of Food Microbiology*, *166*(3), 349–355.

Kim, S. W., Ko, M. J., & Chung, M. S. (2019). Extraction of the flavonol quercetin from onion waste by combined treatment with intense pulsed light and subcritical water extraction. *Journal of Cleaner Production*, *231*, 1192–1199.

Kim, M. J., Mikš-Krajnik, M., Kumar, A., & Yuk, H. G. (2016). Inactivation by 405 ± 5 nm light emitting diode on Escherichia coli O157:H7, Salmonella Typhimurium, and Shigella sonnei under refrigerated condition might be due to the loss of membrane integrity. *Food Control*, *59*, 99–107.

Kim, T., Silva, J. L., & Chen, T. C. (2002). Effects of UV irradiation on selected pathogens in peptone water and on stainless steel and chicken meat. *Journal of Food Protection*, *65*(7), 1142–1145.

Kim, M. J., Tang, C. H., Bang, W. S., & Yuk, H. G. (2017b). Antibacterial effect of 405 ± 5 nm light emitting diode illumination against Escherichia coli O157:H7, Listeria monocytogenes, and Salmonella on the surface of fresh-cut mango and its influence on fruit quality. *International Journal of Food Microbiology*, *244*, 82–89.

Koh, P. C., Noranizan, M. A., Karim, R., & Nur Hanani, Z. A. (2016). Repetitive pulsed light treatment at certain interval on fresh-cut cantaloupe (Cucumis melo L. reticulatus cv. *Glamour*). *Innovative Food Science and Emerging Technologies*, *36*, 92–103.

Koh, P. C., Noranizan, M. A., Nur Hanani, Z. A., Karim, R., & Rosli, S. Z. (2017). Application of edible coatings and repetitive pulsed light for shelf life extension of fresh-cut cantaloupe (Cucumis melo L. reticulatus cv. *Glamour*). *Postharvest Biology and Technology*, *129*, 64–78.

Koutchma, T. (2008). UV light for processing foods. *Ozone: Science and Engineering*, *30*(1), 93–98.

Koutchma, T. (2009). Advances in ultraviolet light technology for non-thermal processing of liquid foods. *Food and Bioprocess Technology*, *2*(2), 138–155.

Koutchma, T. (2019). Pulsed light as a new treatment to maintain physical and nutritional quality of food. In F. Chemat & E. Vorobiev (Eds), *Green Food Processing Techniques* (pp. 391–401). New York: Elsevier.

Koutchma, T., Parisi, B., & Patazca, E. (2007). Validation of UV coiled tube reactor for fresh juices. *Journal of Environmental Engineering and Science, 6*(3), 319–328.

Koyyalamudi, S. R., Jeong, S. C., Pang, G., Teal, A., & Biggs, T. (2011). Concentration of vitamin D2 in white button mushrooms (Agaricus bisporus) exposed to pulsed UV light. *Journal of Food Composition and Analysis, 24*(7), 976–979.

Kramer, B., & Muranyi, P. (2014). Effect of pulsed light on structural and physiological properties of Listeria innocua and Escherichia coli. *Journal of Applied Microbiology, 116*(3), 596–611.

Kramer, B., Wunderlich, J., & Muranyi, P. (2017a). Pulsed light decontamination of endive salad and mung bean sprouts in water. *Food Control, 73*, 367–371.

Kramer, B., Wunderlich, J., & Muranyi, P. (2017b). Recent findings in pulsed light disinfection. *Journal of Applied Microbiology, 122*(4), 830–856.

Krishnamurthy, K., Tewari, J. C., Irudayaraj, J., & Demirci, A. (2008). Microscopic and spectroscopic evaluation of inactivation of Staphylococcus aureus by pulsed UV light and infrared heating. *Food and Bioprocess Technology, 3*(1), 93.

Kumar, A., Ghate, V., Kim, M. J., Zhou, W., Khoo, G. H., & Yuk, H. G. (2017). Inactivation and changes in metabolic profile of selected foodborne bacteria by 460 nm LED illumination. *Food Microbiology, 63*, 12–21.

Kwaw, E., Ma, Y., Tchabo, W., Apaliya, M. T., Sackey, A. S., Wu, M., & Xiao, L. (2018). Impact of ultrasonication and pulsed light treatments on phenolics concentration and antioxidant activities of lactic-acid-fermented mulberry juice. *LWT - Food Science and Technology, 92*(February), 61–66.

Lante, A., Tinello, F., & Nicoletto, M. (2016). UV-A light treatment for controlling enzymatic browning of fresh-cut fruits. *Innovative Food Science and Emerging Technologies, 34*, 141–147.

Lavilla, M., & Gayán, E. (2018). Consumer acceptance and marketing of foods processed through emerging technologies. In F. Barba, A. Sant'Ana, V. Orlien, & M. Koubaa (Eds.), *Innovative technologies for food preservation: Inactivation of spoilage and pathogenic microorganisms* (pp. 233–253). New York: Elsevier.

Lee, B., & Bahneth, W. P. (2013). Effects of installation location on performance and economics of in-duct ultraviolet germicidal irradiation systems for air disinfection. *Building and Environment, 67*, 193–201.

Lee, H. S., Park, H. H., & Min, S. C. (2020a). Microbial decontamination of red pepper powder using pulsed light plasma. *Journal of Food Engineering, 284*(January), 110075.

Lee, S. Y., Park, H. H., & Min, S. C. (2020b). Pulsed light plasma treatment for the inactivation of Aspergillus flavus spores, Bacillus pumilus spores, and Escherichia coli O157:H7 in red pepper flakes. *Food Control, 118*, 107401.

Leivar, P., Tepperman, J. M., Monte, E., Calderon, R. H., Liu, T. L., & Quail, P. H. (2009). Definition of early transcriptional circuitry involved in light-induced reversal of PIF-imposed repression of photomorphogenesis in young Arabidopsis seedlings. *Plant Cell, 21*(11), 3535–3553.

Leng, J., Mukhopadhyay, S., Sokorai, K., Ukuku, D. O., Fan, X., Olanya, M., & Juneja, V. (2020). Inactivation of Salmonella in cherry tomato stem scars and quality preservation by pulsed light treatment and antimicrobial wash. *Food Control, 110*, 107005.

Lester, G. E., Makus, D. J., & Hodges, D. M. (2010). Relationship between fresh-packaged spinach leaves exposed to continuous light or dark and bioactive contents: Effects of cultivar, leaf size, and storage duration. *Journal of Agriculture and Food Chemistry, 58*(5), 2980–2987.

Lim, W., & Harrison, M. A. (2016). Effectiveness of UV light as a means to reduce Salmonella contamination on tomatoes and food contact surfaces. *Food Control, 66*, 166–173.

Liu, J., Bi, J., McClements, D. J., Liu, X., Yi, J., Lyu, J., … Liu, D. (2020). Impacts of thermal and non-thermal processing on structure and functionality of pectin in fruit- and vegetable-based products: A review. *Carbohydrate Polymers, 250*(2), 116890.

Liu, C.-H., Cai, L.-Y., Lu, X.-Y., Han, X.-X., & Ying, T.-J. (2012). Effect of postharvest UV-C irradiation on phenolic compound content and antioxidant activity of tomato fruit during storage. *Journal of Integrative Agriculture, 11*(1), 159–165.

Liu, L. H., Zabaras, D., Bennett, L. E., Aguas, P., & Woonton, B. W. (2009). Effects of UV-C, red light and sun light on the carotenoid content and physical qualities of tomatoes during post-harvest storage. *Food Chemistry, 115*(2), 495–500.

Lopes, M. M. A., Silva, E. O., Canuto, K. M., Silva, L. M. A., Gallão, M. I., Urban, L., … Miranda, M. R. A. (2016). Low fluence pulsed light enhanced phytochemical content and antioxidant potential of "Tommy Atkins" mango peel and pulp. *Innovative Food Science and Emerging Technologies, 33*, 216–224.

Luksiene, Z., Buchovec, I., & Viskelis, P. (2013). Impact of high-power pulsed light on microbial contamination, health promoting components and shelf life of strawberries. *Food Technology and Biotechnology, 51*(2), 284.

Luksiene, Z., & Paskeviciute, E. (2011). Novel approach to the microbial decontamination of strawberries: Chlorophyllin-based photosensitization. *Journal of Applied Microbiology, 110*(5), 1274–1283.

Luo, Y. (2007). Fresh-cut produce wash water reuse affects water quality and packaged product quality and microbial growth in romaine lettuce. *Hortscience Horts, 42*(6), 1413–1419.

Lu, Y., Yang, B., Zhang, H., & Lai, A. C. K. (2021). Inactivation of foodborne pathogenic and spoilage bacteria by single and dual wavelength UV-LEDs: Synergistic effect and pulsed operation. *Food Control, 125*(August 2020), 107999.

Maftei, N. A., Ramos-Villarroel, A. Y., Nicolau, A. I., Martín-Belloso, O., & Soliva-Fortuny, R. (2014). Influence of processing parameters on the pulsed-light inactivation of Penicillium expansum in apple juice. *Food Control, 41*(1), 27–31.

Mahendran, R., Ramanan, K. R., Barba, F. J., Lorenzo, J. M., López-Fernández, O., Munekata, P. E. S., … Tiwari, B. K. (2019). Recent advances in the application of pulsed light processing for improving food safety and increasing shelf life. *Trends in Food Science and Technology, 88*(December 2018), 67–79.

Makinde, O. M., Ayeni, K. I., Sulyok, M., Krska, R., Adeleke, R. A., & Ezekiel, C. N. (2020). Microbiological safety of ready-to-eat foods in low-and middle-income countries: A comprehensive 10-year (2009 to 2018) review. *Comprehensive Reviews in Food Science and Food Safety, 19*(2), 703–732.

Manzocco, L., Da Pieve, S., & Maifreni, M. (2011). Impact of UV-C light on safety and quality of fresh-cut melon. *Innovative Food Science and Emerging Technologies, 12*(1), 13–17.

Manzocco, L., Ignat, A., Bartolomeoli, I., Maifreni, M., & Nicoli, M. C. (2015). Water saving in fresh-cut salad washing by pulsed light. *Innovative Food Science and Emerging Technologies, 28*, 47–51.

Manzocco, L., Panozzo, A., & Nicoli, M. C. (2013). Inactivation of polyphenoloxidase by pulsed light. *Journal of Food Science, 78*(8), E1183–E1187.

Manzocco, L., Quarta, B., & Dri, A. (2009). Polyphenoloxidase inactivation by light exposure in model systems and apple derivatives. *Innovative Food Science and Emerging Technologies, 10*(4), 506–511.

Marquenie, D., Geeraerd, A. H., Lammertyn, J., Soontjens, C., Van Impe, J. F., Michiels, C. W., & Nicolaï, B. M. (2003). Combinations of pulsed white light and UV-C or mild heat treatment to inactivate conidia of Botrytis cinerea and Monilia fructigena. *International Journal of Food Microbiology, 85*(1–2), 185–196.

Martín-Sómer, M., Pablos, C., van Grieken, R., & Marugán, J. (2017). Influence of light distribution on the performance of photocatalytic reactors: LED vs mercury lamps. *Applied Catalysis B: Environmental, 215*, 1–7.

Massa, G. D., Kim, H.-H., Wheeler, R. M., & Mitchell, C. A. (2008). Plant productivity in response to LED lighting. *HortScience, 43*(7), 1951–1956.

Maurice, J. (1994). The rise and rise of food poisoning. *New Scientist, 144*(1956), 28–33.

Ma, L., Zhang, M., Bhandari, B., & Gao, Z. (2017). Recent developments in novel shelf life extension technologies of fresh-cut fruits and vegetables. *Trends in Food Science and Technology, 64*, 23–38.

Menezes, N. M. C., Tremarin, A., Junior, A. F., & de Aragão, G. M. F. (2019). Effect of soluble solids concentration on Neosartorya fischeri inactivation using UV-C light. *International Journal of Food Microbiology, 296*, 43–47.

Mitchell, C. A., Both, A.-J., Bourget, C. M., Burr, J. F., Kubota, C., Lopez, R. G., … Runkle, E. S. (2012). LEDs: The future of greenhouse lighting! *Chronica Horticulturae, 52*(1), 6–12.

Montgomery, N. L., & Banerjee, P. (2015). Inactivation of Escherichia coli O157:H7 and Listeria monocytogenes in biofilms by pulsed ultraviolet light. *BMC Research Notes, 8*, 235.

Moreau, M., Lescure, G., Agoulon, A., Svinareff, P., Orange, N., & Feuilloley, M. (2013). Application of the pulsed light technology to mycotoxin degradation and inactivation. *Journal of Applied Toxicology, 33*(5), 357–363.

Moreíra, M. R., Tomadoni, B., Martín-Belloso, O., & Fortuny, R. S. (2015). Preservation of fresh-cut apple quality attributes by pulsed light in combination with gellan gum-based prebiotic edible coatings. *LWT - Food Science and Technology, 64*(2), 1130–1137.

Mukhopadhyay, S., Sudarsan, K., Sokorai, K., Ukuku, D. O., Jin, T., Fan, X., ... Juneja, V. (2021). Effects of direct and in-package pulsed light treatment on inactivation of E. coli O157:H7 and reduction of microbial loads in Romaine lettuce. *LWT, 139*, 110710.

Mukhopadhyay, S., Ukuku, D. O., Juneja, V., & Fan, X. (2014). Effects of UV-C treatment on inactivation of Salmonella enterica and Escherichia coli O157:H7 on grape tomato surface and stem scars, microbial loads, and quality. *Food Control, 44*, 110–117.

Muñoz, A., Caminiti, I. M., Palgan, I., Pataro, G., Noci, F., Morgan, D. J., ... Lyng, J. G. (2012). Effects on Escherichia coli inactivation and quality attributes in apple juice treated by combinations of pulsed light and thermosonication. *Food Research International, 45*(1), 299–305.

Muramoto, Y., Kimura, M., & Nouda, S. (2014). Development and future of ultraviolet light-emitting diodes: UV-LED will replace the UV lamp. *Semiconductor Science and Technology, 29*(8), 084004.

Nair, G. B., & Dhoble, S. J. (2021). Current trends and innovations. *Fundamentals and Applications of Light-Emitting Diodes, 253*, 253–270.

Nájera, C., Guil-Guerrero, J. L., Enríquez, L. J., Álvaro, J. E., & Urrestarazu, M. (2018). LED-enhanced dietary and organoleptic qualities in postharvest tomato fruit. *Postharvest Biology and Technology, 145*, 151–156.

Nelson, K. L., Boehm, A. B., Davies-Colley, R. J., Dodd, M. C., Kohn, T., Linden, K. G., ... Zepp, R. G. (2018). Sunlight-mediated inactivation of health-relevant microorganisms in water: A review of mechanisms and modeling approaches. *Environmental Science Processes & Impacts Critical Review, 20*, 1089–1122.

Noci, F., Riener, J., Walkling-Ribeiro, M., Cronin, D. A., Morgan, D. J., & Lyng, J. G. (2008). Ultraviolet irradiation and pulsed electric fields (PEF) in a hurdle strategy for the preservation of fresh apple Juice. *Journal of Food Engineering, 85*(1), 141–146.

Noichinda, S., Bodhipadma, K., Mahamontri, C., Narongruk, T., & Ketsa, S. (2007). Light during storage prevents loss of ascorbic acid, and increases glucose and fructose levels in Chinese kale (Brassica oleracea var. alboglabra). *Postharvest Biology and Technology, 44*(3), 312–315.

Nyangaresi, P. O., Qin, Y., Chen, G., Zhang, B., Lu, Y., & Shen, L. (2019). Comparison of the performance of pulsed and continuous UVC-LED irradiation in the inactivation of bacteria. *Water Research, 157*, 218–227.

Ogihara, H., Morimura, K., Uruga, H., Miyamae, T., Kogure, M., & Furukawa, S. (2013). Inactivation of food-related microorganisms in liquid environment by pulsed xenon flash light treatment system. *Food Control, 33*(1), 15–19.

Olaimat, A. N., & Holley, R. A. (2012). Factors influencing the microbial safety of fresh produce: A review. *Food Microbiology, 32*(1), 1–19.

Oms-Oliu, G., Aguiló-Aguayo, I., Martín-Belloso, O., & Soliva-Fortuny, R. (2010). Effects of pulsed light treatments on quality and antioxidant properties of fresh-cut mushrooms (Agaricus bisporus). *Postharvest Biology and Technology, 56*(3), 216–222.

Onwude, D. I., Hashim, N., Janius, R., Abdan, K., Chen, G., & Oladejo, A. O. (2017). Non-thermal hybrid drying of fruits and vegetables: A review of current technologies. *Innovative Food Science and Emerging Technologies, 43*(January), 223–238.

Page, T., Griffiths, G., & Buchanan-Wollaston, V. (2001). Molecular and biochemical characterization of postharvest senescence in broccoli. *Plant Physiol., 125*(2), 18–27.

Palgan, I., Caminiti, I. M., Muñoz, A., Noci, F., Whyte, P., Morgan, D. J., ... Lyng, J. G. (2011). Effectiveness of high intensity light pulses (HILP) treatments for the control of Escherichia coli and Listeria innocua in apple juice, orange juice and milk. *Food Microbiology, 28*(1), 14–20.

Pankaj, S. K., Shi, H., & Keener, K. M. (2018). A review of novel physical and chemical decontamination technologies for aflatoxin in food. *Trends in Food Science and Technology, 71*, 73–83.

Pataro, G., Gianpiero, M., Sinik, M., Capitoli, M. M., Donsì, G., & Ferrari, G. (2015). The influence of postharvest UV-C and pulsed light treatments on quality and antioxidant properties of tomato fruits during storage. *Innovative Food Science and Emerging Technologies, 30*, 103–111.

Pataro, G., Muñoz, A., Palgan, I., Noci, F., Ferrari, G., & Lyng, J. G. (2011). Bacterial inactivation in fruit juices using a continuous flow pulsed light (PL) system. *Food Research International, 44*(6), 1642–1648.

Patras, A., Brunton, N. P., Da Pieve, S., & Butler, F. (2009). Impact of high pressure processing on total anti-oxidant activity, phenolic, ascorbic acid, anthocyanin content and color of strawberry and black-berry purées. *Innovative Food Science and Emerging Technologies, 10*(3), 308–313.

Pellicer, J. A., & Gómez-López, V. M. (2017). Pulsed light inactivation of horseradish peroxidase and associ-ated structural changes. *Food Chemistry, 237,* 632–637.

Pellicer, J. A., Navarro, P., & Gómez-López, V. M. (2020). Pectin methylesterase inactivation by pulsed light. *Innovative Food Science and Emerging Technologies, 62*(January), 102366.

Pellicer, J. A., Navarro, P., Hernández Sánchez, P., & Gómez-López, V. M. (2019). Structural changes associ-ated with the inactivation of lipoxygenase by pulsed light. *LWT, 113,* 108332.

Phimphilai, S., Maimamuang, S., & Phimphilai, K. (2012). Application of ultraviolet radiation in the dry-ing process of longan (Dimocarpus longan'daw'). *IV international symposium on Lychee, Longan and other Sapindaceae Fruits 1029,* 385–391.

Pisanello, D., & Caruso, G. (2018). *Novel foods in the European Union.* Berlin: Springer.

Pombo, M. A., Dotto, M. C., Martínez, G. A., & Civello, P. M. (2009). UV-C irradiation delays strawberry fruit softening and modifies the expression of genes involved in cell wall degradation. *Postharvest Biology and Technology, 51*(2), 141–148.

Pospíšil, J., & Nešpůrek, S. (2008). Polymer additives. In O. G. Piringer & A. L. Baner (Eds.), *Plastic packaging: Interactions with food and pharmaceuticals* (pp. 63–88). Germany: Wiley.

Prasad, A., Du, L., Zubair, M., Subedi, S., Ullah, A., & Roopesh, M. S. (2020). Applications of light-emitting diodes (LEDs) in food processing and water treatment. *Food Engineering Reviews, 12*(3), 268–289.

Preetha, P., Pandiselvam, R., Varadharaju, N., Kennedy, Z. J., Balakrishnan, M., & Kothakota, A. (2021). Effect of pulsed light treatment on inactivation kinetics of Escherichia coli (MTCC 433) in fruit juices. *Food Control, 121*(August 2020), 107547.

Proulx, J., Sullivan, G., Marostegan, L. F., VanWees, S., Hsu, L. C., & Moraru, C. I. (2017). Pulsed light and antimicrobial combination treatments for surface decontamination of cheese: Favorable and antag-onistic effects. *Journal of Dairy Science, 100*(3), 1664–1673.

Qian, H., Liu, T., Deng, M., Miao, H., Cai, C., Shen, W., & Wang, Q. (2016). Effects of light quality on main health-promoting compounds and antioxidant capacity of Chinese kale sprouts. *Food Chemistry, 196,* 1232–1238.

Rajkovic, A., Tomasevic, I., De Meulenaer, B., & Devlieghere, F. (2017). The effect of pulsed UV light on Escherichia coli O157:H7, Listeria monocytogenes, Salmonella typhimurium, Staphylococcus aureus and staphylococcal enterotoxin A on sliced fermented salami and its chemical quality. *Food Control, 73,* 829–837.

Ramesh, M., Valérie, O., & Mark, L. (2012). Effect of pulsed ultraviolet light on the total phenol content of elderberry (Sambucus nigra) fruit. *Food and Nutrition Sciences, 2012,* 774–783.

Ramos, B., Miller, F. A., Brandão, T. R. S., Teixeira, P., & Silva, C. L. M. (2013). Fresh fruits and vegetables - An overview on applied methodologies to improve its quality and safety. *Innovative Food Science and Emerging Technologies, 20,* 1–15.

Ramos-Villarroel, A. Y., Aron-Maftei, N., Martín-Belloso, O., & Soliva-Fortuny, R. (2012a). Influence of spec-tral distribution on bacterial inactivation and quality changes of fresh-cut watermelon treated with intense light pulses. *Postharvest Biology and Technology, 69,* 32–39.

Ramos-Villarroel, A. Y., Aron-Maftei, N., Martín-Belloso, O., & Soliva-Fortuny, R. (2012b). The role of pulsed light spectral distribution in the inactivation of Escherichia coli and Listeria innocua on fresh-cut mushrooms. *Food Control, 24*(1–2), 206–213.

Ramos-Villarroel, A. Y., Martín-Belloso, O., & Soliva-Fortuny, R. (2011). Bacterial inactivation and qual-ity changes in fresh-cut avocado treated with intense light pulses. *European Food Research and Technology, 233*(3), 395–402.

Ramos-Villarroel, A. Y., Martín-Belloso, O., & Soliva-Fortuny, R. (2013). Intense light pulses: Microbial inac-tivation in fruits and vegetables. *CyTA - Journal of Food, 11*(3), 234–242.

Ramos-Villarroel, A. Y., Martín-Belloso, O., & Soliva-Fortuny, R. (2015). Combined effects of malic acid dip and pulsed light treatments on the inactivation of Listeria innocua and Escherichia coli on fresh-cut produce. *Food Control, 52,* 112–118.

Ribeiro, C., Canada, J., & Alvarenga, B. (2012). Prospects of UV radiation for application in postharvest technology. *Emirates Journal of Food and Agriculture*, *24*(6), 586–597.

Rice, J. K., & Ewell, M. (2001). Examination of peak power dependence in the UV inactivation of bacterial spores. *Applied and Environmental Microbiology*, *67*(12), 5830–5832.

Rodov, V., Vinokur, Y., & Horev, B. (2012). Brief postharvest exposure to pulsed light stimulates coloration and anthocyanin accumulation in fig fruit (Ficus carica L.). *Postharvest Biology and Technology*, *68*, 43–46.

Romero Bernal, A. R., Contigiani, E. V., González, H. H. L., Alzamora, S. M., Gómez, P. L., & Raffellini, S. (2019). Botrytis cinerea response to pulsed light: Cultivability, physiological state, ultrastructure and growth ability on strawberry fruit. *International Journal of Food Microbiology*, *309*, 108311.

Rosenfeldt, E. J., Linden, K. G., Canonica, S., & von Gunten, U. (2006). Comparison of the efficiency of OH radical formation during ozonation and the advanced oxidation processes O_3/H_2O_2 and UV/H_2O_2. *Water Research*, *40*(20), 3695–3704.

Ross, A. I. V., Griffiths, M. W., Mittal, G. S., & Deeth, H. C. (2003, December 31). Combining nonthermal technologies to control foodborne microorganisms. *International Journal of Food Microbiology*, *89*(2–3), 125–138.

Rowan, N. J. (2011). Defining established and emerging microbial risks in the aquatic environment: Current knowledge, implications, and outlooks. *International Journal of Microbiology*, *2011*, 462832.

Rowan, N. J. (2019). Pulsed light as an emerging technology to cause disruption for food and adjacent industries – Quo vadis? *Trends in Food Science and Technology*, *88*(November 2018), 316–332.

Rowan, N. J., Valdramidis, V. P., & Gómez-López, V. M. (2015). A review of quantitative methods to describe efficacy of pulsed light generated inactivation data that embraces the occurrence of viable but non culturable state microorganisms. *Trends in Food Science and Technology*, *44*(1), 79–92.

Sahoo, S. K., Tomar, M. S., & Pradhan, R. C. (2021). Disinfecting agents for controlling fruits and vegetable diseases after harvest. *Food Losses, Sustainable Postharvest and Food Technologies*, 103–151.

Salinas-Roca, B., Soliva-Fortuny, R., Welti-Chanes, J., & Martín-Belloso, O. (2016). Combined effect of pulsed light, edible coating and malic acid dipping to improve fresh-cut mango safety and quality. *Food Control*, *66*, 190–197.

Samuoliene, G., Sirtautas, R., Brazaityte, A., & Duchovskis, P. (2012). LED lighting and seasonality effects antioxidant properties of baby leaf lettuce. *Food Chemistry*, *134*(3), 1494–1499.

Sanchez-Vizuete, P., Orgaz, B., Aymerich, S., Le Coq, D., & Briandet, R. (2015). Pathogens protection against the action of disinfectants in multispecies biofilms. *Frontiers in Microbiology*, *6*, 705.

Santhirasegaram, V., Razali, Z., George, D. S., & Somasundram, C. (2015). Comparison of UV-C treatment and thermal pasteurization on quality of Chokanan mango (Mangifera indica L.) juice. *Food and Bioproducts Processing*, *94*, 313–321.

Sanzani, S. M., Nigro, F., Mari, M., & Ippolito, A. (2009). Innovations in the control of postharvest diseases of fresh fruits and vegetables. *Arab Journal of Plant Protection*, *27*(2), 240–244.

Sauceda-Gálvez, J. N., Tió-Coma, M., Martinez-Garcia, M., Hernández-Herrero, M. M., Gervilla, R., & Roig-Sagués, A. X. (2020). Effect of single and combined UV-C and ultra-high pressure homogenisation treatments on inactivation of Alicyclobacillus acidoterrestris spores in apple juice. *Innovative Food Science and Emerging Technologies*, *60*, 102299.

Schmid, F.-X. (2001). *Biological macromolecules: UV-visible spectrophotometry* (pp. 1–4). London, UK: Encyclopedia of Life Sciences.

Schmid, J., Hoenes, K., Vatter, P., & Hessling, M. (2019). Antimicrobial effect of visible light—Photoinactivation of Legionella rubrilucens by irradiation at 450, 470, and 620 nm. *Antibiotics*, *8*(4), 187.

Schottroff, F., Fröhling, A., Zunabovic-Pichler, M., Krottenthaler, A., Schlüter, O., & Jäger, H. (2018). Sublethal injury and viable but non-culturable (VBNC) state in microorganisms during preservation of food and biological materials by non-thermal processes. *Frontiers in Microbiology*, *9*, 2773.

Schweiggert, R. M., & Carle, R. (2017). Critical reviews in food science and nutrition carotenoid deposition in plant and animal foods and its impact on bioavailability carotenoid deposition in plant and animal foods and its impact on bioavailability. *Crit Rev Food Sci Nutr.*, *57*(9), 1807–1830.

Scott, G., Dickinson, M., Shama, G., & Rupar, M. (2018). A comparison of the molecular mechanisms underpinning high-intensity, pulsed polychromatic light and low-intensity UV-C hormesis in tomato fruit. *Postharvest Biology and Technology*, *137*(April 2017), 46–55.

Scott, G., Rupar, M., Fletcher, A. G. D., Dickinson, M., & Shama, G. (2017). A comparison of low intensity UV-C and high intensity pulsed polychromatic sources as elicitors of hormesis in tomato fruit. *Postharvest Biology and Technology, 125*, 52–58.

Shahbaz, H. M., Ryoo, H., Kim, J. U., Kim, S., Lee, D. U., Ghafoor, K., & Park, J. (2017). Effects of UV-C in a Teflon-coil and high hydrostatic pressure combined treatment for maintenance of the characteristic quality of dongchimi (watery radish kimchi) during room temperature storage. *Journal of Food Processing and Preservation, 41*(4), e13057.

Sholtes, K., & Linden, K. G. (2019). Pulsed and continuous light UV LED: Microbial inactivation, electrical, and time efficiency. *Water Research, 165*, 114965.

Solomon, E. B., Yaron, S., & Matthews, K. R. (2002). Transmission of Escherichia coli O157: H7 from contaminated manure and irrigation water to lettuce plant tissue and its subsequent internalization. *Applied and Environmental Microbiology, 68*(1), 397–400.

Solovchenko, A. E., & Merzlyak, M. N. (2008). Screening of visible and UV radiation as a photoprotective mechanism in plants. *Russian Journal of Plant Physiology, 55*(6), 719–737.

Søltoft-Jensen, J., & Hansen, F. (2005). New chemical and biochemical hurdles. In Da-Wen Sun (Ed.), *Emerging technologies for food processing* (pp. 387–416). New York: Elsevier.

Song, K., Mohseni, M., & Taghipour, F. (2016, May 1). Application of ultraviolet light-emitting diodes (UV-LEDs) for water disinfection: A review. *Water Research, 94*, 341–349.

Song, Y., Qiu, K., Gao, J., & Kuai, B. (2020). Molecular and physiological analyses of the effects of red and blue LED light irradiation on postharvest senescence of pak choi. *Postharvest Biology and Technology, 164*(February), 111155.

Song, K., Taghipour, F., & Mohseni, M. (2018). Microorganisms inactivation by continuous and pulsed irradiation of ultraviolet light-emitting diodes (UV-LEDs). *Chemical Engineering Journal, 343*, 362–370.

Stack, H. M., Hill, C., & Gahan, C. G. M. (2008). Stress responses. In D. Liu (Ed.), *Handbook of Listeria monocytogenes* (pp. 61–96). Boca Raton, FL: CRC Press.

Subedi, S., & Roopesh, M. S. (2020). Simultaneous drying of pet food pellets and Salmonella inactivation by 395 nm light pulses in an LED reactor. *Journal of Food Engineering, 286*(December 2019), 110110.

Surjadinata, B. B., Jacobo-Velázquez, D. A., & Cisneros-Zevallos, L. (2017). UVA, UVB and UVC light enhances the biosynthesis of phenolic antioxidants in fresh-cut carrot through a synergistic effect with wounding. *Molecules, 22*(4), 668.

Takeshita, K., Shibato, J., Sameshima, T., Fukunaga, S., Isobe, S., Arihara, K., & Itoh, M. (2003). Damage of yeast cells induced by pulsed light irradiation. *International Journal of Food Microbiology, 85*(1–2), 151–158.

Tao, T., Ding, C., Han, N., Cui, Y., Liu, X., & Zhang, C. (2019). Evaluation of pulsed light for inactivation of foodborne pathogens on fresh-cut lettuce: Effects on quality attributes during storage. *Food Packaging and Shelf Life, 21*, 100358.

Tawema, P., Han, J., Vu, K. D., Salmieri, S., & Lacroix, M. (2016). Antimicrobial effects of combined UV-C or gamma radiation with natural antimicrobial formulations against Listeria monocytogenes, Escherichia coli O157: H7, and total yeasts/molds in fresh cut cauliflower. *LWT - Food Science and Technology, 65*, 451–456.

Thakkar, S. K., Huo, T., Maziya-Dixon, B., & Failla, M. L. (2009). Impact of style of processing on retention and bioaccessibility of β-carotene in cassava (manihot esculenta, crantz). *Journal of Agricultural and Food Chemistry, 57*(4), 1344–1348.

Tikekar, R. V., Anantheswaran, R. C., & LaBorde, L. F. (2011). Ascorbic acid degradation in a model apple juice system and in apple juice during ultraviolet processing and storage. *Journal of Food Science, 76*(2), H62–H71.

Toledo, M. E. A., Ueda, Y., Imahori, Y., & Ayaki, M. (2003). L-ascorbic acid metabolism in spinach (Spinacia oleracea L.) during postharvest storage in light and dark. *Postharvest Biology and Technology, 28*(1), 47–57.

Uesugi, A. R., Hsu, L. C., Worobo, R. W., & Moraru, C. I. (2016). Gene expression analysis for Listeria monocytogenes following exposure to pulsed light and continuous ultraviolet light treatments. *LWT - Food Science and Technology, 68*, 579–588.

Uesugi, A. R., & Moraru, C. I. (2009). Reduction of Listeria on ready-to-eat sausages after exposure to a combination of pulsed light and nisin. *Journal of Food Protection, 72*(2), 347–353.

Urrestarazu, M., Nájera, C., & Gea, M. M. (2016). Effect of the spectral quality and intensity of light-emitting diodes on several horticultural crops. *Hortscience, 51*(3), 268–271.

Vicente, A. R., Pineda, C., Lemoine, L., Civello, P. M., Martinez, G. A., & Chaves, A. R. (2005). UV-C treatments reduce decay, retain quality and alleviate chilling injury in pepper. *Postharvest Biology and Technology, 35*(1), 69–78.

Vincente, A. R., Manganaris, G. A., Ortiz, C. M., Sozzi, G. O., & Crisosto, C. H. (2014). Nutritional quality of fruits and vegetables. In N. Banks, R. L. Shewfelt, S. E. Prussia, & W. J. Florkowski (Eds.), *Postharvest Handling: A Systems Approach* (pp. 69–122). New York: Elsevier.

Wadamori, Y., Gooneratne, R., & Hussain, M. A. (2016). Outbreaks and factors influencing microbiological contamination of fresh produce. *J Sci Food Agric, 97*(5), 1396–1403.

Wang, B., Mahoney, N. E., Pan, Z., Khir, R., Wu, B., Ma, H., & Zhao, L. (2016). Effectiveness of pulsed light treatment for degradation and detoxification of aflatoxin B1 and B2 in rough rice and rice bran. *Food Control, 59*, 461–467.

Wang, B., Zhang, Y., Venkitasamy, C., Wu, B., Pan, Z., & Ma, H. (2017). Effect of pulsed light on activity and structural changes of horseradish peroxidase. *Food Chemistry, 234*, 20–25.

Watada, A. E., & Qi, L. (1999). Quality of fresh-cut produce. *Postharvest Biology and Technology, 15*(3), 201–205.

Wekhof, A. (2000). Disinfection with flash lamps. *PDA Journal of Pharmaceutical Science and Technology, 54*(3), 264–276. Retrieved from http://europepmc.org/abstract/MED/10927918.

Whipps, J. M., Hand, P., Pink, D. A. C., & Bending, G. D. (2008). Chapter 7. Human pathogens and the phyllosphere. *Advances in Applied Microbiology, 64*, 183–221.

Williamson, V. G., Frisina, C., Tareen, M. N., & Stefanelli, D. (2018). Storage performance of two 'Pink Lady®' clones differs, but 1-MCP treatment is beneficial, regardless of maturity at harvest. *Scientia Horticulturae, 235*, 142–151.

World Health Organization (WHO). (2003). *Diet, nutrition, and the prevention of chronic diseases: Report of a joint WHO/FAO expert consultation, 916*. Geneva: World Health Organization.

Wu, J., Zhang, D., Chen, Q., Feng, J., Li, Q., Yang, F., … Cheng, X. (2018). Shifts in soil organic carbon dynamics under detritus input manipulations in a coniferous forest ecosystem in subtropical China. *Soil Biology and Biochemistry, 126*, 1–10.

Xiang, Q., Fan, L., Zhang, R., Ma, Y., Liu, S., & Bai, Y. (2020). Effect of UVC light-emitting diodes on apple juice: Inactivation of Zygosaccharomyces rouxii and determination of quality. *Food Control, 111*, 107082.

Xiao, Z., Lester, G. E., Luo, Y., Xie, Z., Yu, L., & Wang, Q. (2014). Effect of light exposure on sensorial quality, concentrations of bioactive compounds and antioxidant capacity of radish microgreens during low temperature storage. *Food Chemistry, 151*, 472–479.

Xu, W., Chen, H., Huang, Y., & Wu, C. (2013). Decontamination of Escherichia coli O157: H7 on green onions using pulsed light (PL) and PL-surfactant-sanitizer combinations. *International Journal of Food Microbiology, 166*(1), 102–108.

Xu, W., Chen, H., & Wu, C. (2015). Application of pulsed light (PL)-surfactant combination on inactivation of Salmonella and apparent quality of green onions. *LWT - Food Science and Technology, 61*(2), 596–601.

Xu, F., Wang, B., Hong, C., Telebielaigen, S., Nsor-Atindana, J., Duan, Y., & Zhong, F. (2019). Optimization of spiral continuous flow-through pulse light sterilization for Escherichia coli in red grape juice by response surface methodology. *Food Control, 105*, 8–12.

Xu, W., & Wu, C. (2014). Decontamination of Salmonella enterica Typhimurium on green onions using a new formula of sanitizer washing and pulsed UV light (PL). *Food Research International, 62*, 280–285.

Xu, W., & Wu, C. (2016). The impact of pulsed light on decontamination, quality, and bacterial attachment of fresh raspberries. *Food Microbiology, 57*, 135–143.

Yi, J. Y., Bae, Y. K., Cheigh, C. I., & Chung, M. S. (2017). Microbial inactivation and effects of interrelated factors of intense pulsed light (IPL) treatment for Pseudomonas aeruginosa. *LWT - Food Science and Technology, 77*, 52–59.

Yi, J. Y., Lee, N. H., & Chung, M. S. (2016). Inactivation of bacteria and murine Norovirus in untreated groundwater using a pilot-scale continuous-flow intense pulsed light (IPL) system. *LWT - Food Science and Technology*, *66*, 108–113.

Zhang, L., Ma, G., Yamawaki, K., Ikoma, Y., Matsumoto, H., Yoshioka, T., … Kato, M. (2015). Regulation of ascorbic acid metabolism by blue LED light irradiation in citrus juice sacs. *Plant Science*, *233*, 134–142.

Zhan, L., Hu, J., Ai, Z., Pang, L., Li, Y., & Zhu, M. (2013a). Light exposure during storage preserving soluble sugar and L-ascorbic acid content of minimally processed romaine lettuce (Lactuca sativa L.var. longifolia). *Food Chemistry*, *136*(1), 273–278.

Zhan, L., Hu, J., Li, Y., & Pang, L. (2012). Combination of light exposure and low temperature in preserving quality and extending shelf-life of fresh-cut broccoli (Brassica oleracea L.). *Postharvest Biology and Technology*, *72*, 76–81.

Zhu, Y., Koutchma, T., Warriner, K., & Zhou, T. (2014). Reduction of patulin in apple juice products by uv light of different wavelengths in the uvc range. *Journal of Food Protection*, *77*(6), 963–971.

Zhu, Y., Li, C., Cui, H., & Lin, L. (2019). Antimicrobial mechanism of pulsed light for the control of Escherichia coli O157:H7 and its application in carrot juice. *Food Control*, *106*, 106751.

Zhuang, R. Y., Beuchat, L. R., & Angulo, F. J. (1995). Fate of Salmonella montevideo on and in raw tomatoes as affected by temperature and treatment with chlorine. *Applied and Environmental Microbiology*, *61*(6), 2127–2131.

Chapter 9

Commercial Feasibility and Viability of Non-Thermal Processing Technologies for the Fruit and Vegetable Processing Industry

M. Selvamuthukumaran

CONTENTS

9.1 INTRODUCTION

Non-thermal processing technologies can be widely adopted for implementation in the fruit and vegetable processing industry (Figure 9.1). These techniques can minimize the destruction of valuable nutrients, especially vitamins and minerals during the processing of fruits and vegetables and can prevent spoilage which causes microorganisms thereby ensuring the safety of the processed foods. There are a lot of techniques available, but each technique has its own merits and demerits. The various advantages and disadvantages associated with specific non-thermal processing technologies for their application in the fruit and vegetable processing industry are detailed below.

DOI: 10.1201/9781003222170-9

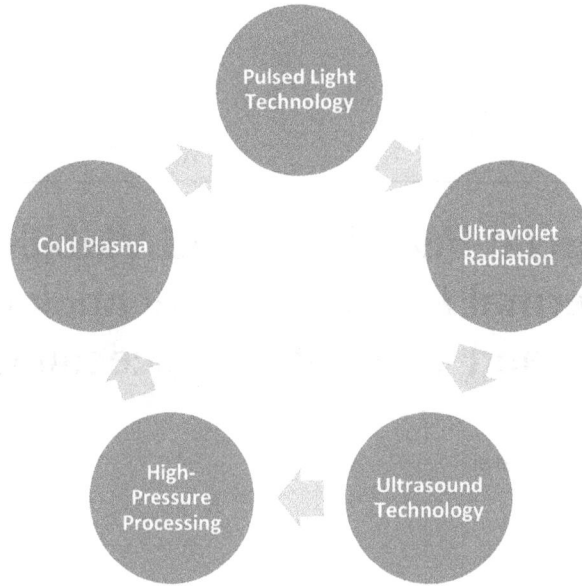

Figure 9.1   Non-thermal processing technologies for the fruit and vegetable processing industry.

9.2 PROS OF THE SCALING UP OF NON-THERMAL PROCESSING TECHNOLOGIES

9.2.1 Pulsed Light Technology

This is considered to be one of the quick approaches for the disinfection of various fruits and vegetable products. Fewer nutrients are degraded during pulse light applications (Pan et al., 2017). Organoleptic characteristics of the products were highly achieved with maximum inactivation of microbes, and microorganisms' presence may lead to significant quality losses (Bhavya and Hebbar, 2017). This technology is environmentally friendly and has wide adaptability because of the significant bacterial reduction in a shorter time (Bhavya and Hebbar, 2017; Oms-Oliu et al., 2010). This technique provides an assured economic return thereby reducing the microorganism and also by enhancing the stability of the final product (Bhavya and Hebbar, 2017; Oms-Oliu et al., 2010; Chaine et al., 2012; Ozer and Demirci, 2006). The pulse light can be applied successfully to the packed fruit and vegetable products, which can avoid contamination (Junior et al., 2020).

9.2.2 Ultrasound Technology

A major advantage laying behind the use of ultrasound technology is the consumption of lower amounts of energy with both higher material output as well as input (Hernández-Hernández et al., 2019; Rawson et al., 2011). Treatment time with food products is greatly reduced. Ultrasound technology can be scaled up successfully for the fruit and vegetable product processing industry, as it is deemed to be safe as well as eco-friendly (Dong et al., 2021). In addition to these advantages, the technique is quite feasible, easy to adopt because of its simplicity, and cheaper when compared with traditional heat processing techniques (Bahrami et al., 2020; Gabriel, 2015). This technology can prevent microbial attacks that can interfere with the fermentation process of juices.

9.2.3 Ultraviolet Radiation

The application of ultraviolet radiation for treating fresh fruit juice especially watermelon results in the prevention of nutrition degradation, thereby enhancing the quality to a higher extent. Antioxidants viz. vitamin C, lycopene, and phenols were preserved in this way (Bhattacharjee et al., 2019). Therefore, this UV radiation can be scaled up for the juice processing industry, thereby strictly adhering to government regulations. It has already been shown that the application of such radiation techniques can enhance the microbial lethality effect, when compared to the use of chemical agents like chlorine and hydrogen peroxide (Bahrami et al., 2020; Mikš-Krajnik et al., 2017). The techniques can be used for several packed fruit and vegetable products, as they possess excellent permeation properties (Hernández-Hernández et al., 2019; Khan et al., 2018).

Ultraviolet radiation is relatively cost-effective (Bhattacharjee et al., 2019; Gayan et al., 2012) and can be user-friendly, thereby impacting less on the food quality. It also increases organoleptic characteristics, especially the taste of the product, and this can be applied both on solid as well as liquid food products.

9.2.4 High-Pressure Processing

High-pressure processing is a non-thermal technique that can be applied for either solid based or liquid based fruit or vegetable products (Barba et al., 2015). The major advantages of using these non-thermal processing techniques include zero waste emission (Table 9.1) (Jung and Tonello-Samson, 2018; Boye and Arcand, 2013; Rodriguez-Gonzalez et al., 2015) with very limited use of preservatives (Hernández-Hernández et al., 2019; Chawla et al., 2011). Even pressurization water can be recycled for its use thereby saving energy. The quality of fruit and vegetable products can be retained by enhancing the organoleptic acceptability, preserving the nutrients, and increasing the stability. This technique reduces the rate of spoilage, which can increase the economic value (Dong et al., 2021; Huang et al., 2020; Junior et al., 2019).

TABLE 9.1   MERITS AND DEMERITS OF USING HIGH-PRESSURE PROCESSING FOR THE FRUIT AND VEGETABLE PROCESSING INDUSTRY

Non-thermal processing technique	Merits	Demerits
High-Pressure Processing	• Applied on either solid based or liquid based fruits or vegetable products • Zero waste emission • Needs to be used with very few preservatives • Energy is saved • Stability increased • Acceptability improved • Nutrients retained	• Cannot be applied on dried and porous food products • Needs big investment • May require large space • Needs skilled persons to operate it effectively • Used for only handling food products, which are packed in plastic only

The commercial application of high-pressure processing shows that this technique helps to enhance the higher rate of bioactive constituents extraction, especially anthocyanin in the raw materials as a result of applying pressure during the extraction process (Barba et al., 2015). In the wine industry such techniques can be adopted to extract bioactive constituents from by-products and waste. The main advantage of using this method lies in the saving of energy and it is also energy efficient (Jung and Tonello-Samson, 2018).

9.2.5 Cold Plasma

The cold plasma technique is widely used to inactivate microorganisms like fungi, bacteria, and viruses (Ma et al., 2015; Charoux et al., 2021). The major advantages of using this technique are both the authentic taste of foods and the nutrients are retained (Sonawane and Patil, 2020). The entire process is carried out at lower temperatures even at room temperature (Pan et al., 2017), therefore the quality of the final product is greatly enhanced (Kim and Min, 2018). This technique can be commercially adopted as this will save the use of more water, as this approach will lead to the use of natural gas and electricity (Hernández-Hernández et al., 2019; Ziuzina et al., 2015). By using this technique successfully, the microbes, which are present on the external surface of the food, can be controlled to a greater extent, and surface food decontamination is also possible by using this technique. In animal foods, this technique has shown excellent results in killing and destroying a wide variety of microbial species especially in fresh meat and fish (Mishra et al., 2016).

9.3 CONS OF THE SCALING UP OF NON-THERMAL PROCESSING TECHNOLOGIES

9.3.1 Pulsed Light Technology

It requires more installation cost (Bhavya and Hebbar, 2017). This method seems to not be an option for handling fruits and vegetables, which are irregular in shape and opaque in appearance; it can cause heating to products because of the high treatment time period and finally it will affect the reduction power of the bacterial species (Bhavya and Hebbar, 2017; Huang and Chen, 2014).

9.3.2 Ultrasound Technology

Fruit and vegetable product quality can be reduced as a result of the application of ultrasound due to the formation of free radicals; the cavitation process may lead to the generation of many free radicals, which can reduce the quality of food products (Bahrami et al., 2020; Pingret et al., 2013). The quality changes brought about in foods include protein denaturation, lipid peroxidation, ascorbic acid reduction, and organoleptic acceptability reduction (Bhargava et al., 2021). It also reduces nutrients (Hernández-Hernández et al., 2019; Harder et al., 2016). Even if higher ultrasound wave is used at 20 kHz, it cannot even destroy specific microbes like *Listeria monocytogenes*. For reducing such microbes, higher sonication power and pressure are required (Chemat et al., 2017). The commercial application of ultrasound can pose a threat to the workers or operators, as it can potentially result in occupational risks.

9.3.3 Ultraviolet Radiation

Consumer acceptability is the drawback of the commercial application of UV radiation techniques. Consumers have doubts as to whether food processed out of UV radiation may be harmful to our health or not (Hernández-Hernández et al., 2019; Khan et al., 2018). A major consumer worry is whether this processing may leave some radioactive materials in the foods, which may even pose some health risks to the public (Bahrami et al., 2020; Xuan et al., 2017). It was observed that the repeated and continuous use of UV radiation may affect the properties of food materials; antioxidants like lycopene are oxidized as a result of the enhanced use of UV radiation concentration for a prolonged time (Bhattacharjee et al., 2019). UV radiation is applied to liquid foods and may affect the turbidity (Pinto et al., 2016). Both a prolonged time as well as higher manpower use in a fruit and vegetable processing unit will reduce the commercial application of UV radiation on a larger scale. Being its lower penetration power, they become less effective when they are applied to food particles that are of an irregular shape. This process requires higher capital investment, which can further limit its commercial applications (Hernández-Hernández et al., 2019)

9.3.4 High-Pressure Processing

High-pressure processing cannot be used for dried and porous food products (La Peña et al., 2019). The treated food products need to be kept in chilled conditions, since pressure in combination with a cold temperature can successfully destroy microbial vegetative cells and therefore this process deems to be a very cost-effective and labor-oriented process (Huang et al., 2017). It can be used for only handling food products which are packed in plastic as this process leads the packed materials to compress by about 15% (Huang et al., 2017). Product quality is also altered especially while processing dairy products i.e. the free fatty acid level is enhanced, and the magnitude of casein micelle is reduced (Liepa et al., 2016). This process may also require big primary investment and in addition to that it may require a large space and skilled people to operate it effectively.

9.3.5 Cold Plasma

This process results in the generation of free radicals; as a result the foods may undergo oxidation, particularly lipids may deteriorate and antioxidants may be destroyed, and it also affects the organoleptic attributes, especially aroma and taste (Sonawane and Patil, 2020; Yong et al., 2015). It also requires higher investment costs, and it needs sound knowledge for the skilled people to understand plasma chemistry (Roobab et al., 2018; Atik and Gümü, 2021). This process has its limitations, particularly when being implemented at a commercial level especially by adopting this technique on pilot scale level to industrial level, as there is no process optimization at a larger scale (Hernández-Hernández et al., 2019; Priyadarshini et al., 2019; Knorr et al., 2013). In dairy products, this process may enhance the lipid oxidation rate, thereby affecting the organoleptic attributes (Coutinho et al., 2018). The application of plasma will not be effective for foods which have an irregular structure, as the plasma cannot penetrate food particles (Coutinho et al., 2018; Song et al., 2009). This process can be implemented either at a lab or on a pilot plant scale level only (Coutinho et al., 2018; Rathod et al., 2021; Asaithambi et al., 2021).

9.4 CONCLUSIONS

The application of non-thermal technologies especially for fruits and vegetable processing may ensure that by using this technique it is possible to obtain nutritious, wholesome, and safe food. However, these techniques are available only at a pilot scale level and for commercialization there is a need for large infrastructural facilities, huge capital investment, and skilled people for its realization. Such an implementation at a high level is quite a big challenge. Fruit and vegetable processors can even adopt any specific technique, which can give good results with respect to product quality as well as commercial feasibility.

REFERENCES

Asaithambi, N.; Singh, S.K.; Singha, P. Current status of non-thermal processing of probiotic foods: A review. *J. Food Eng.* 2021, 303, 110567.

Atik, A.; Gümüs, T. The effect of different doses of UV-C treatment on microbiological quality of bovine milk. *LWT* 2021, 136, 110322.

Bahrami, A.; Baboli, Z.M.; Schimmel, K.; Jafari, S.M.; Williams, L. Efficiency of novel processing technologies for the control of Listeria monocytogenes in food products. *Trends Food Sci. Technol.* 2020, 96, 61–78.

Barba, F.J.; Parniakov, O.; Pereira, S.A.; Wiktor, A.; Grimi, N.; Boussetta, N.; Saraiva, J.A.; Raso, J.; Martin-Belloso, O.; Witrowa-Rajchert, D.; et al. Current applications and new opportunities for the use of pulsed electric fields in food science and industry. *Food Res. Int.* 2015, 77, 773–798.

Bhargava, N.; Mor, R.S.; Kumar, K.; Sharanagat, V.S. Advances in application of ultrasound in food processing: A review. *Ultrason. Sonochem.* 2021, 70, 105293.

Bhattacharjee, C.; Saxena, V.; Dutta, S. Novel thermal and non-thermal processing of watermelon juice. *Trends Food Sci. Technol.* 2019, 93, 234–243.

Bhavya, M.L.; Hebbar, H.U. Pulsed light processing of foods for microbial safety. *Food Qual. Saf.* 2017, 1(3), 187–202.

Boye, J.I.; Arcand, Y. Current trends in green technologies in food production and processing. *Food Eng. Rev.* 2013, 5(1), 1–17.

Chaine, A.; Levy, C.; Lacour, B.; Riedel, C.; Carlin, F. Decontamination of sugar syrup by pulsed light. *J. Food Prot.* 2012, 75(5), 913–917.

Charoux, C.M.G.; Patange, A.; Lamba, S.; O'Donnell, C.P.; Toward, B.K.; Scannell, A.G.M. Application of non thermal plasma technology on safety and quality of dried food ingredients. *J. Appl. Microbiol.* 2021, 130(2), 325–340.

Chawla, R.; Patil, G.R.; Singh, A.K. High hydrostatic pressure technology in dairy processing: A review. *J. Food Sci. Technol.* 2011, 48(3), 260–268.

Chemat, F.; Rombaut, N.; Sicaire, A.-G.; Meullemiestre, A.; Fabiano-Tixier, A.-S.; Abert-Vian, M. Ultrasound assisted extraction of food and natural products. Mechanisms, techniques, combinations, protocols and applications. A review. *Ultrason. Sonochem.* 2017, 34, 540–560.

Coutinho, N.M.; Silveira, M.R.; Rocha, R.S.; Moraes, J.; Ferreira, M.V.S.; Pimentel, T.C.; Freitas, M.Q.; Silva, M.C.; Raices, R.S.; Ranadheera, C.S.; et al. Cold plasma processing of milk and dairy products. *Trends Food Sci. Technol.* 2018, 74, 56–68.

Dong, X.; Wang, J.; Raghavan, V. Critical reviews and recent advances of novel non-thermal processing techniques on the modification of food allergens. *Crit. Rev. Food Sci. Nutr.* 2021, 61(2), 196–210.

Gabriel, A.A. Inactivation of Listeria monocytogenes in milk by multifrequency power ultrasound. *J. Food Process. Preserv.* 2015, 39(6), 846–853.

Gayán, E.; Serrano, M.J.; Raso, J.; Álvarez, I.; Condón, S. Inactivation of Salmonella enterica by UV-C light alone and in combination with mild temperatures. *Appl. Environ. Microbiol.* 2012, 78(23), 8353–8361.

Harder, M.N.C.; Arthur, V.; Arthur, P.B. Irradiation of foods: Processing technology and effects on nutrients: Effect of ionizing radiation on food components. In: *Encyclopedia of food and health*; Caballero, B., Finglas, P.M., Toldrá, F., Eds.; Academic Press: Oxford, 2016; pp. 476–481.

Hernández-Hernández, H.; Moreno-Vilet, L.; Villanueva-Rodríguez, S. Current status of emerging food processing technologies in Latin America: Novel non-thermal processing. *Innov. Food Sci. Emerg. Technol.* 2019, 58, 102233.

Huang, H.-W.; Hsu, C.-P.; Wang, C.-Y. Healthy expectations of high hydrostatic pressure treatment in food processing industry. *J. Food Drug Anal.* 2020, 28(1), 1–13.

Huang, H.-W.; Wu, S.-J.; Lu, J.-K.; Shyu, Y.-T.; Wang, C.-Y. Current status and future trends of high-pressure processing in food industry. *Food Control* 2017, 72, 1–8.

Huang, Y.; Chen, H. A novel water-assisted pulsed light processing for decontamination of blueberries. *Food Microbiol.* 2014, 40, 1–8.

Jung, S.; Tonello-Samson, C. High hydrostatic pressure food processing: Potential and limitations. In: *Alternatives to conventional food processing*; Proctor, A. Ed.; Royal Society of Chemistry: London, 2018; pp. 251–315.

Júnior, L.M.; Cristianini, M.; Anjos, C.A.R. Packaging aspects for processing and quality of foods treated by pulsed light. *J. Food Process. Preserv.* 2020, 44, 14902.

Júnior, L.M.; Cristianini, M.; Padula, M.; Anjos, C.A.R. Effect of high-pressure processing on characteristics of flexible packaging for foods and beverages. *Food Res. Int.* 2019, 119, 920–930.

Khan, M.K.; Ahmad, K.; Hassan, S.; Imran, M.; Ahmad, N.; Xu, C. Effect of novel technologies on polyphenols during food processing. *Innov. Food Sci. Emerg. Technol.* 2018, 45, 361–381.

Kim, J.H.; Min, S.C. Moisture vaporization-combined helium dielectric barrier discharge-cold plasma treatment for microbial decontamination of onion flakes. *Food Control* 2018, 84, 321–329.

Knorr, D.; Froehling, A.; Jaeger, H.; Reineke, K.; Schlueter, O.; Schoessler, K. Emerging technologies for targeted food processing. In: *Advances in food process engineering research and applications*; Yanniotis, S., Taoukis, P., Stoforos, N.G., Karathanos, V.T., Eds.; Springer: Boston, MA, 2013; pp. 341–374.

La Peña, M.M.-D.; Welti-Chanes, J.; Martín-Belloso, O. Novel technologies to improve food safety and quality. *Curr. Opin. Food Sci.* 2019, 30, 1–7.

Liepa, M.; Zagorska, J.; Galoburda, R. High-pressure processing as novel technology in dairy industry: A review. *Res. Rural Dev.* 2016, 1, 76–83. Available online: https://llufb.llu.lv/conference/Research-for-Rural-Development/2016/LatviaResRuralDev_22 nd_vol1-76-83.pdf (accessed on 15 March 2021).

Ma, R.; Wang, G.; Tian, Y.; Wang, K.; Zhang, J.; Fang, J. Non-thermal plasma-activated water inactivation of food-borne pathogen on fresh produce. *J. Hazard. Mater.* 2015, 300, 643–651.

Mishra, R.; Bhatia, S.; Pal, R.; Visen, A.; Trivedi, H. Cold plasma: Emerging as the new standard in food safety. *Res. Inv. Int. J. Eng. Sci.* 2016, 6, 15–20.

Mikš-Krajnik, M.; Feng, L.X.J.; Bang, W.S.; Yuk, H.-G. Inactivation of Listeria monocytogenes and natural microbiota on raw salmon fillets using acidic electrolyzed water, ultraviolet light or/and ultrasounds. *Food Control* 2017, 74, 54–60.

Oms-Oliu, G.; Martín-Belloso, O.; Soliva-Fortuny, R. Pulsed light treatments for food preservation. A review. *Food Bioprocess Technol.* 2010, 3(1), 13–23.

Ozer, N.P.; Demirci, A. Inactivation of Escherichia coli O157:H7 and Listeria monocytogenes inoculated on raw salmon fillets by pulsed UV-light treatment. *Int. J. Food Sci. Technol.* 2006, 41(4), 354–360.

Pan, Y.; Sun, D.-W.; Han, Z. Applications of electromagnetic fields for nonthermal inactivation of microorganisms in foods: An overview. *Trends Food Sci. Technol.* 2017, 64, 13–22.

Pingret, D.; Fabiano-Tixier, A.-S.; Chemat, F. Degradation during application of ultrasound in food processing: A review. *Food Control* 2013, 31(2), 593–606.

Pinto, L.; Baruzzi, F.; Ippolito, A. Recent advances to control spoilage microorganisms in washing water of fruits and vegetables: The use of electrolyzed water. In: *III international symposium on postharvest pathology: Using science to increase food availability, Bari, Italy*; Ippolito, A., Sanzani, S.M., Wisniewski, M., Droby, S., Eds.; International Society for Horticultural Science (ISHS): Leuven, Belgium, 2016, p. 1144.

Priyadarshini, A.; Rajauria, G.; O'Donnell, C.P.; Tiwari, B.K. Emerging food processing technologies and factors impacting their industrial adoption. *Crit. Rev. Food Sci. Nutr.* 2019, 59(19), 3082–3101.

Rathod, N.B.; Kahar, S.P.; Ranveer, R.C.; Annapure, U.S. Cold plasma an emerging non thermal technology for milk and milk products: A review. *Int. J. Dairy Technol.* 2021, 74. 10.1111/1471-0307.12771

Rawson, A.; Patras, A.; Tiwari, B.; Noci, F.; Koutchma, T.; Brunton, N. Effect of thermal and non thermal processing technologies on the bioactive content of exotic fruits and their products: Review of recent advances. *Food Res. Int.* 2011, 44(7), 1875–1887.

Rodriguez-Gonzalez, O.; Buckow, R.; Koutchma, T.; Balasubramaniam, V. Energy requirements for alternative food processing technologies-principles, assumptions, and evaluation of efficiency. *Compr. Rev. Food Sci. Food Saf.* 2015, 14(5), 536–554.

Roobab, U.; Aadil, R.M.; Madni, G.M.; Bekhit, A.E.-D. The impact of nonthermal technologies on the microbiological quality of juices: A review. *Compr. Rev. Food Sci. Food Saf.* 2018, 17(2), 437–457.

Sonawane, S.K.; Patil, S. Non-thermal plasma: An advanced technology for food industry. *Food Sci. Technol. Int.* 2020, 26(8), 727–740.

Song, H.P.; Kim, B.; Choe, J.H.; Jung, S.; Moon, S.Y.; Choe, W.; Jo, C. Evaluation of atmospheric pressure plasma to improve the safety of sliced cheese and ham inoculated by 3-strain cocktail Listeria monocytogenes. *Food Microbiol.* 2009, 26(4), 432–436.

Xuan, X.-T.; Ding, T.; Li, J.; Ahn, J.-H.; Zhao, Y.; Chen, S.-G.; Ye, X.-Q.; Liu, D.-H. Estimation of growth parameters of Listeria monocytogenes after sublethal heat and slightly acidic electrolyzed water (SAEW) treatment. *Food Control* 2017, 71, 17–25.

Yong, H.I.; Kim, H.-J.; Park, S.; Kim, K.; Choe, W.; Yoo, S.J.; Jo, C. Pathogen inactivation and quality changes in sliced cheddar cheese treated using flexible thin-layer dielectric barrier discharge plasma. *Food Res. Int.* 2015, 69, 57–63.

Ziuzina, D.; Han, L.; Cullen, P.; Bourke, P. Cold plasma inactivation of internalised bacteria and biofilms for Salmonella enterica serovar typhimurium, Listeria monocytogenes and Escherichia coli. *Int. J. Food Microbiol.* 2015, 210, 53–61.

Chapter 10

Packaging Criteria for Non-Thermally Processed Fruit and Vegetable Products

Gurbuz Gunes, Ayse Merve Cellat, and Feride Sonverdi

CONTENTS

10.1 INTRODUCTION

Consumers have become more conscious and educated on the health aspect of their diet. Demand for food products that are minimally processed with fresh-like quality, and that are safe and convenient has been increasing over a long time. Preservation of fresh-like quality and safety of foods with minimal processing and additives has been a major challenge for scientists and industry. This has stimulated research on novel non-thermal food processing and preservation methods. Conventional thermal food processing technologies allow the production of shelf-stable and safe food products but have a significant negative impact on the sensory and nutritional quality of foods. Fruits and vegetables constitute an important part of the human diet with high nutritional value. They contain high levels of bioactive compounds associated with protection against various diseases including cancer and cardiovascular diseases. It is well established that thermal processing methods cause significant loss of these bioactive compounds in fruit and

DOI: 10.1201/9781003222170-10

vegetable products as well as losses in organoleptic quality. The loss in organoleptic quality of the products would also adversely affect their consumption by consumers. This would indirectly result in the under-utilization of the health-protective effects of fruit and vegetable products on the consumer diet. As for all processed products, the packaging is an inevitable part of non-thermally processed fruit and vegetable products. Packaging would strongly affect the degree of success obtained in the application of such goods.

Food packaging has four basic functions: containment, protection, communication, and convenience (Robertson, 2013). Among these, the protection function is important in keeping the food products' quality and safety during their shelf life. Food packaging presents a physical barrier between its content and the environment. It prevents physical contamination or loss of its content to the environment in the first place. Small molecules including O_2, CO_2, water vapor, and aroma components can pass through many of the packaging materials except glass and metal packages. Transport of the gases and the small molecules through the package can take place from the foods to the environment or in the opposite direction. Thus, the gas and aroma barrier properties or migration of small molecules of packaging materials to foods must also be considered for the selection of proper packaging materials. Food products prone to oxidation require packaging under a vacuum or a reduced O_2 environment. Some products are packaged under modified atmospheres with elevated CO_2 to control microbial growth. The moisture content of a food product must be preserved during its shelf life. All these types of requirements necessitate the use of packaging material with good/appropriate gas and water vapor barrier properties. The thermal and mechanical properties of the packaging materials must also be suitable for specific food products. The selected packaging materials must be resistant to processing, packaging, logistic, and storage environment so that the basic function of the package can be fulfilled properly. The thermal properties of a package are important because they affect its performance during processes such as hot-filling, heat-sealing, in-package thermal processing, freezing, and frozen storage. The mechanical properties of a package must be sufficient to resist mechanical stresses imposed during high-speed filling and sealing operations, product weight, handling, and storage operations. Overall, the selection of proper packaging materials requires careful evaluations of food-package interactions and package-food processing interactions on top of other environmental conditions to which the packaged products are exposed until consumption.

10.2 PACKAGING REQUIREMENTS FOR NON-THERMALLY PROCESSED FOODS

The basic principles of food packaging are also applicable to non-thermally processed foods. Non-thermal processes are applied to foods at either the pre-packaging or post-packaging stage depending on the nature of the product and processes. For those processes that are applied before packaging the general principles and requirements of packaging are applicable since the packaging materials are not exposed to non-thermal processes. However, for the non-thermal processes that are applied after the packaging of foods, the interaction of packaging materials and the non-thermal processes must be considered because the processes may affect the properties and functions of the packaging materials. It is important to maintain the critical functional properties of a packaging material after a non-thermal treatment is applied so that the package would function properly for the food product. The packaging requirements for each non-thermal process are discussed in the following sections.

10.3 HIGH-PRESSURE PROCESSING

High-pressure processing (HPP) is one of the most commonly employed non-thermal technology in the industry. HPP is a non-thermal process that is applied between 100 and 800 MPa from a few seconds to minutes, and effectively inactivates microorganisms (Farkas and Hoover, 2000; Knorr, 1995). The mechanisms of microbial inactivation include pressure damage to the primary structure of the cell membrane, changes in membrane permeability, separation of the cell membrane from the cytoplasm, leakage of cell components, denaturation of key enzymes, and disruption of DNA replication and transcription (Yamamoto, 2017). HPP can activate or inactivate enzymes that affect the quality and nutritional values of the food (Chakraborty et al., 2014). The efficiency of the process relies on different factors such as pressure, temperature, process time, type of microorganism, and food composition (Sehrawat et al., 2021). HPP has been studied extensively and applied industrially to fruit and vegetable juice, puree, jams, and marmalades, but research on the application of fresh produce is rather limited. But the existing literature shows that it has also some potential in preserving fresh fruits and vegetables. Application of HPP (300–350 MPa) to fresh produce such as lettuce and tomatoes resulted in up to 6-log inactivation in inoculated *E. coli* without adversely affecting organoleptic quality (Arroyo et al., 1997). A 5-log inactivation of natural bacterial count in fresh green beans was obtained upon HPP at 500 MPa and 20°C with no loss of quality (Krebbers et al., 2002). Retention of texture and quality of fresh blueberries treated by HPP at 200–600 MPa was noted during 28-day storage (Scheidt & Silva, 2018).

High-pressure processing can be applied to foods either before or after packaging. For liquid foods such as fruit juices, HPP can be applied to the juice in a series of treatment chambers called isolators in a continuous manner, and then the treated juice can be aseptically packaged as typically done in conventional aseptic packaging systems (Farkas and Hoover, 2000). This system offers advantages in operation at high capacity, but it applies to fluid foods and probably requires higher investment costs. Alternatively, fruit and vegetable products including fluid and solid forms can also be treated by HPP after packaging. In this case, the products are first packaged in their final packages and then subjected to HPP. This system offers flexibility in treating various types of food products in various package formats and is dominated in industrial use. Types and characteristics of packaging material are substantially important in this application. The packaging material must be flexible so that the high-pressure applied can be transmitted to the product of pressure (Marangoni Júnior et al., 2019).

In the meantime, the package and its seal must be strong enough to resist the applied pressure and volume fluctuation during the process. Considering these basic requirements, rigid packages such as metal cans, glass bottles, and paper-based packages are not suitable for foods to be treated by high-pressure post-packaging (Caner et al., 2004a). Flexible packages made of single or multilayer polymeric materials are suitable if they resist the HPP without losing their integrity and critical characteristics. Thus, special emphasis must be given to the critical properties such as gas barriers, migration, sealing, and integrity of the structure after HPP treatments. Headspace volume is another factor to be considered. Gases like air in headspace can be compressed during HPP and this may adversely affect the applied pressure, package integrity, and energy cost during treatment. The presence of CO_2 in the headspace may increase microbial inactivation during treatment (Zhou et al., 2019). So the amount and composition of headspace gases in the packages need to be evaluated for a given food product and package format in HPP. The interaction of HPP with flexible food packaging characteristics was critically reviewed by Caner et al. (2004a, 2004b). Many of the studies reported in the literature involved HPP at moderate temperatures. However, a combination of high temperature and high pressure has special potential in the inactivation of bacterial spores and enzymes. Thus, more investigation of the effects of HPP at elevated temperatures on the properties of packaging materials is needed.

One of the important characteristics of a food package is the gas (O_2, CO_2, water vapor) barrier property because it adds to the protection of the initial quality of the foods from oxidation, moisture loss or gain, and microbial growth. The published research showed that HPP can increase, decrease, or have no effect on the gas barrier properties of the packaging films. Caner et al. (2000) found that the barrier properties of the metallized-PET/EVA/LLDPE film increased substantially by the HPP while less than 11% increase was noted in the barrier properties of the other materials tested. Lambert et al. (2000) found that the oxygen transmission rate (OTR) of a PA/medium density PE film increased by 25 while OTR of PA/PE surlyn:ionomer Zn film decreased by 16% out of six packaging films they tested. The authors also noted different trends in the water vapor transmission rate (WVTR) of the films in response to HPP. An increase of WVTR of BB4L (Cryovac-Grace packaging) film out of several other films was observed after HPP treatments in another study (Le-Bail et al., 2006). A slight reduction of WVTR of PET/PE and PA/PE films by HPP at 300–500 MPa was reported by others (Zhao & Tang, 2017). HPP caused no significant modification of gas barrier properties of various films in other studies (Masuda et al., 1992; Pastorelli, 1997). Al-Ghamdi et al. (2020) treated various vegetable purees in two different EVOH containing high barrier packages (PE/PA6//EVOH//PA6/PE and PP/PA//EVOH//PA/PP) with HPP combined with high temperatures (pressure-assisted thermal sterilization) and found no visual defects on the packages but OTR and WVTR of the PE/PA6//EVOH//PA6/PE film increased substantially by the treatment. The mechanisms of the changes in OTR or WVTR upon HPP are unclear but could be associated with structural modifications and delamination. So, the gas barrier properties of the packaging films used for HPP foods should be studied to make sure the potential changes induced by the HPP are in the acceptable range for the product. A few studies investigated the effect of HPP on aroma barrier or packaging films and reported that HPP does not have an adverse effect on aroma barrier properties unless some structural deformation on delamination takes place (Caner et al., 2003; Kübel et al., 1996). Absorption of food components such as p-cymene and acetophenone by packaging films was reported to be lower during HPP treatment, but no difference in loss of aroma from foods during storage was detected between the HPP-treated and untreated samples (Kübel et al., 1996). Less absorption of compounds by the packaging materials during HPP may be associated with a reversible increase of the density of the packaging film under pressure. The chemical characteristics of the aroma compounds and the packaging materials would also affect the degree of the mass transport through the packaging film as affected by HPP (Sheung et al., 2004). This was shown in minced beef and chicken meat vacuum packaged with or without Al-foil wrapping before the HPP treatment, where the Al-foil wrapping resulted in higher retention of flavor profile after the treatment (Rivas-Canedo et al., 2009).

Mechanical and structural properties and heat sealing strength are also important and they need to be investigated as affected by HPP treatments. Structural deformation such as cracks, wrinkles, pores, delamination, and loss of integrity of the seal could be potential problems induced by HPP and these defects would fail the containment and protection functions of the food package. Lambert et al. (2000) studied several coextruded packaging films with PE, PA, PVDC, PP, PET, and adhesives as affected by HPP treatment at pressures up to 500 MPa in different food simulants. The HPP treatment caused some increase in tensile strength of the packages resulting in a reduction in their flexibility, but the changes were not substantial enough (below 25% as the industrial norm) to adversely affect the package functions. Increased tensile strength of PET/PE and PA/PE films by HPP was also reported in another work (Zhao & Tang, 2017). No significant effects of

HPP on mechanical properties and heat-sealing strength were reported in other studies (Caner et al., 2000; Masuda et al., 1992; Pastorelli, 1997; Mertens, 1993; Ochiai & Nakagawa, 1992). However, HPP caused delamination in one of the packages with PA/linear PE free radical. Delamination and some structural deformation in packaging films upon HPP treatment were also reported by others (Caner et al., 2000; Goetz & Weisser, 2002; Schauwecker et al., 2002). Delamination induced by HPP may be associated with the adhesive used in lamination and headspace volume (Lambert et al., 2000). Dobiáš et al. (2004) investigated the effect of HPP on different packaging materials. PP cast 40 showed loss of sealability whereas sealability increased in Surlyn 1605 film. Changes in structure and properties of the packaging film by HPP may also be caused by a rapid expansion of the gases within the film during decompression (Sterr et al., 2018).

The consequences of HPP treatment on migration must also be considered in the HPP of packaged products. HPP treatment did not affect migration from various packaging films (Pastorelli, 1997; Ting et al., 2002; Caner, 2004b; Schauwecker et al., 2002). On the contrary, Galotto et al. (2010) found that HPP at 400 MPa resulted in a higher overall migration in olive oil but a lower migration in water in various packaging films such as laminates of PE/EVOH/PE, metalized polyester/PE, PET/PE, and PPSiOx. The total migration in fatty food simulants exceeded the 10 mg/dm^2 limit value in some of the packages. HPP treatment with headspace gases like CO_2 may result in the extraction of polymer additives that higher migration to foods during treatment can take place, thus it must be investigated (Fleckenstein et al., 2014).

Biodegradable packaging materials have also been considered in the HPP of foods. Cubeddu et al. (2021) showed that PLA bottles for apple juice performed equally well under HPP as compared to conventional PET bottles. The suitability of PLA for HPP was also studied and shown in other works (Mauricio-Iglesias et al., 2010; Sansone et al., 2012). Cellulose acetate films were evaluated for HPP at 200–400 MPa and it was found that HPP had minimal effects on functional properties including reduction of tensile strength, increase in elongation at break and heat sealing strength, delamination, changes in color and clarity, and reduction in WVTR (Gonçalves et al., 2020). It was reported in another work that HPP induced some changes in mechanical properties of PLASiOx-PLA film, and that these changes were higher when the film was in the water as compared to being in contact with oil (Galotto et al., 2009). It was also reported that HPP caused cracks and pinholes in PETALOx film in the same study. The impacts of all these changes in the film properties on their performance in specific food packaging need to be investigated for evaluation of their use in the HPP treatment of foods.

When the samples were treated with HPP before packaging, the packaging materials have no interaction with the HPP and should be selected based on general requirements dictated by the product, storage, logistics, and other conditions. Fernández-García et al. (2001) studied the effect of different packaging on the quality of HPP-treated orange juices and orange-carrot-lemon juice during storage. In this study glass, polypropylene, Barex® (modified acetonitrile-methyl acrylate copolymers), teflon flasks, PE foil were used as packaging material. No significant differences in antioxidative capacity, carotene content, vitamin C, and sensory quality were detected due to the different packaging treatments. In another study vacuum-skin and vacuum-nylon packaging were investigated in the sensory properties of jackfruit bulbs treated at 500 MPa for 5 min. The HPP treatment resulted in a shelf-life extension in both packagings with no difference in quality due to packaging type (Ng et al., 2020). Kaushik et al. (2018) studied conventional thermal treatment and thermal-assisted HPP of mango pulp in different packaged products. Sensory quality was not affected by the packaging and the process but the product in the aluminum-based retort pouch showed a lower loss of ascorbic acid than other PE-nylon and EVOH-based packaging during storage.

10.4 PULSED ELECTRIC FIELD

Pulsed electric field (PEF) processing involves exposing foods to an electrical field typically 2–4 kV/mm for an extremely short time (fractions of a millisecond) in the form of pulses by placing the foods between two electrodes (Barbosa-Cánovas et al., 1998). It is a non-thermal technology that inactivates microorganisms and enzymes in foods and facilitates mass transfer and extraction of bioactive components (Zhang et al., 2020). PEF is one of the most widely used commercial non-thermal processes in food preservation and processing worldwide (Kempkes, 2017).

The mechanism of the effects of PEF is based on the phenomena of electroporation and dielectric breakdown, which creates pores in the cell membrane and causes increased permeability and in some cases ruptures (Zimmermann et al., 1974; Weaver & Chizmadzhev, 1996; Wouters et al., 2001). The efficiency of the process is affected by product composition and process parameters, including pulse duration, pulse shape, pulse number, and initial temperature (Wouters et al., 2001).

The creation of an electrical field in foods during PEF treatment requires direct contact of the foods with the two electrodes. Thus, the food products are treated with PEF before packaging in current applications. Aseptic packaging is required for the product treated with PEF to prevent recontamination. The packaging materials required are dictated solely by the treated products' requirements for preservation since no interaction of the PEF and package takes place in this way of the treatment. For example, a high gas barrier material is needed to prevent oxidative degradations and extend the shelf life of the treated product. Ayhan et al. (2001) studied the effect of glass, polyethylene terephthalate (PET), high-density polyethylene (HDPE), and low-density polyethylene (LDPE) bottles on the quality of orange juice treated with PEF and found that glass and PET packages resulted in higher aroma compounds, vitamin C and color quality than the other packages. This was associated with higher O_2 barrier properties of the glass and PET as compared to the HDPE and LDPE. Similarly, the high O_2 permeability of polypropylene packages used for PEF-treated tomato juice resulted in a reduction in lycopene content at the end of storage (Min et al., 2003).

The application of the PEF treatment to pre-packaged products, if successful, would be favorable in exclusion of aseptic packaging system thereby reducing the overall cost of investment and operation. This requires integration of the electrodes with the packages and/or conductive packaging materials, which can be very challenging. Plastic packaging materials are not conductive but some can be made so by the incorporation of conductive fillers. Roodenburg et al. (2010) designed an in-package PEF treatment system ('PEF-in-pack') which utilized a plastic conductive film (composite Ethylene Vinyl Acetate (EVA) copolymer film) and tested its efficiency in microbial inactivation. They found that *L. plantarum* was inactivated by up to 2.1 log which was only slightly lower than the levels that would be obtained by metal electrodes. These researchers later showed that a higher level of microbial inactivation (5.9 log) by PEF-in-pack treatment at 2.3 kV/mm with treatment time up to 600 μs was achievable using a commercial conductive film (EVA incorporated with 30% carbon black particles) used for packaging of electronic components (Roodenburg et al., 2013). The commercial success of the PEF-in-pack treatment concept would depend on the suitability of the conductive plastic films used in the process of packaging the treated foods. Further research on the development of conductive packaging films with desirable barrier, mechanical, and optical properties for food packaging is needed.

10.5 IRRADIATION

Irradiation is a well-studied food preservation technology that has great potential in providing extended shelf life and safe food thereby reducing food loss (Prakash, 2020). Food products are exposed to ionizing radiations such as gamma rays, x-rays, or electron beams either before or after packaging at the required dose levels which are highly regulated. The ionizing radiation from these sources has high energies and captures electrons from atoms resulting in chemical modifications in the materials. The ionization process leads to both direct and indirect impacts on microorganisms and product quality. The direct effects are those that damage the cell components like DNA, lipids, and carbohydrates while indirect effects come from free radicals and reactive species which are formed in the radiolysis of water and other compounds. Considering those effects, gamma irradiation with high frequency and energy can be used in sterilization and microbial decontamination owing to its high penetration into the materials (Bisht et al., 2021). The typically approved doses of irradiation for various purposes include 0.2–0.5 kGy for insect disinfestation, 0.05-0.1 kGy for inhibition of sprouting, 1.5–7 kGy for fresh and frozen meat products, 10–30 kGy for microbial disinfection of dried spices, and sterilization in aseptic packaging (Min et al., 2014). New packaging material developed in the concept of active and intelligent packaging or improving properties using nano-compounds, novel layers, or coating must be evaluated for irradiation-induced changes and compliance with current regulations before being used in the irradiation process. In some cases, irradiation may control the release of an active compound (for example naphthalene) to foods in active packages and this could be advantageous, but irradiation may also disrupt the active compounds (for example trans-cinnamaldehyde) in such system disrupting the active packaging functions in a dose-dependent manner (Han et al., 2008; Han et al., 2007; Arvayitoyannis & Tsarouhas, 2010).

Foods are usually irradiated in their final packages and therefore the packaging materials are also exposed to irradiation during the treatment. Besides, food packages are sterilized by irradiation before being used in the packaging of foods, as in aseptic food packaging. Irradiation may cause radiolytic products in the packaging materials which may affect the quality and safety of foods and also functional properties of the packaging materials. The irradiation-induced free radicals and ions can cause cross-linking or chain scission and result in off-odors by radiolysis of volatile compounds (Welle et al., 2002). One of the common interactions of plastic films with irradiation is cross-linking which can affect mechanical properties like tensile strength and elongation, migration, and gas barrier properties of the packaging materials. Chain scission induced by irradiation results in a reduction of molecular weight of the polymer and release of volatile compounds such as hydrogen, hydrogen chloride, methane, etc. depending on the type of polymer in the main body of the plastic films like PE, PP, PET, PVC, and PA. Atmospheric gas composition during irradiation has effects on the type of reactions induced in packaging films: treatment delivered under high O_2 causes predominantly chain scission while the treatment applied under vacuum or low O_2 results in cross-link to a greater extent (Guillard et al., 2010). A higher degree of migration is expected when chain-scission type degradation that produces small molecular weight compounds takes place upon irradiation. Besides, the treatment may interact with the polymer additives causing migration of these compounds or their decomposition products to foods, which may affect the quality and safety of the packaged foods (Arvanitoyannis & Tsarouhas, 2010). Thus, the packaging materials intended to be used for irradiated foods are also regulated by FDA or similar authorities around the world. Seal integrity and strength of two multilayer films (EVA/EVOH/EVA and PE/EVOH/PE) were not affected by irradiation at doses up to 270 kGy, and the tensile properties of the films were unaffected by irradiation up to 45 kGy (Girard-Perier et al., 2020). Irradiation up to 45 kGy caused no significant increase in the overall migration of commercial packaging films in both polar and nonpolar food simulants (Welle et al., 2000). On the other hand, it was reported in other research that irradiation above 10 kGy caused an increase in overall migration from PVC in all food simulants (Lirio et al., 2017). Increased migration of acetyl tributyl citrate (ATBC) in PVDC/PVC film was observed upon irradiation up to 25 kGy (Zygoura et al., 2011). Specific migration of plastic additives was decreased by irradiation (Marque et al., 1998), but the migration of the decomposition products of the plastic additives must also be considered. It was shown that specific migration of degradation products of Irgafos 168, a plastic additive, was increased by irradiation, but remained below 0.1 mg/dm^2 (Welle et al., 2000). It was also found that irradiation can induce volatile formation in packaging films such as LDPE, PET, and this may adversely affect the sensory quality of foods being packaged with these materials (Welle et al., 2002). FDA regulations on plastic films that are allowed to be irradiated consider the migration of irradiation-induced volatiles, decomposition products in food simulants, and approvals are given based on how the results of migration tests comply with the regulation. Thus, it should be known that the packaging materials to be exposed to irradiation must be approved in the regulations and also the use of these materials in the packaging of a specific food product must be studied for the consequences of their impact on especially sensory quality of the foods.

More studies with emerging packaging materials such as bioplastics, active packaging films, etc. need to be conducted to determine their suitability for irradiation. Irradiation of LDPE/PA coated with antioxidant and antimicrobials resulted in enhancement of mechanical and barrier properties, possibly due to cross-linking, and maintenance of the antioxidant and antimicrobial activity (Han et al., 2007, 2006). Improvement of mechanical and barrier properties of protein and carbohydrate films by irradiation in a dose-dependent manner has been reported in many other pieces of research (Lacroix et al., 2002; Ressouay et al., 1998; Abd El-Mohdy, 2007; Zhai et al., 2003).

Irradiation not only inactivates microorganisms in foods but also induces biochemical changes in foods. Some of these are chemical changes such as inhibition of sprouting and ethylene synthesis, but some are undesirable such as oxidative reactions, increased respiration, and loss of firmness and color in fruits and vegetables. The effects of irradiation on these changes depend on irradiation dose, dose rate, temperature, and atmospheric gas composition around the product during irradiation. The presence of O_2 around the product causes a higher level of irradiation-induced chemical reactions which cause undesirable effects on quality. Modified atmosphere packaging under reduced O_2 can be applied before irradiation to reduce the unwanted chemical changes and quality loss due to irradiation.

10.6 ULTRAVIOLET LIGHT

Ultraviolet light (UV) treatment involves exposure of foods to electromagnetic radiation in the UV region. The types of UV radiations are categorized depending on their wavelength as UV-A (315–400 nm), UV-B (280–315 nm), UV-C (200–280 nm), and Vacuum-UV (100–200 nm). UV-C at 254 nm is known as germicide with great lethal effects on several microorganisms, and thus it is one of the most commonly used in food preservation. UV treatment leaves no residues of toxic compounds while preserving the quality parameters of foods such as flavor, taste, pH, and odor. However, the treatment is confined to surfaces only since it has practically no penetration in foods that contain a lot of organic molecules. The efficiency of the treatment depends on factors such as treatment time, wavelength, dose, and treatment chamber design (Bisht et al., 2021).

Besides the sterilization effect of UV-C, studies showed that it can also increase the content of bioactive compounds and may prevent chilling injuries (Choi, et al., 2015; Formica-Oliveira et al., 2017). UV-C treatment also controls postharvest ripening and physiology of fresh produce and thus positively contributing to their storage life (Ma et al., 2021). However, a limitation of its use is that surface properties of material should be smooth, free from pores and roughness as in tomatoes, peppers, pears, and apples to obtain a higher microbial inactivation (Abdussamad et al., 2016). Packaging of food material is generally done after the treatment to avoid shadowing and screening effects of packages. For fresh fruits and vegetable products, a combination of modified atmosphere packaging (MAP) and UV-C can be applied for better maintenance of product quality. Tomás-Callejas et al. (2012) studied the effects of pretreatment of UV-C at 4.54 kJ/m^{-2} combined with high O_2 MAP on the quality of fresh-cut tatsoi baby leaves and found positive effects of the combined treatment on microbial inactivation and quality maintenance. The positive effects of UV-C pretreatments at 15 J/cm^2 applied to fresh-cut paprika packaged in a modified atmosphere were reported in another work in which a lower microbial count, better antioxidant capacity, and extended shelf life were reported (Choi et al., 2015). Manzocco et al. (2016) investigated the UV-C treatment both on pre-packaged (by PA/PE pouches) and post-packaged (by PET/EVOH/PE trays) fresh-cut pineapples to observe their potential in preserving the quality and eliminating contamination risks after treatment. While the samples treated before packaging were exposed to a dose of 200 J/m^2, the samples treated after packaging received 160 J/m^2 with the same setup because of the screening effect of plastic material. It was found that UV-C treatment did not cause any reduction in the initial microbial load, yet together with the packaging, it slowed down the bacterial growth during storage. According to the results, the treatment applied to pre-packaged samples would be favorable in this case as it would avoid contamination after the treatment. However, the application of UV to pre-packaged foods would have limitations on the form and amount of samples in the packages and the type of packaging materials since they would have shadowing effects limiting the efficiency of UV-C exposure. Besides, the effect of UV on the properties of the packaging materials needs also be evaluated carefully if the food is to be treated after packaging. Cucumbers packaged in PE films (24.7 μm) were treated with UV-C light at doses up to 560 mJ/cm^2, and a 1.6 log reduction in *E. coli* count was obtained (Abdussamad et al., 2016). The treatment resulted in a significant delay in the quality loss while it caused no effects on the gas and moisture barrier properties of the package. The UV transmission rate of the PE film was measured as 82% in the study.

Determination of suitable packaging or the systems for UV-C treated fruits and vegetable products is done by considering the product requirements as done with untreated products. On the other hand, potential effects of the treatment on the product characteristics, for example, respiration and ethylene production of fresh produce, must also be considered because these changes may affect the packaging requirements. If the UV treatment is applied after packaging the product, the UV transmission rate of the film and the effect of UV on film properties must also be considered in addition to the effects on product characteristics.

UV-light treatment is also used in the sterilization of packaging materials before food packaging as in aseptic packaging. It has been shown that exposure of plastic films such as PE, PP, and PET to UV light caused oxidative surface modifications such as increased hydrophilicity, surface tension, and adhesion properties (Ozen & Floros, 2001; Luo & Wong, 2001). Formation of low molecular weight product upon UV treatment of plastic films was reported and potential migration of these compounds to food can be an issue that needs to be investigated.

10.7 PULSED LIGHT

Pulsed light (PL) processing is based on exposure of food products to intense broad-spectrum white light (200–1000 nm) that includes ultraviolet (UV), visible (VL) and infrared (IR) regions in the form of pulses. The application can be done by a single pulse, a burst of pulses, or a continuous pulse either in an array or random mode depending on the process requirements. The efficiency of PL treatment depends on many factors such as pulse fluence, input voltage of PL emission, amount of UV, number of pulses, the distance between product and lamp, type of microorganisms, and environmental factors like packaging, food material, and surface characteristics (John & Ramaswamy, 2018). Although pulsed light (PL) is applied directly to foods before packaging there have been studies investigating its applications on pre-packaged foods. PL treatment through PP packaging resulted in a 2.6–3.0 log inactivation of *L. inoccua* and *E. coli* in fresh-cut mushrooms and watermelon (Ramos-Villarroel et al., 2012a, 2012b). The microbial inactivation levels obtained in pre-packaged foods are comparable with the ones in unpackaged food according to a limited number of studies in the literature, although this may be dependent on the type of packaging materials used as discussed below.

PL is used mainly for surface sterilization, inactivation of spoilage and pathogenic microorganisms, and extending shelf life. The technology is mostly applicable to the surface of solid foods and also liquid foods. Many studies investigated the effects of PL on fresh produce such as berries, apple juice, minimally processed vegetables, alfalfa seeds, and salmon fillets. Although PL could achieve higher microbial inactivation than UV-light due to its broad spectrum and high peak power, the exposed surface is still a limiting factor that should be chemically stable against depolymerization and transparent so that light is allowed to pass into the food (Abida et al., 2014; Huang & Chen, 2014). PL has also potential use in the sterilization of packaging materials. It has been shown that PL treatment at 6–8 J/cm^2 caused 3–7 log inactivation of various microorganisms on different packaging films (Marangoni Júnior et al., 2020).

For in-package applications of PL, only glass or some polymer materials having high transmissivity to respective lights can be used. Paper, metal, multilayered opaque packaging materials, or packages with a high degree of printing would not be suitable for PL treatment of packaged foods. Among various types of polymers, selection can be done by considering the factors which affect light transmittances such as the incorporation of additives and crystallinity. For example, polystyrene (PS), polyvinyl chloride (PVC), polyethylene terephthalate (PET), and polypropylene (PP) are considered as non- or low- crystalline structured polymers. Also, the absorption of light is insignificant in amorphous and homogenous polymers which allows high transparency (Heinrich et al., 2015). The efficiency of PL treatment is highly correlated with the inclusion of UV-C wavelength, thus the degree of transparency of packaging materials to light, especially in the UV-C range, is important for the selection of packaging materials in in-package treatments. Among the commercial polymeric films PE, PP, PVS, PA, PS, PC has relatively good UV-C transparency while PET has poor transparency (Heinrich et al., 2015). On the other hand, specific packaging films must be evaluated for their light transmission properties since their formulation and processing technologies may affect them. Besides, the impact of PL treatment on other functional properties of the packaging materials such as mechanical, barrier, and migration should be considered (Andrady, 2007). Some alteration of mechanical properties of packaging films upon PL treatment has been reported although the applied intensities were much higher than the ones used for treating foods (Keklik et al., 2009, 2010). It has been shown that PL treatment caused degradation of some polymer additives some of which migrated over the threshold limits (Castillo et al., 2013). Depending on the spectrum of light used in PL, photothermal and photochemical effects can be seen on the treated foods or the packages. Resistance or response of the packaging materials to the photothermal effects of the PL treatment must also be evaluated if applicable in the applied process since overheating of packaging film can happen and affect its properties negatively.

10.8 COLD PLASMA

Among other non-thermal methods that increase the shelf life of fresh products but struggle with the structure and shape of food or process defects like shadowing, cold plasma (CP) stands out as an alternative method. The technology not only ensures the preservation of food quality and safety but also offers some advantages such as low processing temperature and convenience with in-package applications. Besides, cold plasma, obtained from different gases, can be used in the surface modification, deposition, and sterilization of packaging materials, decontamination of foods before packaging, and in-package decontamination of foods without causing a serious effect on food quality parameters like color, texture, and sensory properties (Pankaj & Thomas, 2016; Pan et al., 2019).

Cold plasma is the fourth state of matter, basically supplied from a neutral gas exposed to a strong electromagnetic field that results in ionization and the occurrence of the plasma phase. The electrical discharges can be applied in various forms such as corona discharge, glow discharge, dielectric barrier discharge (DBD), plasma jet, and plasma needle in CP treatment. Among these, DBD and plasma jet are mainly used for environmental and biological applications since they have a simple design and are suitable for different types of materials and treatment requirements (Bourke et al., 2017). However, the selection of the type of CP system depends on the energy source, type of electrode, treatment factors, the composition of the gas, and plasma chemistry.

From the packaging point of view, DBD and corona discharge plasma are commonly used in various modifications of polymer material surfaces. The packaging materials mostly used are polyethylene (PE), polypropylene (PP), and polyethylene terephthalate (PET) (Pankaj et al., 2014a). The surface modification by CP includes: pretreatments for cleaning and sterilization, increasing the functionality of the surface, deposition on the surface to supplement coating materials on the film surface, and etching to obtain a certain roughness (Pankaj & Thomas, 2016). The modification of polymer materials can be done either by chemical or physical methods that may be obtained from CP treatment. The process stimulates various effects between plasma and polymer interface which benefits surface functionalization such as increased adhesion, ease in printing and sealing, preventing fogging, and enhancing the resistance of polymer to physical damage. Another way of modification, called plasma deposition, benefits the polymer materials' barrier properties both for gases and chemical solvents. Lamination or coating of a layer can be enhanced by surface modification of the main film before the process to obtain a film with enhanced properties such as increased gas barrier, active packaging with antimicrobial or anti-oxidant activities, etc. Plasma species may react with packaging material components resulting in cross-linking, formation of functional groups, chain scission, and formation of low molecular weight compounds (Pankaj et al., 2014b). These reactions can affect crystallinity, contact angle, roughness, and surface energy resulting in modification of the functional properties of the packaging materials. The nature and intensity of these reactions depend on the polymer, plasma processing conditions, and especially the gases used for plasma formation. Cross-liking is expected to decrease migration from the treated films while chain-scission and degradation reactions would increase migration. The impact of CP on overall and specific migration should also be investigated because the reactions and their consequences can affect it. Such investigations are published in very limited numbers. Cold plasma did not increase migration to the excess of the permitted levels in PLA films (Pankaj et al., 2014a). CP treatment resulted in a reduction in migration of di-2-ethylhexyladipate (DEHA) and epoxidized soybean oil (ESO) from plasticized PVC, and this reduction was much higher when the plasma was generated from Ar as compared with other gases like N_2, CO_2, H_2, and He (Audic et al., 2001). Studies on the effects of CP on biopolymers are relatively few but similar types of reactions and effects on polymer characteristics would be expected. CP treatment caused increased roughness and hydrophilicity, improved mechanical properties, decreased water vapor permeability in zein films (Dong et al., 2018).

Although fruit and vegetable products can be treated with CP before packaging, in-package treatments are also possible with the advantage that there would be no re-contamination after the treatment. This process, called in-package CP treatment, gained interest in the food industry since it has the potential to be performed on a commercial scale. At the same time, the treatment can be compatible with currently used processing lines and is proposed to be applied to foods in both rigid and flexible packages when atmospheric dielectric barrier discharge cold plasma (ADCP) is used (Min et al., 2018).

The treatment parameters such as applied voltage, treatment time, the distance between electrodes, and air composition can be adjusted depending on the food material and its packaging. Besides, the efficiency is affected by the amount and composition of CP species and the packaging materials. The packaging parameters include the type of material, shape, gas composition of the package headspace, headspace ratio (HSR), and secondary packaging (Kim et al., 2019). For instance, in a study, the microbial and organoleptic and physiological quality and stability of fresh-cut pears packaged in polypropylene were evaluated as affected by CP treatments (Zhang et al., 2021). The results showed that among different ADCP treatment times and voltages, higher voltage (65kV) with a short time (1min) resulted in a better microbial and enzyme inactivation and better maintenance of the quality. Although all the treatment conditions showed considerable effects, the effectiveness of ADCP varied with the target microorganism, type of enzymes, and treatment conditions.

The CP technology is also applicable with MAP. After plasma formation is achieved inside the package by applying a strong electric field the unreacted plasma species transforms back to its original gas and the MAP condition is preserved (Misra et al., 2019). Although a high level of O_2 may be beneficial for microbial inactivation by cold plasma, the O_2 level used in MAP is generally low in fruit and vegetable packaging (Min et al., 2016). However, according to a study where atmospheric and modified atmosphere conditions were compared on mixed vegetable packaging, there was no significant difference in microbial inactivation by the change of headspace gas composition in the CP treatment (Kim et al., 2019). This may be associated with the limited volume of gases in the package headspace. Further studies that investigate the use of various gases including novel MAP gases like Ar, N_2O, He, etc., and the package headspace volume on the effects of CP treatments are needed. In-package plasma treatment of foods would have the advantage that the process also sterilizes packaging material.

10.9 ULTRASOUND TECHNOLOGY

Ultrasound (US) is a novel technology that utilizes sound waves at high frequencies (São José et al., 2014). Ultrasound equipment generates frequencies from 20 kHz to 10 MHz (Piyasena et al., 2003). Ultrasound treatment can be applied in washing and disinfection of fruits and vegetables, extraction of value-added compounds, filtration of juices, dehydration, and to a lesser extent in pasteurization of juices when combined with other treatments (Povey & Mason, 1998). It has also the potential for non-invasive quality monitoring of the products. Solid target products are exposed to ultrasound energy in a liquid medium where the cavitation is formed. The system typically consists of a power supply, a transducer, and a delivery system usually in the form of a probe. Ultrasonic bath systems are also available for extraction and cleaning operations as well. Liquid food products like juices are directly treated with ultrasound for microbial inactivation.

The inactivation mechanism by ultrasound technology is based on a phenomenon called 'cavitation' that forms the dynamics of gas bubbles in the medium (Piyasena et al., 2003). The target cells or enzymes can be destroyed physically with the action of shear forces generated or chemically with the actions of reactive species formed during sonication (Dolas et al., 2019). The efficiency of this technology depends on factors including food composition, temperature, processing time, ultrasound power, frequency, and amplitude (Rojas et al., 2017).

Although ultrasound technology has many advantages, the commercialization of ultrasound for microbial inactivation processes is limited. This is mainly due to the limited level of microbial inactivation (around 2-log) obtained by ultrasound alone at low temperatures. Besides, this technology requires high energy utilization (Bhargava et al., 2021).

Ultrasound treatment is usually applied to foods before packaging so that the target gets exposed to ultrasound energy more efficiently. There are very limited investigations on the in-package application of ultrasound to foods, possibly because it is thought that it may not be feasible as the effects of ultrasound may not pass through the packaging materials effectively without causing degradations in the material. However, some improvements in the in-package pasteurization of sausages using an ultrasound-aided system over a conventional system were reported in the literature (Cichoski et al., 2015). But no studies on in-package ultrasound treatment of fruits and vegetable products were found in the literature. As the microbial inactivation achieved by ultrasound is relatively low the treatment is applied in combination with others such as elevated temperature and pressure (manothermosonication). The treated products can also be packaged with special packaging techniques such as aseptic, modified atmosphere (MA), and active packaging.

Fresh-cut cucumber processed with ultrasound at 20 kHz for up to 15 min and packaged with MA with a gas composition of 5% O_2, 5% CO_2, and 90% N_2 (Fan et al., 2019). It was found that the US treatment significantly inhibited microbial growth and quality loss and extended the storage life of the product. In another study, Zhang et al. (2019) treated pakchoi with ultrasound at 30 kHz for up to 15 min combined with MAP containing 5% O_2 and 10% CO_2 and showed that the US treatment resulted in a better microbial and organoleptic quality than MAP alone and the control without ultrasound and MAP. Benefits of ultrasound in the improvement of mushrooms were found when combined with chemicals in washing treatments before MAP in another piece of work (Lagnika et al., 2014).

Ultrasound treatment has also been used in the synthesis of new packaging materials containing active ingredients (active packaging), improving the degradation of their waste, and improving their properties (Liu et al., 2018, 2019). Exposure of packaging films caused various modifications in their properties such as barrier properties and surface characteristics depending on ultrasound treatment parameters and types of film (Ščetar et al., 2017). As stated earlier, there is a lack of studies on ultrasound treatment applied to in-package fruits and vegetables, and the impact of ultrasound on packaging materials' properties must be evaluated very carefully for potential in-package applications.

10.10 DENSE PHASE CARBON DIOXIDE

Dense phase carbon dioxide (DPCD) is a novel technology in which the food is exposed to pressurized CO_2 (up to 30 MPa) for a sufficient time (Ferrentino & Spilimbergo, 2017). CO_2 offers some advantages including being inexpensive, non-toxic, non-flammable, and recyclable compared to other solvents commonly used in food processing (Clifford & Williams, 2000). CO_2 is soluble in both hydrophilic and hydrophobic media, and it has both antimicrobial activity and solvating power for various value-added components. The activities of CO_2 increase when it is pressurized. Dense phase carbon dioxide is a non-thermal technology that inactivates microorganisms and enzymes in different ways including pH lowering effect, the inhibitory effect of molecular CO_2 and bicarbonate ion, physical disruption of cells, modification of cell membrane and extraction of cellular components, conformational changes (Damar & Balaban, 2006). Many of the microorganisms causing spoilage and safety concerns in fruits and vegetable products can be effectively inactivated by $DPCO_2$. Polyphenol oxidase, pectinesterase, peroxidase, and lipoxygenase are considered important in the fruit and vegetable industry and can be inactivated by DPCD. (Balaban et al., 1994; Chen et al., 2010; Ferrentino & Spilimbergo, 2017; Tang et al., 2021). $DPCO_2$ is effective in microbial and enzyme inactivation, especially at the supercritical stage. Although DPCD technology has significant benefits and accordingly, various efforts have been made for its commercialization, it currently has no commercial pasteurization application (Yuan & Novak, 2012).

Technically, DPCD needs to come into direct contact with the product to achieve microbial and enzyme inactivation. In this respect, food products are treated with $DPCO_2$ before packaging. Continuous $DPCO_2$ treatment systems on pilot scales have been developed for the treatment of liquid foods (REF). Integration of $DPCO_2$ processing with aseptic packaging is considered appropriate for the prevention of recontamination and providing shelf-stability of the treated product. Several studies show effective inactivation of various microorganisms in different fruit juices with $DPCO_2$ treatment (Gunes et al., 2005, 2006; Damar & Balaban, 2006).

There is a lack of information about the use of the DPCD process alone applied to in-packaged foods. On the other hand, a combination of high-pressure processing (HPP) and $DPCO_2$ on some products has been tested. This combined treatment involves pre-carbonation of the food product before packaging and application of HPP to the packaged pre-carbonated product. During the HPP the dissolved CO_2 in the carbonated foods transforms to $DPCO_2$ and thus the result obtained is a combined effect of $DPCO_2$ and HPP. For example, the effect of a combination of HPP and DPCD on pectin methylesterase (PME), polyphenol oxidase (PPO), *L. plantarum*, and *E. coli* was investigated in orange juice (Corwin & Shellhammer, 2002). The orange juice was carbonated before packaging in PET or LDPE bottles and it was subjected to HPP at 500 and 800 MPa. The treatment resulted in improved inactivation of PME, PPO, and *L. plantarum* in the product. Similarly, feijoa puree samples with three different CO_2 levels (no CO_2, pre-carbonated, or pre-carbonated plus the addition of 8.5 mL CO_2/g into the package headspace) were treated with HPP (Ortuño et al., 2013). They found that more enzyme inactivation occurred in the samples with CO_2 added to the headspace than in the other samples.

10.11 MAGNETIC FIELD

Magnetic field processing is another non-thermal treatment that can be applied to foods for microbial inactivation (Minano et al., 2020). Either direct or alternating electric currents or permanent magnets form a magnetic field. Magnetic field (MF) can be applied as static magnetic field (SMF), oscillating magnetic field (OMF), or pulsed magnetic field (PMF) to foods. MF has effects on charged molecules in the target. The microbial inactivation mechanisms of MF are not clearly understood but are thought to be associated with the interaction of the MF with the ions in the cells and the further consequences of this interaction in the biological system (Barbosa-Cánovas et al., 2000; Minano et al., 2020). These ions may have specific cellular functions which may be disrupted by MF treatment. The efficiency of the inactivation depends on factors such as type and characteristics of the magnetic field, number of pulses, frequency, electrical resistivity, treatment time, microbial growth stage, and food composition (Barbosa-Cánovas et al., 2000). The treatment can be applied to foods in their final packages. However, metal packages cannot be used in magnetic field operation because the metal prevents the product to be affected by the magnetic field and high-intensity magnetic field damages metal containers. (Hofmann, 1985; Pothakamury, et al., 1993). There is a lack of published research on the effects of MF on the properties of food packaging materials. Due to the limited degree of microbial inactivation (around 2-log) by MF shown so far and the lack of understanding of the microbial inactivation mechanisms, interest in this novel food preservation process is relatively low among the researchers and industry.

10.12 MODIFIED ATMOSPHERE

Modified atmospheres can also be considered as a non-thermal food preservation method that involves the use of a gas mixture other than air to preserve the quality of foods. Typically, the food product is exposed to the special gas mixture during its shelf life in a package and this is the aforementioned MAP. The exposure of the food to the MA takes place in-package at ambient, refrigerated, or frozen temperatures depending on the specific storage temperature requirements. Thus, there is an interaction of MA with the food and the package that needs to be considered in its application. Different types of fruits and vegetables require different modified atmospheres (MA) depending on their characteristics. For example, fully processed fruits and vegetable products such as dehydrated, cooked, canned, and juices are not physiologically active and would require a headspace gas usually free from O_2 with some CO_2 to control oxidative and microbiological deteriorations. Fresh and fresh-cut fruits and vegetables require some O_2 and CO_2 in the headspace to allow their aerobic respiration to continue at reduced rates for a shelf-life extension because they are physiologically active. The MA and package interaction takes place through permeation of the gases through the packaging films, so correct gas barrier properties of the package are required to achieve the target MA exposure of the foods.

MA in the headspace can be established by directly flushing packages with the desired gas mixture (called active modification) before sealing for fully processed products. Flexible plastic packaging materials are the most common type of packaging used in the MAP of foods. The packaging materials used for the fully processed product must have a very high gas barrier property so that the desired gas in the headspace can be maintained during storage. Using a high gas barrier plastic film prevents diffusion of O_2 from air into the package headspace during storage. Various plastic packaging materials used in the MAP of foods include low-density polyethylene (LDPE), linear low-density polyethylene (LLDPE), high-density polyethylene (HDPE), polypropylene (PP), polyvinyl chloride (PVC), ethylene-vinyl alcohol (EVOH), polyamide, polychlorotrifluoroethylene (PCTFE), ethylene-vinyl acetate (EVA), ionomers, polycarbonate films, polystyrene and biodegradable polymers (Mangaraj et al., 2009). For non-respiring products, plastic films with high gas barrier properties are needed. They are usually multilayered films with a high O_2 barrier layer such as EVOH, PVDC, or AL layers.

The MAP of fresh and fresh-cut produce has a complicated dynamic system because of their active respiratory activities through which they consume O_2 and release CO_2 in the package headspace. MA in the headspace of a fresh produce package can be created through active modification as described above or passive modification (Gunes & Kirkin, 2012). In the case of passive modified packaging, the product is packaged and sealed without initial gas flushing to the headspace. In both approaches, an equilibrium MA will be created through the respiration and permeation process. Active modification is desirable for fresh products that are very perishable such as berries, and fresh-cut products (Charles et al., 2003; Day, 2003). For fresh produce having a relatively long storage life such as whole apples, pears, etc. passive modifications would be advantageous since they do not require the use of special packaging equipment and gases. The gas barrier properties of the packaging materials must be compatible with the product respiration rate so that the desired equilibrium MA forms in the headspace. Because each fresh or fresh-cut product has a different respiration rate there is a need for packaging materials with different gas permeabilities. The equilibrium MA in a fresh produce package depends mainly on respiration rate, gas permeability, amount of product, package surface area for permeation, and storage temperature. Thus, the design of MAP for fresh produce requires consideration of all these factors (Gunes & Kirkin, 2012). Among the available commercial plastic packaging films LDPE, OPP, and PVC are the ones that have suitable gas permeabilities for at least some fresh produce products. However, these and many other commercial plastic films are not suitable for all fresh produce products; some products require higher gas permeabilities while others require a lower permeability (Exama et al., 1993). Besides, the MAP of fresh produce products requires certain optimum O_2 and CO_2 levels in the headspace of a package, and this requires special CO_2 permeability as well as O_2 permeability. It is a big challenge to get both the desired O_2 and CO_2 permeabilities in a single packaging film. Because of this, the design of MAP for fresh produce is made in a way to get the most appropriate equilibrium MA that would have safe levels of O_2 and CO_2 for the specific product. Micro-perforation of existing commercial packaging film is considered as a tool to obtain the desired gas barrier properties, especially for fresh produce packaging. The perforation of an existing packaging film can be made with different sizes and amounts on a unit surface by laser, cold/hot needle, or tube perforations. Besides enhancing the gas transmission rate, this method facilitates moisture transfer, shortens cooling time, prevents the water vapor condensation inside the package, and maintains an equilibrium pressure inside the package (Hussein et al., 2015). Several studies with various fresh fruits and vegetables such as strawberries (Matar et al., 2021), cherry tomatoes and peaches (Mistriotis et al., 2016), jalapeño peppers (de Jesús Ornelas-Paz et al., 2015), pomegranate arils (Hussein et al., 2015), and broccoli (Serrano et al., 2006) have shown its potential in fresh produce packaging. However, increased moisture loss from the product and potential microbial contamination through the pores can be a serious concern that needs to be resolved through the development of technologies and hence make perforation at a smaller size and enable controlled geometries on various types of commercial packaging films.

Another challenge in the MAP of fresh produce is due to the different temperature sensitivity of respiration rates of fresh produce and gas permeability of plastic films. In case of temperature abuse during logistics and storage of the MA-packaged product, the equilibrium MA is disrupted causing the formation of unfavorable MA in the headspace which causes quality loss. It is desirable to have a packaging film with a similar temperature sensitivity of gas permeability to the temperature sensitivity of the respiration rate of fresh produce. Under these conditions, temperature abuse will not result in disruption of equilibrium MA in the package headspace. Recently, a polymeric material with a high temperature sensitivity has been commercialized and was claimed to solve this problem (Kim & Seo, 2018). Similarly, a polyurethane-based polymer film was developed in other research and this film was shown to have higher temperature sensitivity than commercial plastic films (Turan et al., 2016; Ilhan et al., 2021).

In addition to the conventional gases like O_2, CO_2, and N_2, various novel gases such as argon (Ar), nitrogen oxide (N_2O), helium (He), Xenon (Xe), and carbon monoxide (CO) have been studied in the MAP of fruits and vegetable products with certain benefits for quality maintenance and shelf life (Rocculi et al., 2004; Silveira et al., 2014). The potential use of these gases in MAP requires an extensive investigation of the barrier properties of packaging films to these novel gases in order to assess the suitability of the existing packaging materials.

10.13 CONCLUSIONS

Packaging of fruits and vegetable products treated by non-thermal preservation technologies must be evaluated and conducted carefully to have a successful result. If the non-thermal process is applied before packaging there is no interaction of the treatment with the package and thus, mainly the product characteristics determine the packaging requirements. On the other hand, if the treatment is applied to in-package foods then the interaction of the treatment with the packaging material must also be considered and evaluated for proper packaging materials and format for the product. The applied non-thermal treatments may affect the characteristics, integrity, and safety of the packaging materials used or may require special characteristics of the packaging materials to obtain successful results from the treatment. Further studies on the treatment-packaging material interactions, and the development of suitable packaging materials for each of these non-thermal technologies, are required.

REFERENCES

Abd El-Mohdy, H. L. (2007). Synthesis of starch-based plastic films by electron beam irradiation. *Journal of Applied Polymer Science, 104*(1), 504–513. https://doi.org/10.1002/app.25524.

Abdussamad, T., Rasco, B., & Sablani, S. (2016). Ultraviolet-C light sanitization of English cucumber (cucumis sativus) packaged in polyethylene film. *Journal of Food Science, 81*(6), 1419–1430. https://doi.org/10.1111/1750-3841.13314.

Abida, J., Rayees, B., & Masoodi, F. (2014). Pulsed light technology: A novel method for food preservation. *International Food Research Journal, 21*(3), 839–848. Retrieved from https://eds.p.ebscohost.com/eds/pdfviewer/pdfviewer?vid=0&sid=e00489d2-b41c-4874-af51-7461c180a510%40redis.

Al-Ghamdi, S., Sonar, C. R., Patel, J., Albahr, Z., & Sablani, S. S. (2020). High pressure-assisted thermal sterilization of low-acid fruit and vegetable purees: Microbial safety, nutrient, quality, and packaging evaluation. *Food Control, 114*, 107233. https://doi.org/10.1016/j.foodcont.2020.107233.

Andrady, A. (2007). Ultraviolet radiation and polymers. In J. E. Mark (Ed.), *Physical properties of polymers handbook* (pp. 857–866), Springer. https://doi.org/10.1007/978-0-387-69002-5_51.

Arroyo, G., Sanz, P., & Préstamo, G. (1997). Effect of high pressure on the reduction of microbial populations in vegetables. *Journal of Applied Microbiology, 82*(6), 735–742. https://doi.org/10.1046/j.1365-2672.1997.00149.x.

Arvanitoyannis, I., & Tsarouhas, P. (2010). Food packaging materials for irradiation. In I. Arvanitoyannis (Ed.), *Irradiation of food commodities*. London: Elsevier. https://doi.org/10.1016/B978-0-12-374718-1.10003-3.

Audic, J., Poncin-Epaillard, F., Reyx, D., & Brosse, J. (2001). Cold plasma surface modification of conventionally and nonconventionally plasticized poly(vinyl chloride)-based flexible films: Global and specific migration of additives into isooctane. *Journal of Applied Polymer Science, 79*(8), 1384–1393. https://doi.org/10.1002/1097-4628(20010222)79:8<1384::aid-app50>3.0.co;2-h.

Ayhan, Z., Yeom, H. W., Howard Zhang, Q., & Min, D. B. (2001). Flavor, color, and vitamin C retention of pulsed electric field processed orange juice in different packaging materials. *Journal of Agricultural and Food Chemistry, 49*(2), 669–674. https://doi.org/10.1021/jf000984b.

Balaban, M. O., Pekyardimci, S., Chen, C. S., Arreola, A., & Marshall, M. R. (1994). Enzyme inactivation by pressurized carbon dioxide. In T. Yano, R. Matsuno, & K. Nakamura (Eds.), *Developments in food engineering* (pp. 855–857), Springer. https://doi.org/10.1007/978-1-4615-2674-2_279.

Barbosa-Cánovas, G. V., Pothakamury, U. R., Palou, E., Enrique, E., & Swanson, B. G. (1998). *Non-thermal preservation of foods*. New York: Marcel Dekker Inc.

Barbosa-Cánovas, G. V., Schaffner, D. W., Pierson, M. D., & Zhang, Q. H. (2000). Oscillating magnetic fields. *Journal of Food Science, 65*(4), 86–89. https://doi.org/10.1111/j.1750-3841.2000.tb00622.x.

Bhargava, N., Mor, R. S., Kumar, K., & Sharanagat, V. S. (2021). Advances in application of ultrasound in food processing: A review. *Ultrasonics Sonochemistry, 70,* 105293. https://doi.org/10.1016/j.ultsonch.2020.105293.

Bisht, B., Bhatnagar, P., Gururani, P., Kumar, V., Tomar, M., Sinhmar, R., … Kumar, S. (2021). Food irradiation: Effect of ionizing and non-ionizing radiations on preservation of fruits and vegetables – A review. *Trends in Food Science and Technology, 114,* 372–385. https://doi.org/10.1016/j.tifs.2021.06.002.

Bourke, P., Ziuzina, D., Han, L., Cullen, P., & Gilmore, B. (2017). Microbiological interactions with cold plasma. *Journal of Applied Microbiology, 123*(2), 308–324. https://doi.org/10.1111/jam.13429.

Caner, C., Hernandez, R. J., & Harte, B. R. (2004b). High-pressure processing effects on the mechanical, barrier and mass transfer properties of food packaging flexible structures: A critical review. *Packaging Technology and Science, 17*(1), 23–29. https://doi.org/10.1002/pts.635.

Caner, C., Hernandez, R. J., & Pascal, M. A. (2000). Effect of high-pressure processing on the permeance of selected high-barrier laminated films. *Packaging Technology and Science, 13*(5), 183–195. https://doi.org/10.1002/1099-1522(200009)13:5<183::AID-PTS514>3.0.CO;2-U.

Caner, C., Hernandez, R. J., Pascall, M. A., & Riemer, J. (2003). The use of mechanical analyses, scanning electron microscopy and ultrasonic imaging to study the effects of high pressure processing on multilayer films. *Journal of the Science of Food and Agriculture, 83*(11), 1095–1103. https://doi.org/10.1002/jsfa.1437.

Caner, C., Hernandez, R. J., Pascall, M. A., Balasubramaniam, V. M., & Horte, B. R. (2004a). The effect of high-pressure food processing on the sorption behaviour of selected packaging materials. *Packaging Technology and Science, 17*(3), 139–153. https://doi.org/10.1002/pts.650.

Castillo, R., Biedermann, M., Riquet, A., & Grob, K. (2013). Comprehensive on-line HPLC-GC for screening potential migrants from polypropylene into food: The effect of pulsed light decontamination as an example. *Polymer Degradation and Stability, 98*(9), 1679–1687. https://doi.org/10.1016/j.polymdegradstab.2013.06.007.

Chakraborty, S., Kaushik, N., Rao, P. S., & Mishra, H. N. (2014). High-pressure inactivation of enzymes: A review on its recent applications on fruit purees and juices. *Comprehensive Reviews in Food Science and Food Safety, 13*(4), 578–596. https://doi.org/10.1111/1541-4337.12071.

Charles, F., Sanchez, J., & Gontard, N. (2003). Active modified atmosphere packaging of fresh fruits and vegetables: Modeling with tomatoes and oxygen absorber. *Journal of Food Science, 68*(5), 1736–1742. https://doi.org/10.1111/j.1365-2621.2003.tb12321.x.

Chen, J. L., Zhang, J., Song, L., Jiang, Y., Wu, J., & Hu, X. S. (2010). Changes in microorganism, enzyme, aroma of hami melon (cucumis melo L.) juice treated with dense phase carbon dioxide and stored at 4°C. *Innovative Food Science and Emerging Technologies, 11*(4), 623–629. https://doi.org/10.1016/j.ifset.2010.05.008.

Choi, I., Yoo, T., & Kang, H. (2015). UV-C treatments enhance antioxidant activity, retain quality and microbial safety of fresh-cut paprika in MA storage. *Horticulture, Environment and Biotechnology, 56*(3), 324–329. https://doi.org/10.1007/s13580-015-0141-y.

Cichoski, A. J., Rampelotto, C., Silva, M. S., De Moura, H. C., Terra, N. N., Wagner, R., … Barin, J. S. (2015). Ultrasound-assisted post-packaging pasteurization of sausages. *Innovative Food Science and Emerging Technologies, 30,* 132–137. https://doi.org/10.1016/j.ifset.2015.04.011.

Clifford, A. A., & Williams, J. R. (2000). Introduction to supercritical fluids and their applications. In J. R. Williams & A. A. Clifford (Eds.), *Supercritical fluid methods and protocols. Methods in biotechnology™* (Vol. 13, pp. 1–16). New York: Humana Press. https://doi.org/10.1385/1-59259-030-6:1.

Corwin, H., & Shellhammer, T. H. (2002). Combined carbon dioxide and high pressure inactivation of pectin methylesterase, polyphenol oxidase, lactobacillus plantarum and Escherichia coli. *Journal of Food Science, 67*(2), 697–701. https://doi.org/10.1111/j.1365-2621.2002.tb10661.x.

Cubeddu, A., Fava, P., Pulvirenti, A., Haghighi, H., & Licciardello, F. (2021). Suitability assessment of PLA bottles for high-pressure processing of apple juice. *Foods, 10*(2), 1–11. https://doi.org/10.3390/foods10020295.

Damar, S., & Balaban, M. O. (2006). Review of dense phase CO_2 technology: Microbial and enzyme inactivation, and effects on food quality. *Journal of Food Science, 71*(1), 1–11. https://doi.org/10.1111/j.1365-2621.2006.tb12397.x.

Day, B. (2003). Novel MAP applications for fresh prepared produce. In R. Ahvenainen (Ed.), *Novel food packaging techniques* (pp. 189–207). Woodhead Publishing Limited. https://doi.org/10.1533/9781855737020.2.189.

de Jesús Ornelas-Paz, J., Castañeda-Jiménez, A., Estrada-Alvarado, M., Ramos-Aguilar, O., Ibarra-Junquera, V., Pérez-Martínez, J., … Ruiz-Cruz, S. (2015). Effect of the perforation level of recycled-LDPE bags on the modification of the atmosphere development, bioactive compounds content, and other qualities of Jalapeño peppers during storage. *Journal of Food Science and Technology, 52*(10), 6415–6424. https://doi.org/10.1007/s13197-015-1749-8.

de São José, J. F. B., de Andrade, N. J., Ramos, A. M., Vanetti, M. C. D., Stringheta, P. C., & Chaves, J. B. P. (2014). Decontamination by ultrasound application in fresh fruits and vegetables. *Food Control, 45*, 36–50. https://doi.org/10.1016/j.foodcont.2014.04.015.

Dobiáš, J., Voldřich, M., Marek, M., & Chudáčková, K. (2004). Changes of properties of polymer packaging films during high pressure treatment. *Journal of Food Engineering, 61*(4), 545–549. https://doi.org/10.1016/S0260-8774(03)00214-0.

Dolas, R., Saravanan, C., & Kaur, B. P. (2019). Emergence and era of ultrasonic's in fruit juice preservation: A review. *Ultrasonics Sonochemistry, 58*, 104609. https://doi.org/10.1016/j.ultsonch.2019.05.026.

Dong, S., Guo, P., Chen, Y., Chen, G., Ji, H., Ran, Y., … Chen, Y. (2018). Surface modification via atmospheric cold plasma (ACP): Improved functional properties and characterization of zein film. *Industrial Crops and Products, 115*, 124–133. https://doi.org/10.1016/j.indcrop.2018.01.080.

Exama, A., Arul, J., Lencki, R. W., Lee, L. Z., & Toupin, C. (1993). Suitability of plastic films for modified atmosphere packaging of fruits and vegetables. *Journal of Food Science, 58*(6), 1365–1370. https://doi.org/10.1111/j.1365-2621.1993.tb06184.x.

Fan, K., Zhang, M., & Jiang, F. (2019). Ultrasound treatment to modified atmospheric packaged fresh-cut cucumber: Influence on microbial inhibition and storage quality. *Ultrasonics Sonochemistry, 54*, 162–170. https://doi.org/10.1016/j.ultsonch.2019.02.003.

Farkas, D. F., & Hoover, D. G. (2000). High pressure processing. *Journal of Food Science, 65*, 47–64. https://doi.org/10.1111/j.1750-3841.2000.tb00618.x.

Fernández-García, A., Butz, P., Bognàr, A., & Tauscher, B. (2001). Antioxidative capacity, nutrient content and sensory quality of orange juice and an orange-lemon-carrot juice product after high pressure treatment and storage in different packaging. *European Food Research and Technology, 213*(4), 290–296. https://doi.org/10.1007/s002170100332.

Ferrentino, G., & Spilimbergo, S. (2017). Non-thermal pasteurization of apples in syrup with dense phase carbon dioxide. *Journal of Food Engineering, 207*, 18–23. https://doi.org/10.1016/j.jfoodeng.2017.03.014.

Fleckenstein, B. S., Sterr, J., & Langowski, H.-C. (2014). The effect of high pressure processing on the integrity of polymeric packaging – Analysis and categorization of occurring defects. *Packaging Technology and Science, 27*(2), 83–103. https://doi.org/10.1002/pts.2018.

Formica-Oliveira, A., Martínez-Hernández, G., Díaz-López, V., Artés, F., & Artés-Hernández, F. (2017). Effects of UV-B and UV-C combination on phenolic compounds biosynthesis in fresh-cut carrots. *Postharvest Biology and Technology, 127*, 99–104. https://doi.org/10.1016/j.postharvbio.2016.12.010.

Galotto, M. J., Ulloa, P., Escobar, R., Guarda, A., Gavara, R., & Miltz, J. (2010). Effect of high-pressure food processing on the mass transfer properties of selected packaging materials. *Packaging Technology and Science, 23*(5), 253–266. https://doi.org/10.1002/pts.893.

Galotto, M. J., Ulloa, P. A., Guarda, A., Gavara, R., & Miltz, J. (2009). Effect of high-pressure food processing on the physical properties of synthetic and biopolymer films. *Journal of Food Science, 74*(6), E304–E311. https://doi.org/10.1111/j.1750-3841.2009.01212.x.

Girard-Perier, N., Gaston, F., Dupuy, N., Marque, S., Delaunay, L., & Dorey, S. (2020). Study of the mechanical behavior of gamma-irradiated single-use bag seals. *Food Packaging and Shelf Life, 26*, 100582. https://doi.org/10.1016/j.fpsl.2020.100582.

Goetz, J., & Weisser, H. (2002). Permeation of aroma compounds through plastic films under high pressure: In-situ measuring method. *Innovative Food Science and Emerging Technologies, 3*(1), 25–31. https://doi.org/10.1016/S1466-8564(01)00061-3.

Gonçalves, S. M., Chávez, D. W. H., de Oliveira, L. M., Sarantópoulos, C. I. G. de L., de Carvalho, C. W. P., de Melo, N. R., & Rosenthal, A. (2020). Effects of high hydrostatic pressure processing on structure and functional properties of biodegradable film. *Heliyon, 6*(10), e05213. https://doi.org/10.1016/j.heliyon.2020.e05213.

Guillard, V., Mauricio-Iglesias, M., & Gontard, N. (2010). Effect of novel food processing methods on packaging: Structure, composition, and migration properties. *Critical Reviews in Food Science and Nutrition, 50*(10), 969–988. https://doi.org/10.1080/10408390903001768.

Gunes, G., Blum, L. K., & Hotchkiss, J. H. (2005). Inactivation of yeasts in grape juice using a continuous dense phase carbon dioxide processing system. *Journal of the Science of Food and Agriculture, 85*(14), 2362–2368. https://doi.org/10.1002/jsfa.2260.

Gunes, G., Blum, L. K., & Hotchkiss, J. H. (2006). Inactivation of Escherichia coli (ATCC 4157) in diluted apple cider by dense-phase carbon dioxide. *Journal of Food Protection, 69*(1), 12–16. https://doi.org/10.4315/0362-028X-69.1.12.

Gunes, G., & Kirkin, C. (2012). Design of modified and controlled atmospheres. In J. Ahmed & M. S.Rahman (Eds.), *Handbook of food process design* (pp. 1340–1368). Wiley-Blackwell. https://doi.org/10.1002/9781444398274.ch43.

Han, J., Castell-Perez, M., & Moreira, R. (2006). The influence of electron beam irradiation on the effectiveness of trans-cinnamaldehyde-coated LDPE/polyamide films. *Journal of Food Science, 71*(5), E245–E251. https://doi.org/10.1111/j.1750-3841.2006.00049.x.

Han, J., Castell-Perez, M., & Moreira, R. (2007). The influence of electron beam irradiation of antimicrobial-coated LDPE/polyamide films on antimicrobial activity and film properties. *LWT - Food Science and Technology, 40*(9), 1545–1554. https://doi.org/10.1016/j.lwt.2006.11.012.

Han, J., Castell-Perez, M., & Moreira, R. (2008). Effect of food characteristics, storage conditions, and electron beam irradiation on active agent release from polyamide-coated LDPE films. *Journal of Food Science, 73*(2), E37–E43. https://doi.org/10.1111/j.1750-3841.2007.00616.x.

Heinrich, V., Zunabovic, M., Bergmair, J., Kneifel, W., & Jäger, H. (2015). Post-packaging application of pulsed light for microbial decontamination of solid foods: A review. *Innovative Food Science and Emerging Technologies, 30*, 145–156. https://doi.org/10.1016/j.ifset.2015.06.005.

Hofmann, G. A. (1985). *Deactivation of microorganisms by an oscillating magnetic field* (U.S. Patent No. 4,524,079). Washington, DC: U.S. Patent and Trademark Office. Retrieved from https://patents.google.com/patent/US4524079A/en.

Huang, Y., & Chen, H. (2014). A novel water-assisted pulsed light processing for decontamination of blueberries. *Food Microbiology, 40*, 1–8. https://doi.org/10.1016/j.fm.2013.11.017.

Hussein, Z., Caleb, O., & Opara, U. (2015). Perforation-mediated modified atmosphere packaging of fresh and minimally processed produce—A review. *Food Packaging and Shelf Life, 6*, 7–20. https://doi.org/10.1016/j.fpsl.2015.08.003.

Ilhan, I., Kaya, M., Turan, D., Gunes, G., Guner, F. S., & Kılıç, A. (2021). Thermoresponsive polyurethane films for packaging applications: Effects of film formulation on their properties. *Food Packaging and Shelf Life, 29*, 100695. https://doi.org/10.1016/j.fpsl.2021.100695.

John, D., & Ramaswamy, H. (2018). Pulsed light technology to enhance food safety and quality: A mini-review. *Current Opinion in Food Science, 23*, 70–79. https://doi.org/10.1016/j.cofs.2018.06.004.

Kaushik, N., Rao, P. S., & Mishra, H. N. (2018). Comparative analysis of thermal-assisted high pressure and thermally processed mango pulp: Influence of processing, packaging, and storage. *Food Science and Technology International, 24*(1), 15–34. https://doi.org/10.1177/1082013217724578.

Keklik, N., Demirci, A., & Puri, V. (2009). Inactivation of Listeria monocytogenes on unpackaged and vacuum-packaged chicken frankfurters using pulsed UV-light. *Journal of Food Science, 74*(8), M431–M439. https://doi.org/10.1111/j.1750-3841.2009.01319.x.

Keklik, N., Demirci, A., & Puri, V. (2010). Decontamination of unpackaged and vacuum-packaged boneless chicken breast with pulsed ultraviolet light. *Poultry Science, 89*(3), 570–581. https://doi.org/10.3382/ps.2008-00476.

Kempkes, M. A. (2017). Industrial pulsed electric field systems. In D. Miklavčič (Ed.), *Handbook of electroporation*. Cham: Springer. https://doi.org/10.1007/978-3-319-32886-7_211.

Kim, D., & Seo, J. (2018). A review: Breathable films for packaging applications. *Trends in Food Science and Technology, 76*, 15–27. https://doi.org/10.1016/j.tifs.2018.03.02.

Kim, S., Bang, I., & Min, S. (2019). Effects of packaging parameters on the inactivation of Salmonella contaminating mixed vegetables in plastic packages using atmospheric dielectric barrier discharge cold plasma treatment. *Journal of Food Engineering, 242*, 55–67. https://doi.org/10.1016/j.jfoodeng.2018.08.020.

Knorr, D. (1995). Hydrostatic pressure treatment of food: Microbiology. In G. W. Gould (Ed.), *New methods of food preservation* (pp. 159–175). Springer. https://doi.org/10.1007/978-1-4615-2105-1_8.

Krebbers, B., Matser, A. M., Koets, M., & Van Den Berg, R. W. (2002). Quality and storage-stability of high-pressure preserved green beans. *Journal of Food Engineering, 54*(1), 27–33. https://doi.org/10.1016/S0260-8774(01)00182-0.

Kübel, J., Ludwig, H., Marx, H., & Tauscher, B. (1996). Diffusion of aroma compounds into packaging films under high pressure. *Packaging Technology and Science, 9*(3), 143–152. https://doi.org/10.1002/pts.2770090304.

Lacroix, M., Le, T., Ouattara, B., Yu, H., Letendre, M., Sabato, S., … Patterson, G. (2002). Use of γ-irradiation to produce films from whey, casein and soy proteins: Structure and functional characteristics. *Radiation Physics and Chemistry, 63*(3–6), 827–832. https://doi.org/10.1016/s0969-806x(01)00574-6.

Lagnika, C., Zhang, M., Nsor-Atindana, J., & Bashari, M. (2014). Effects of ultrasound and chemical treatments on white mushroom (Agaricus bisporus) prior to modified atmosphere packaging in extending shelf-life. *Journal of Food Science and Technology, 51*(12), 3749–3757. https://doi.org/10.1007/s13197-012-0904-8.

Lambert, Y., Demazeau, G., Largeteau, A., Bouvier, J. M., Laborde-Croubit, S., & Cabannes, M. (2000). Packaging for high-pressure treatments in the food industry. *Packaging Technology and Science, 13*(2), 63–71.https://doi.org/10.1002/1099-1522(200003/04)13:2<63::AID-PTS495>3.0.CO;2-6.

Le-Bail, A., Hamadami, N., & Bahuaud, S. (2006). Effect of high-pressure processing on the mechanical and barrier properties of selected packagings. *Packaging Technology and Science, 19*(4), 237–243. https://doi.org/10.1002/pts.727.

Lirio, F., Souza, M., Soares, B., Borges, V., & Xavier, M. (2017). Quantification of overall migration levels from poly (vinyl chloride) films: Effects of gamma and electron beam radiation. *Journal of Bioenergy and Food Science, 4*(2), 90–98. https://doi.org/10.18067/jbfs.v4i2.135.

Liu, P., Wang, R., Kang, X., Cui, B., & Yu, B. (2018). Effects of ultrasonic treatment on amylose-lipid complex formation and properties of sweet potato starch-based films. *Ultrasonics Sonochemistry, 44*, 215–222. https://doi.org/10.1016/j.ultsonch.2018.02.029.

Liu, Y., Wang, S., Lan, W., & Qin, W. (2019). Development of ultrasound treated polyvinyl alcohol/tea polyphenol composite films and their physicochemical properties. *Ultrasonics Sonochemistry, 51*, 386–394. https://doi.org/10.1016/j.ultsonch.2018.07.043.

Luo, S., & Wong, C. (2001). Effect of UV/ozone treatment on surface tension and adhesion in electronic packaging. *IEEE Transactions on Components and Packaging Technologies, 24*(1), 43–49. https://doi.org/10.1109/6144.910801.

Ma, L., Wang, Q., Li, L., Grierson, D., Yuan, S., Zheng, S., … Zuo, J. (2021). UV-C irradiation delays the physiological changes of bell pepper fruit during storage. *Postharvest Biology and Technology, 180*, 111506. https://doi.org/10.1016/j.postharvbio.2021.111506.

Mangaraj, S., Goswami, T. K., & Mahajan, P. V. (2009). Applications of plastic films for modified atmosphere packaging of fruits and vegetables: A review. *Food Engineering Reviews, 1*(2), 133–158. https://doi.org/10.1007/s12393-009-9007-3.

Manzocco, L., Plazzotta, S., Maifreni, M., Calligaris, S., Anese, M., & Nicoli, M. (2016). Impact of UV-C light on storage quality of fresh-cut pineapple in two different packages. *LWT - Food Science and Technology, 65*, 1138–1143. https://doi.org/10.1016/j.lwt.2015.10.007.

Marangoni Júnior, L., Cristianini, M., & Anjos, C. (2020). Packaging aspects for processing and quality of foods treated by pulsed light. *Journal of Food Processing and Preservation, 44*(11). https://doi.org/10.1111/jfpp.14902.

Marangoni Júnior, L., Cristianini, M., Padula, M., & Anjos, C. A. R. (2019). Effect of high-pressure processing on characteristics of flexible packaging for foods and beverages. *Food Research International, 119,* 920–930. https://doi.org/10.1016/j.foodres.2018.10.078.

Marque, D., Feigenbaum, A., Dainelli, D., & Riquet, A. (1998). Safety evaluation of an ionized multilayer plastic film used for vacuum cooking and meat preservation. *Food Additives and Contaminants, 15*(7), 831–841. https://doi.org/10.1080/02652039809374717.

Masuda, M., Saito, Y., Iwanami, T., & Hirai, Y. (1992). Effect of hydrostatic pressure on packaging materials for food. In C. Balny, R. Hayashi, K. Heremans, & P. Masson (Eds.), *High pressure and biotechnology* (pp. 545–547). London: Colloque INSERM/John Libbey.

Matar, C., Salou, T., Hélias, A., Pénicaud, C., Gaucel, S., Gontard, N., … Guillard, V. (2021). Benefit of modified atmosphere packaging on the overall environmental impact of packed strawberries. *Postharvest Biology and Technology, 177,* 111521. https://doi.org/10.1016/j.postharvbio.2021.111521.

Mauricio-Iglesias, M., Jansana, S., Peyron, S., Gontard, N., & Guillard, V. (2010). Effect of high-pressure/temperature (HP/T) treatments of in-package food on additive migration from conventional and bio-sourced materials. *Food Additives and Contaminants, 27*(1), 118–127. https://doi.org/10.1080/19440040903268054.

Mertens, B. (1993). Packaging aspects of high-pressure food processing technology. *Packaging Technology and Science, 6*(1), 31–36. https://doi.org/10.1002/pts.2770060106.

Min, S., Jin, Z. T., & Zhang, Q. H. (2003). Commercial scale pulsed electric field processing of tomato juice. *Journal of Agricultural and Food Chemistry, 51*(11), 3338–3344. https://doi.org/10.1021/jf0260444.

Min, S., Roh, S., Niemira, B., Boyd, G., Sites, J., Fan, X., … Jin, T. Z. (2018). In-package atmospheric cold plasma treatment of bulk grape tomatoes for microbiological safety and preservation. *Food Research International, 108,* 378–386. https://doi.org/10.1016/j.foodres.2018.03.033.

Min, S., Roh, S., Niemira, B., Sites, J., Boyd, G., & Lacombe, A. (2016). Dielectric barrier discharge atmospheric cold plasma inhibits Escherichia coli O157:H7, Salmonella, Listeria monocytogenes, and Tulane virus in Romaine lettuce. *International Journal of Food Microbiology, 237,* 114–120. https://doi.org/10.1016/j.ijfoodmicro.2016.08.025.

Min, S., Zhang, H., & Han, J. (2014). Packaging for non-thermal food processing. *Innovations in Food Packaging,* 515–535. https://doi.org/10.1016/b978-0-12-394601-0.00021-7.

Minano, H. L. A., Silva, A. C. D. S., Souto, S., & Costa, E. J. X. (2020). Magnetic fields in food processing perspectives, applications and action models. *Processes, 8*(7), 814. https://doi.org/10.3390/pr8070814.

Misra, N., Yepez, X., Xu, L., & Keener, K. (2019). In-package cold plasma technologies. *Journal of Food Engineering, 244,* 21–31. https://doi.org/10.1016/j.jfoodeng.2018.09.019.

Mistriotis, A., Briassoulis, D., Giannoulis, A., & D'Aquino, S. (2016). Design of biodegradable bio-based equilibrium modified atmosphere packaging (EMAP) for fresh fruits and vegetables by using micro-perforated poly-lactic acid (PLA) films. *Postharvest Biology and Technology, 111,* 380–389. https://doi.org/10.1016/j.postharvbio.2015.09.022.

Ng, S. K., Tan, T. B., Tan, P. F., Chong, G. H., & Tan, C. P. (2020). Effect of selected high pressure processing parameters on the sensory attributes and shelf life of jackfruit (Artocarpus heterophyllus L.) bulb packed using different packaging materials. *International Food Research Journal, 27*(4), 675–682.

Ochiai, S., & Nakagawa, Y. (1992). Packaging for high pressure food processing. In C. Balny, R. Hayashi, K. Heremans, & P. Masson (Eds.), *High pressure and biotechnology* (pp. 515–519). London: Colloque INSERM/John Libbey.

Ortuño, C., Duong, T., Balaban, M., & Benedito, J. (2013). Combined high hydrostatic pressure and carbon dioxide inactivation of pectin methylesterase, polyphenol oxidase and peroxidase in feijoa puree. *Journal of Supercritical Fluids, 82,* 56–62. https://doi.org/10.1016/j.supflu.2013.06.005.

Ozen, B. F., & Floros, J. D. (2001). Effects of emerging food processing techniques on the packaging materials. *Trends in Food Science and Technology, 12*(2), 60–67. https://doi.org/10.1016/S0924-2244(01)00053-X.

Pan, Y., Cheng, J., & Sun, D. (2019). Cold plasma-mediated treatments for shelf life extension of fresh produce: A review of recent research developments. *Comprehensive Reviews in Food Science and Food Safety, 18*(5), 1312–1326. https://doi.org/10.1111/1541-4337.12474.

Pankaj, S., Bueno-Ferrer, C., Misra, N., Milosavljević, V., O'Donnell, C., Bourke, P., … Cullen, P. J. (2014a). Applications of cold plasma technology in food packaging. *Trends in Food Science and Technology*, *35*(1), 5–17. https://doi.org/10.1016/j.tifs.2013.10.009.

Pankaj, S., Bueno-Ferrer, C., Misra, N., O'Neill, L., Jiménez, A., Bourke, P., & Cullen, P. (2014b). Surface, thermal and antimicrobial release properties of plasma-treated Zein films. *Journal of Renewable Materials*, *2*(1), 77–84. https://doi.org/10.7569/jrm.2013.634129.

Pankaj, S., & Thomas, S. (2016). Cold plasma applications in food packaging. In *Cold plasma in food and agriculture* (pp. 293–307). Amsterdam: Elsevier. https://doi.org/10.1016/b978-0-12-801365-6.00012-3.

Pastorelli, S. (1997). Packaging and high pressure technology in the food industry. *Italia Imbaliaggio*, *9*, 98–101.

Piyasena, P., Mohareb, E., & McKellar, R. C. (2003). Inactivation of microbes using ultrasound: A review. *International Journal of Food Microbiology*, *87*(3), 207–216. https://doi.org/10.1016/S0168-1605(03)00075-8.

Pothakamury, U. R., Barbosa-Cánovas, G. V., & Swanson, B. G. (1993). Magnetic-field inactivation of microorganisms and generation of biological changes. *Food Technology*, *47*(12), 85–93.

Povey, M. J., & Mason, T. J. (1998). *Ultrasound in food processing*. London: Springer Science & Business Media.

Prakash, A. (2020). *What is the benefit of irradiation compared to other methods of food preservation? Genetically modified and irradiated food* (pp. 217–231). https://doi.org/10.1016/b978-0-12-817240-7.00013-9.

Ramos-Villarroel, A., Aron-Maftei, N., Martín-Belloso, O., & Soliva-Fortuny, R. (2012a). Influence of spectral distribution on bacterial inactivation and quality changes of fresh-cut watermelon treated with intense light pulses. *Postharvest Biology and Technology*, *69*, 32–39. https://doi.org/10.1016/j.postharvbio.2012.03.002.

Ramos-Villarroel, A., Aron-Maftei, N., Martín-Belloso, O., & Soliva-Fortuny, R. (2012b). The role of pulsed light spectral distribution in the inactivation of Escherichia coli and Listeria innocua on fresh-cut mushrooms. *Food Control*, *24*(1–2), 206–213. https://doi.org/10.1016/j.foodcont.2011.09.029.

Ressouany, M., Vachon, C., & Lacroix, M. (1998). Irradiation dose and calcium effect on the mechanical properties of cross-linked caseinate films. *Journal of Agricultural and Food Chemistry*, *46*(4), 1618–1623. https://doi.org/10.1021/jf970805z.

Rivas-Cañedo, A., Fernández-García, E., & Nuñez, M. (2009). Volatile compounds in fresh meats subjected to high pressure processing: Effect of the packaging material. *Meat Science*, *81*(2), 321–328. https://doi.org/10.1016/j.meatsci.2008.08.008.

Robertson, G. L. (2013). *Food packaging: Principles and practice* (3rd ed.). Boca Raton, FL: CRC Press.

Rocculi, P., Romani, S., & Dalla Rosa, M. (2004).Evaluation of physico-chemical parameters of minimally processed apples packed in non-conventional modified atmosphere. *Food Research International*, *37*(4), 329–335. https://doi.org/10.1016/j.foodres.2004.01.006.

Rojas, M. L., Miano, A. C., & Augusto, P. E. D. (2017). Ultrasound processing of fruit and vegetable juices. In D. Bermudex-Aguirre (Ed.), *Ultrasound: Advances in food processing and preservation* (pp. 181–199). Academic Press. https://doi.org/10.1016/B978-0-12-804581-7.00007-5.

Roodenburg, B., de Haan, S. W. H., Ferreira, J. A., Coronel, P., Wouters, P. C., & Hatt, V. (2013). Toward 6 $\log_{10}$ pulsed electric field inactivation with conductive plastic packaging material. *Journal of Food Process Engineering*, *36*(1), 77–86. https://doi.org/10.1111/j.1745-4530.2011.00655.x.

Roodenburg, B., de Haan, S. W. H., van Boxtel, L. B. J., Hatt, V., Wouters, P. C., Coronel, P., & Ferreira, J. A. (2010). Conductive plastic film electrodes for pulsed electric field (PEF) treatment-a proof of principle. *Innovative Food Science and Emerging Technologies*, *11*(2), 274–282. https://doi.org/10.1016/j.ifset.2010.01.005.

Sansone, L., Aldi, A., Musto, P., Di Maio, E., Amendola, E., & Mensitieri, G. (2012). Assessing the suitability of polylactic acid flexible films for high pressure pasteurization and sterilization of packaged foodstuff. *Journal of Food Engineering*, *111*(1), 34–45. https://doi.org/10.1016/j.jfoodeng.2012.01.034.

Sčetar, M., Kurek, M., Režek Jambrak, A., Debeaufort, F., & Galić, K. (2017). Influence of high power ultrasound on physical–chemical properties of polypropylene films aimed for food packaging: Barrier and contact angle features. *Polymer International*, *66*(11), 1572–1578. https://doi.org/10.1002/pi.5415.

Schauwecker, A., Balasubramaniam, V. M., Sadler, G., Pascall, M. A., & Adhikari, C. (2002). Influence of high-pressure processing on selected polymeric materials and on the migration of a pressure-transmitting fluid. *Packaging Technology and Science, 15*(5), 255–262. https://doi.org/10.1002/pts.595.

Scheidt, T. B., & Silva, F. V. M. (2018). High pressure processing and storage of blueberries: Effect on fruit hardness. *High Pressure Research, 38*(1), 80–89. https://doi.org/10.1080/08957959.2017.1402895.

Sehrawat, R., Kaur, B. P., Nema, P. K., Tewari, S., & Kumar, L. (2021). Microbial inactivation by high pressure processing: Principle, mechanism and factors responsible. *Food Science and Biotechnology, 30*(1), 19–35. https://doi.org/10.1007/s10068-020-00831-6.

Serrano, M., Martinez-Romero, D., Guillén, F., Castillo, S., & Valero, D. (2006). Maintenance of broccoli quality and functional properties during cold storage as affected by modified atmosphere packaging. *Postharvest Biology and Technology, 39*(1), 61–68. https://doi.org/10.1016/j.postharvbio.2005.08.004.

Sheung, K. S. M., Min, S., & Sastry, S. K. (2004). Dynamic head space analyses of orange juice flavor compounds and their absorption into packaging materials. *Journal of Food Science, 69*(7), 549–556. https://doi.org/10.1111/j.1365-2621.2004.tb13649.x.

Silveira, A. C., Araneda, C., Hinojosa, A., & Escalona, V. H. (2014). Effect of non-conventional modified atmosphere packaging on fresh cut watercress (Nasturtium officinale r. Br.) Quality. *Postharvest Biology and Technology, 92*, 114–120. https://doi.org/10.1016/j.postharvbio.2013.12.012.

Sterr, J., Fleckenstein, B. S., & Langowski, H. C. (2018). The theory of decompression failure in polymers during the high-pressure processing of food. *Food Engineering Reviews, 10*(1), 14–33. https://doi.org/10.1007/s12393-017-9171-9.

Tang, Y., Jiang, Y., Jing, P., & Jiao, S. (2021). Dense phase carbon dioxide treatment of mango in syrup: Microbial and enzyme inactivation, and associated quality change. *Innovative Food Science and Emerging Technologies, 70*, 102688. https://doi.org/10.1016/j.ifset.2021.102688.

Ting, E., Balasubramaniam, V. M., & Raghubeer, E. (2002). Determining thermal effects in high-pressure processing. *Food Technology, 56*(2), 31–35. Retrieved from https://pascal-francis.inist.fr/vibad/index.php?action=getRecordDetail&idt=13503438.

Tomás-Callejas, A., Otón, M., Artés, F., & Artés-Hernández, F. (2012). Combined effect of UV-C pretreatment and high oxygen packaging for keeping the quality of fresh-cut Tatsoi baby leaves. *Innovative Food Science and Emerging Technologies, 14*, 115–121. https://doi.org/10.1016/j.ifset.2011.11.007.

Turan, D., Gunes, G., & Seniha Güner, F. (2016). Synthesis, characterization and O_2 permeability of shape memory polyurethane films for fresh produce packaging. *Packaging Technology and Science, 29*(7), 415–427. https://doi.org/10.1002/pts.2222.

Weaver, J. C., & Chizmadzhev, Y. A. (1996). Theory of electroporation: A review. *Bioelectrochemistry and Bioenergetics, 41*(2), 135–160. https://doi.org/10.1016/S0302-4598(96)05062-3.

Welle, F., Haack, G., & Franz, R. (2000). Investigation into migrational and sensorial changes of packaging plastics caused by ionising irradiation. *Deutsche Lebensmittel-Rundschau, 96*, 423–430. Retrieved from http://pascalfrancis.inist.fr/vibad/index.php?action=getRecordDetail&idt=863659.

Welle, F., Mauer, A., & Franz, R. (2002). Migration and sensory changes of packaging materials caused by ionising radiation. *Radiation Physics and Chemistry, 63*(3–6), 841–844. https://doi.org/10.1016/S0969-806X(01)00576-X.

Wouters, P. C., Alvarez, I., & Raso, J. (2001). Critical factors determining inactivation kinetics by pulsed electric field food processing. *Trends in Food Science and Technology, 12*(3–4), 112–121. https://doi.org/10.1016/S0924-2244(01)00067-X.

Yamamoto, K. (2017). Food processing by high hydrostatic pressure. *Bioscience, Biotechnology and Biochemistry, 81*(4), 672–679. https://doi.org/10.1080/09168451.2017.1281723.

Yuan, J. T. C., & Novak, J. S. (2012). Industrial applications using supercritical carbon dioxide for food. In M. O. Balaban & G. Ferrentino (Eds.), *Dense phase carbon dioxide: Food and pharmaceutical applications* (pp. 227–238). Wiley-Blackwell. https://doi.org/10.1002/9781118243350.ch11.

Zhai, M., Yoshii, F., & Kume, T. (2003). Radiation modification of starch-based plastic sheets. *Carbohydrate Polymers, 52*(3), 311–317. https://doi.org/10.1016/s0144-8617(02)00292-8.

Zhang, X. T., Zhang, M., Devahastin, S., & Guo, Z. (2019). Effect of combined ultrasonication and modified atmosphere packaging on storage quality of pakchoi (Brassica chinensis L.). *Food and Bioprocess Technology, 12*(9), 1573–1583. https://doi.org/10.1007/s11947-019-02316-9.

Zhang, Y., Zhang, J., Zhang, Y., Hu, H., Luo, S., Zhang, L., ... Li, P. (2021). Effects of in-package atmospheric cold plasma treatment on the qualitative, metabolic and microbial stability of fresh-cut pears. *Journal of the Science of Food and Agriculture, 101*(11), 4473–4480. https://doi.org/10.1002/jsfa.11085.

Zhang, Z., Zhang, B., Yang, R., & Zhao, W. (2020). Recent developments in the preservation of raw fresh food by pulsed electric field. *Food Reviews International*, 1–19. https://doi.org/10.1080/87559129.2020.1860083.

Zhao, W., & Tang, Y. (2017). Effects of lipid, PET/PE and PA/PE with high-pressure processing. *International Journal of Food Properties, 20*(4), 801–809. https://doi.org/10.1080/10942912.2016.1181650.

Zhou, B., Zhang, L., Wang, X., Dong, P., Hu, X., & Zhang, Y. (2019). Inactivation of Escherichia coli O157:H7 by high hydrostatic pressure combined with gas packaging. *Microorganisms, 7*(6), 1–11. https://doi.org/10.3390/microorganisms7060154.

Zimmermann, U., Pilwat, G., & Riemann, F. (1974). Dielectric breakdown of cell membranes. *Biophysical Journal, 14*(11), 881–899. https://doi.org/10.1016/S0006-3495(74)85956-4.

Zygoura, P., Paleologos, E., & Kontominas, M. (2011). Changes in the specific migration characteristics of packaging–food simulant combinations caused by ionizing radiation: Effect of food simulant. *Radiation Physics and Chemistry, 80*(8), 902–910. https://doi.org/10.1016/j.radphyschem.2011.03.020.

Chapter 11

Safety Aspects of Non-Thermal Processing Applications for Fruit and Vegetable Processing

Dilara Devecioglu, Dilara Nur Dikmetas, Sebahat Oztekin, and Funda Karbancioglu-Guler

CONTENTS

11.1 INTRODUCTION

Vitamins, antioxidants, dietary fiber, minerals, and plant protein are abundant in fruit and vegetables (Wu et al., 2020). There are several studies conducted on the nutrients in fruits and vegetables that have an essential role in preventing some diseases, such as cardiovascular disease, cancer, diabetes, osteoporosis, and visual problems (Ishihara et al., 2018; Zeng et al., 2019; Kuete et al., 2017). Consuming a proper amount of fruits and vegetables may help individuals maintain a healthy lifestyle (Yousuf et al., 2020). However, fruits are highly vulnerable to spoilage due to their carbohydrate content and enzyme activity (Yildiz et al., 2021). There are numerous explanations for fresh food deterioration; in general, it could be due to water loss, microbial growth, oxidation, texture and flavor deterioration, as well as an increase in the respiration rate and ripening process (Yousuf et al., 2020). Besides, fruits and vegetables might be contaminated with pathogens from human, animal, or environmental sources during growth, harvest, transportation, and previous processing and handling (Ramos et al., 2013). As a result of contamination, 20% of fruits and vegetables are lost every year (Mostafidi et al., 2020). Furthermore, according

DOI: 10.1201/9781003222170-11

to the Centers for Disease Control and Prevention 2019 report, the foodborne outbreak is mostly caused by fruits and vegetables (38%), followed by processed foods (20%) (Bermudez-Aguirre, 2020). Fresh fruits and vegetables, as well as their processed products, have a short shelf life. For this reason, various studies are carried out to extend the shelf life and ensure food safety. For example, thermal processes are applied to extend the shelf life of fruit juice (Waghmare, 2021). However, this is very difficult to implement in fresh products and has adverse effects on the product in various aspects such as physicochemical, nutritional, rheological, and sensory deterioration. On the other hand, non-thermal food processing techniques showed a minimal negative impact on the fruit and vegetable industry (Bassey et al., 2021; Silva et al., 2020). Both the food industry and researchers have increased interest in non-thermal processes as an alternative to the preservation of minimally processed or freshly consumed fruits and vegetables. In line with the research, it is seen that it supports the preservation of the food's freshness and maintains food quality and safety. In addition, it has been shown that the number of important components for health increases (Denoya et al., 2021). Therefore, non-thermal technologies including ultrasound, pulsed electric field, ozone, cold plasma, high hydrostatic pressure, and supercritical carbon dioxide have gained importance (Waghmare, 2021). Process conditions should be optimized in the fruit and vegetable industry to maintain food quality and sensory properties. Several studies have been performed on process optimization with disinfection efficiency while preserving food quality (de Souza et al., 2018; Wang et al., 2020). Certain fruits and vegetables have been extensively studied to understand the safety aspects of non-thermal processes, including lettuce (Alexopoulos et al., 2013; Takundwa et al., 2021; Lee et al., 2015), tomato (Fan et al., 2020; He et al., 2021), strawberry (Ding et al., 2015; Gibson et al., 2019), juices (Dasan & Boyaci, 2018; Groot et al., 2019; Pallarés et al., 2021; Panigrahi et al., 2021; Szczepańska et al., 2020). This chapter focuses on the extensively studied non-thermal processes applied to fruits and vegetables and processed products, such as juices. It presents an overview of the removal of contaminants that pose a risk to food safety. In addition, some studies about the detection of the process's effect on quality parameters that threaten food safety have been the focus of the study.

11.2 APPLICATION OF COLD PLASMA

Cold plasma, non-thermal technology, is applied to various food products and ensures food safety from different aspects (Mutlu-Ingok et al., 2021). There are different studies on using cold plasma for fruits and vegetables, most of which are about microbial inactivation, and there are limited studies on other contaminants. Fruits and vegetables are consumed both fresh and processed by making fresh cuts. Several disinfection methods are used for these foods, but cold plasma application provides advantages by reducing chemical residues. Besides, using cold plasma results in microbial inactivation, which is the intended use of disinfection (Bermudez-Aguirre, 2019). Research has not only focused on fresh fruit and vegetables, including mandarin (Bang et al., 2020), cabbage (Lee et al., 2015), lettuce (Min et al., 2016), and radish sprouts (Oh et al., 2017), but also different studies are present which investigated the effect of cold plasma on their processed products such as apple juice (Surowsky et al., 2014), sour cherry nectar (Dasan & Boyaci, 2018), and orange juice (Almeida et al., 2015). Due to the highly variable process conditions, there are varying studies in the literature (Table 11.1). Cold plasma, which basically damages the cell membrane with the movement of charged particles and reactive substances, also shows its antimicrobial effect by damaging the cell and DNA (Pasquali et al., 2016).

TABLE 11.1   STUDIES CONDUCTED WITH COLD PLASMA TREATMENT

Products	Hazard Microorganism	System	Process condition and feed gas	Reduction	References
Apple	Listeria monocytogenes	Plasma jet and nisin antimicrobial solution (AMS)	40 s with 3 min AMS & Air	3.4 log cfu/g	Ukuku et al. (2019)
Apple juice	Escherichia coli	Atmospheric pressure plasma jet	650 W, 120 s & Air	4.02 log cfu/mL	Dasan and Boyaci (2018)
Apple juice	Citrobacter freundii	Plasma jet	65V, 480 s and 24 h storage & Argon/O_2	5 log cfu/mL	Surowsky et al. (2014)
Blueberry	Tulane virus Murine norovirus	Plasma jet	47 kHz, 120 s & Air 90 s & Air	3.5 log pfu[1]/g 5 log pfu/g	Lacombe et al. (2017)
Cherry tomatoes	Aerobic bacteria, coliform bacteria, Yeast and mold	Corona discharge plasma	25°C, 20 kHz, 2–4 A, 0–120 s & Air	0.68–2.00 log cfu/g	Lee et al. (2018)
Cherry tomatoes	Salmonella	Microwave-powered cold plasma	25°C, 400–900W, 2-10 min & He-O_2 gas mixture	0.30–3.50 log cfu/tomato	Kim and Min (2017)
Dried fig	Escherichia coli O157:H7, Listeria monocytogenes	Microwave-powered cold plasma	400 W, 10 min & Nitrogen or the mixture of helium and oxygen	1.3–1.6 log cfu/g	Lee et al. (2015)
Lettuce	Listeria monocytogenes	Microwave-powered cold plasma	400 W, 10 min & Nitrogen or the mixture of helium and oxygen	1.8 log cfu/g	Lee et al. (2015)
Mandarin	Penicillium italicum	Microwave-powered cold plasma	900 W, 10 min & N_2	84% disease incidence	Won et al. (2017)
Melon	Mesophilic bacteria Lactic acid bacteria	Dielectric barrier discharge	19 V, 3 A, 30–30 min & Air	3.4 log cfu/g 2.0 log cfu/g	Tappi et al. (2016)
Onion flakes	Escherichia coli, Listeria monocytogenes, Salmonella Enteritidis	Dielectric barrier discharge	15 kHz, 9 kV, 20 min & Helium	1.1–3.1 log cfu/cm^2	Kim and Min (2018)

(Continued)

TABLE 11.1 (CONTINUED)   STUDIES CONDUCTED WITH COLD PLASMA TREATMENT

Products	Hazard	System	Process condition and feed gas	Reduction	References
Orange juice	*Escherichia coli*	Atmospheric pressure plasma jet	650 W, 120 s & Air	1.59 log cfu/mL	Dasan and Boyaci (2018)
Radicchio	*Listeria monocytogenes*	Atmospheric cold plasma	22°C & Air 30 min	2.2 log cfu/cm²	Pasquali et al. (2016)
	Escherichia coli O157:H7		15 min	1.35 log MPN/cm²	
Radish sprouts	*Salmonella typhimurium*	Cold plasma	900 W, 20 min & N₂	2.6 log cfu/g	Oh et al. (2017)
Red currants	Yeast and mold	Diffuse coplanar surface barrier discharge	300 W & Air	1.28 log cfu/g	Limnaios et al. (2021)
Romaine lettuce	*Escherichia coli* O157:H7, *Listeria monocytogenes*	Dielectric barrier discharge	34.8 kV, 1 A & Air	1.0–1.1 log cfu/g	Min et al. (2016)
	Tulane virus			1.3 log cfu/g	
Sour cherry nectar	*Escherichia coli*	Atmospheric pressure plasma jet	650 W, 120 s & Air	3.34 log cfu/mL	Dasan and Boyaci (2018)
Strawberry	*Escherichia coli, Listeria monocytogenes, Salmonella enterica* Typhimurium	Dielectric barrier discharge	50 Hz, 120 kV, 120–300 s & Air	1.6–4.2 log cfu/sample	Ziuzina et al. (2014)
Tomato juice	*Escherichia coli*	Atmospheric pressure plasma jet	650 W, 120 s & Air	1.43 log cfu/mL	Dasan and Boyaci (2018)
Pesticide					
Blueberry	Boscalid	Atmospheric cold plasma	25°C, 50 Hz, 80 kV, 5 min & Air	80%	Sarangapani et al. (2017)
	Imidacloprid			76%	
Mango	Chlorpyrifos	Gliding arc discharge	5 min, 5 L/min & Ar and water vapour	74%	Phan et al. (2018)
	Cypermethrin			62.9%	

ᵃPlaque forming unit

Cherry tomato is a product on which various studies have been carried out, and successful microbial inactivation results have been observed with a cold plasma application (Kim & Min, 2017; Lee et al., 2018; Ziuzina et al., 2014). As mentioned, process conditions are critical for research to ensure effective microbial reduction. While the most effective conditions against *Salmonella* were found to be 10 min and 900 W in the different treatment time (2–10 min) and power (400–900 W) range (Kim & Min, 2017), the reduction values of the contaminants increased as the current increased in the study of Lee et al. (2018). Similar to Kim and Min (2017), the microbial load decreased more by increasing the treatment time in Ziuzina et al. (2014). For cherry tomatoes, it was observed that as the application time of the method applied at certain times (30–300 s) increased, the load of *Escherichia coli*, *Salmonella enterica* serovar Typhimurium, and *Listeria monocytogenes* remained below the detection limits. Furthermore, in an experiment conducted on kumquat fruit, a result below the detection limit of the microbial count was found with intermittent corona discharge plasma jet treatment at 4.0 A (Puligundla et al., 2018).

Moreover, when cold plasma is used as a postharvest technology, it has been effective in preserving mandarins (Won et al., 2017). A combined application (CaO washing and cold plasma) was also studied for mandarins. It was concluded that it increased its storability by decreasing *Penicillium digitatum* disease incidence (Bang et al., 2020). Atmospheric cold plasma also reduced the natural decay of blueberries and improved postharvest quality (Hu et al., 2021). In terms of processed products, an effective bacterial inactivation was achieved without adversely damaging the product properties (Dasan & Boyaci, 2018). However, there are also cases where the studies do not reach the desired goal and are insufficient to either inactivate or reduce microbial load. Lacombe et al. (2015) showed that cold plasma treatment was inadequate to provide a significant reduction in the number of yeast and mold in blueberries. On the other hand, the treatment showed potential to inactive spoilage microorganisms.

In parallel with reducing microbial load, cold plasma may provide advantages for both the producer and the consumer in terms of shelf life. For example, atmospheric cold plasma has been shown to prolong the shelf life of fresh-cut pears by inhibiting the growth of mesophilic aerobic bacteria, yeast, and mold (Zhang et al., 2021a). Enhanced storage ability was also shown by Bang et al. (2020) for mandarins by testing a combination of atmospheric dielectric barrier discharge cold plasma and calcium oxide. Pasquali et al. (2016) also showed in the study performed by radicchio that cold plasma may be an alternative to chemical disinfection methods because it resulted in an observable reduction in both *Listeria monocytogenes* and *Escherichia coli* with a certain treatment time.

While the process ensures microbial safety, the quality parameters that may be affected should not be ignored and forgotten. These parameters also affect food safety and acceptability. For this reason, when the literature is examined, studies that also examine quality parameters are of note. Bioactive components and sensory properties such as color, texture, and flavor that may be affected by the process are the parameters to be considered. Zhou et al. (2020) showed that the drying time of the wolfberry fruit pre-treated with cold plasma under atmospheric pressure with air (3 L/min, 20 kHz, and 750 W, 15–60 s) decreased, and it is stated that different processing time affects color, and the total phenolic and total flavonoid content. For this reason, the processing time should be optimized not only in terms of microbial safety but also considering that it may adversely affect quality parameters. In the study of Kim and Min (2017), which applied a longer process time, the process was effective, but it was commented that the process conditions should be improved by shortening the application time. Additionally, Lee et al. (2018) showed no change in sensory properties of cherry tomatoes and a decrease in microbial load by using corona discharge plasma. It has been shown that cold plasma may be applied to a processed fruit product by maintaining the pH and color of the orange juice while ensuring safety in terms of physico-chemical properties (Almeida et al., 2015). Besides, anthocyanin content in pomegranate juice increased after cold plasma treatment (Kovačević et al., 2016).

Differently, enzymes, one of the criteria affecting food quality and safety, may also be the subject of study. With the cold plasma application, food preservation may be carried out with the inactivation of critical enzymes in fruits and vegetables, including polyphenol oxidases and pectic enzymes (Pankaj et al., 2013). They showed the efficiency of cold plasma to reduce the activity of peroxidase in tomatoes. Similarly, after cold plasma application, polyphenol oxidase activity decreased, and it has been pointed out that it contributes to the stabilization of freshly cut packaged apples (Tappi et al., 2014). On the other hand, the treatment's inhibition ability of peroxidase (17%) and pectin methylesterase (7%) activity was lower for fresh-cut melons (Tappi et al., 2016). Differences may be observed depending on the applied process conditions, the type of product, and the enzyme.

In addition to microbial safety, the realization of pesticide degradation has been another food safety point that has been the subject of studies. With the cold plasma application, while the blueberry quality parameters were kept within acceptable limits, the pesticides examined were significantly degraded (Sarangapani et al., 2017). It has been proven again that cold plasma treatment provides chemical safety by reducing pesticides in mango (Phan et al., 2018). It has been commented that the reason for this successful result may be the effect of reactive oxygen species formed by cold plasma on the chemical bonds of pesticides (Gavahian et al., 2020).

As a result, although cold plasma is seen as an alternative method to thermal treatments in ensuring food safety, it needs more extended studies for its industrial application.

11.3 APPLICATION OF IRRADIATION

Irradiation is one of the methods that may be applied to fresh food products, and with the advantage of being a non-thermal process, the negative effects of temperature on properties such as color and flavor may be eliminated (Verde et al., 2013). Irradiation covers different types of treatments, including electron beam and gamma irradiation. Since irradiation may be applied to fresh food products, various studies were carried out on different fruits, vegetables, and their processed products such as baby spinach leaves (Gomes et al., 2011), cabbage (Grasso et al., 2011), carrot (Chaudry et al., 2004), frozen crushed garlic (Kim et al., 2014), mango juice (Naresh et al., 2015a,) and pineapple (Hajare et al., 2006).

Similar to other non-thermal processes, the quality and sensory characteristics of the food that may be affected by the treatment must also be considered. Verde et al. (2013) applied an irradiation dose of 1.5 kGy, and no major effect on the sensory and quality of raspberry was observed. Similarly, with the same fruit, the dose of 2 kGy was found highly effective against microbial growth; however, the treatment between 1 and 2 kGy was suggested to eliminate the adverse effect of the dose (Guimarães et al., 2013). Gomes et al. (2011) proposed the treatment at 0.7 kGy to reduce *Salmonella* spp. and *Listeria* spp. by 5 log cfu/g. Change in food color, one of the most important parameters for consumers, is also critical. Therefore, it has been the subject of study by researchers such as mango juice (Naresh et al., 2015b).

After the treatment, Naresh et al. (2015b) determined the total phenolic and flavonoid content and antioxidant activity of mango juice in addition to microbial properties. Although the high antimicrobial property was observed as the dose increased, it was concluded that an optimum dose should be applied considering the other characteristics of the food. In addition to the microbial load reduction, the microbial change of the treated products during storage is also critical in terms of food safety. For example, Tawema et al. (2016) used a combination of gamma irradiation and natural antimicrobial for the cauliflower treatment. A major reduction of *Listeria monocytogenes* and *Escherichia coli* O157:H7 during storage and inhibition of yeast and mold was observed. According to Alighourchi et al. (2008), gamma irradiation of pomegranate juice at 2 kGy completely inhibited the total plate and fungal counts. The dose that was higher than 2 kGy had no suggestion in terms of anthocyanin content. Another study recommended gamma irradiation doses of lower than 3 kGy to achieve the microbial and quality safety of sour cherry juice (Arjeh et al., 2015). While mandarins irradiated at 0.4 or 1 kGy showed dose-dependent microbial safety during storage, it was stated that 0.4 kGy irradiation should be used to ensure safety and acceptability in terms of other parameters (Nam et al., 2019).

It has been shown by Majeed et al. (2014) for strawberries that irradiation may also be used to increase shelf life while maintaining food safety. Microbial safety was also obtained in the study of de Jesus Filho et al. (2018) as a result of gamma irradiation at 2 kGy, and strawberries had no major changes during certain storage. Also, Martins et al. (2004) showed a remarkable increase in the shelf life of watercress (112 days) by gamma irradiation. Electron beam irradiation at 3 kGy reduced natural microbiota, *Salmonella* Typhimurium, *Escherichia coli*, and *Listeria monocytogenes* in raspberry while treatment preserved antioxidant activity (Elias et al., 2020).

Extending the shelf life and maintaining food safety during this period are associated with enzyme activities and the control of microbial load during storage. Therefore, nowadays, studies also focus on the effect of processes on enzymes. For example, the application of irradiation at 0.5 kGy decreased the content of catalase and peroxidase in Ponkan fruit as well as affected the growth of *Penicillium digitatum* (Zhao et al., 2020). The fungal decontamination efficiency of gamma irradiation was also proved by Zarbakhsh and Rastegar (2019) for date fruits.

Although various studies (Table 11.2) show that microbial safety is provided by irradiation, industrial applications should be carried out by determining that all changes in the structure of a food product are safe for consumption. For this reason, extensive studies on the subject continue to be carried out.

11.4 APPLICATION OF THE HIGH PRESSURE PROCESS

Non-thermal processes have become more popular than their conventional counterparts due to rising customer demand for minimally processed clean-label foods with improved safety, sensory, and quality properties (Picart-Palmade et al., 2019). High pressure processing (HPP) is one of the non-thermal food processing methods, also called high hydrostatic pressure (HHP), or pascalization, or ultra high pressure processing (Roobab et al., 2021). The HPP uses hydrostatic pressure (100–1000 MPa) in a pressure transmitting liquid medium at mild temperatures (Aaliya et al., 2021; Pallarés et al., 2021).

The HPP process uses the principle of *Le Chatelier* based on the isostatic principle, where the pressure is applied uniformly and instantly using a pressure transmitting liquid (usually water) on food (solid, fluid, or particulate). During the adiabatic heating and cooling of the vessel, the temperature may increase or decrease about 3°C for each 100MPa, respectively, depending on the nature of the product, temperature, and applied pressure (Lopez et al., 2018) (Figure 11.1).

HPP can be used alone or in combination with other thermal and non-thermal processing techniques leading to a synergistic effect on foodborne pathogens (Aaliya et al., 2021), fungal spores (Merkulow et al., 2000), enzymes (Fernandez et al., 2018), pesticides (Gavahian et

TABLE 11.2   STUDIES CONDUCTED WITH IRRADIATION TREATMENT

Products	Hazard	Process condition	Reduction	References
Ashitaba and kale juices	Total aerobic bacteria and coliform	5 kGy, ^{60}Co gamma	2 log cfu/mL	Jo et al. (2012)
Black pepper	*Escherichia coli* O157:H7, *Salmonella* Typhimurium	1–5 kGy, ^{60}Co gamma	1.2->5.2 log cfu/g	Song et al. (2014)
Cabbage	*Escherichia coli*	1.1–4.1 kGy, electron beam	2.0–7.2 log cfu/g	Grasso et al. (2011)
Carrot	Total aerobic viable cell count	1.5 kGy, ^{60}Co gamma	2 log cfu/g	Mohácsi-Farkas et al. (2014)
Carrot juice	Total aerobic bacteria, coliform	1–3 kGy, ^{60}Co gamma	2.40–5.96 log cfu/mL	Song et al. (2007)
Celery	Total bacteria	0.5–1.5 kGy, gamma	2.11–4.15 log cfu/g	Lu et al. (2005)
	Yeast and mold		0.94–1.34 log cfu/g	
Cucumber	*Escherichia coli, Listeria ivanovii, Salmonella* Typhimurium, *Staphylococcus aureus*	1–3 kGy, ^{60}Co gamma	1.94–7.31 log cfu/g	Lee et al. (2006)
Cucumber	*Salmonella* Poona	0.38 kGy, electron beam	1 log cfu/g	Joshi et al. (2018)
Eggplant	Total bacteria, yeast, and mold	0.25–1 kGy, gamma	0.1–4.7 log cfu/g	Hussain et al. (2014)
Kale juice	Viable counts	3–5 kGy, ^{60}Co gamma	4.46–6.15 log cfu/mL	Kim et al. (2007)
Lettuce	Hepatitis A	2.72 kGy, ^{60}Co gamma	1 log pfu[1]/mL	Bidawid et al. (2000)
Lettuce	Rotavirus	1.03 kGy, electron beam	1 log pfu/g	Espinosa et al. (2012)
Lettuce	Total bacterial count	0.5–1.5 kGy, gamma	0.89–3.48 log cfu/g	Zhang et al. (2006)
Raspberry	Mesophilic counts	1.5 kGy, ^{60}Co gamma	1 log cfu/g	Verde et al. (2013)
Red pepper	*Escherichia coli* O157:H7, *Salmonella* Typhimurium	1–5 kGy, ^{60}Co gamma	0.9->5.2 log cfu/g	Song et al. (2014)
Tomato seeds	*Salmonella*	7 kGy, electron beam	4.36 log cfu/g	Trinetta et al. (2011)
Spinach	Rotavirus	1.29 kGy, electron beam	1 log pfu/g	Espinosa et al. (2012)
Strawberry	Hepatitis A	2.97 kGy, ^{60}Co gamma	1 log pfu/mL	Bidawid et al. (2000)
Watercress	*Salmonella* spp.	1.7 kGy, gamma	4 log cfu/g	Martins et al. (2004)

[1]Plaque forming unit

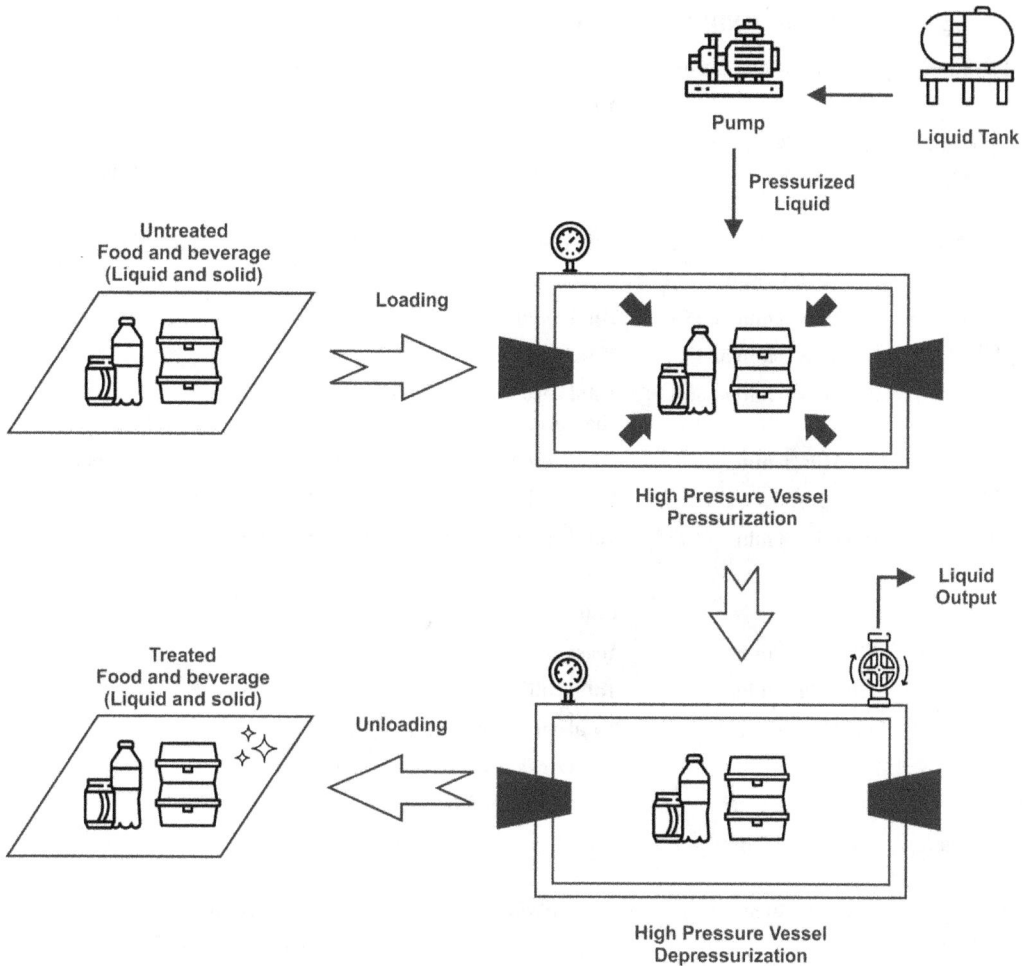

Figure 11.1   Schematic presentation of high pressure processing (HPP).

al., 2020), and mycotoxins (Mirza Alizadeh et al., 2021; Pallarés et al., 2021) (Table 11.3). HPP's mode of action involves disrupting cell membrane integrity and selectivity and inactivating the cytoplasmic or membrane enzymes required for DNA replication and transcription (Rastogi et al., 2007). The formation of electrostatic and hydrophobic interactions and sulfhydryl bond-containing compounds has been associated with the possible cause of mycotoxin degradation in HPP processing (Woldemariam & Emire, 2019). Although the removal of pesticides by HPP is debatable, this technique was mostly recommended to reduce mycotoxin concentrations (Gavahian et al., 2020). When high pressure is applied, hydrophobic pesticides can pass through the pressure transmitting liquid media due to weakening hydrophobic interactions (Iizuka et al., 2013). Likewise, Iizuka and Shimizu (2014) reported using ethanol solution to improve pesticide removal in HPP-treated (0.1–400MPa, 30 min) cherry tomatoes instead of water. Furthermore, HPP with 10% ethanol solution eliminated the chlorpyrifos pesticide from Brussels sprouts at low pressure (Iizuka & Shimizu, 2014). HPP treatment is equivalent to thermal pasteurization when using 600 MPa pressure for 3–5 minutes at room temperature (-25°C) (Augusto et al., 2018).

TABLE 11.3   STUDIES CONDUCTED WITH HIGH-PRESSURE PROCESSING TREATMENT

Products	Process condition	Hazard	Reduction	References
		Microorganism		
Apple juice	600 MPa, 30 min at 75°C + US (24 kHz, 0.33 W/mL)	*Neosartorya fischeri* ascospores	3.3 log	Evelyn et al. (2016)
Apple juice	350 MPa, 15 min	*Penicillium expansum*	6.0 log	Merkulow et al. (2000)
	500 MPa, 15 min	*Eurotium repens*	4.2 log	
Black table olives	250 MPa, 5 min at 25°C	Total mold	90%	Tokuşoğlu et al. (2010)
	250 MPa, 5 min at 4°C	Total mold	100%	
	250 MPa, 5 min at 35°C	Total aerobic-mesophilic bacteria	35–76%	
Cucumber juice	500 MPa, 5 min	Total aerobic bacteria	3.88 log to > 1.00 log	Liu et al. (2016)
		Yeast and mold	2.51 log to > D.L.[1]	
Grape juice	600 MPa, 3 min	Total aerobic bacteria	3.50 log to > 1.00 log	Chang et al. (2017)
		Yeast and mold	2.20 log to > D.L.	
		Coliforms	2.10 log to > D.L.	
Liquorice root sherbet	355 MPa, 1 min	Yeast and mold	2.55 log to > D.L.	Aday et al. (2018)
	250 MPa, 5 min	Total coliforms	3.51 log to > D.L.	
	450 MPa, 5min	Total aerobic bacteria	>2.7 log	
		Escherichia coli	>6 log	
		Salmonella Typhimurium		
Orange juice	500 MPa, 1.1 min	*Saccharomyces cerevisiae*	6.0 log	Zook et al. (1999)
Pineapple fruit juice	500 MPa, 10 min	Total aerobic bacteria	5.60 log to > 2.02 log	Wu et al. (2021)
		Yeast and mold	4.68 log to > 1.45 log	
		Coliforms	3.81 log to > 0.64 log	
Strawberry juice	300 MPa, 1 min at 4°C	Total aerobic bacteria	3.11 log to > 1.7 log	Yildiz et al. (2021)
		Yeast and mold	3.4 log to > 1.97 log cfu/mL	
Strawberry puree	600 MPa/ 30 min at 75°C + US (24 kHz/0.33 W/mL)	*Byssochlamys nivea* ascospores	2.7 log	Evelyn and Silva (2016)
		Neosartorya fischeri ascospores	2 log	
Vegetable smoothie	630 MPa, 6 min	*Listeria monocytogenes*	6 log up to D.L.	Fernández et al. (2019)
		Escherichia coli O157:H7		
		Total aerobic-mesophilic bacteria	1.2 log to < D.L.	
		Yeast and mold	1.96 log to < D.L.	
		Enterobacteriaceae	1.9 log to < D.L.	

(*Continued*)

TABLE 11.3 (CONTINUED)   STUDIES CONDUCTED WITH HIGH-PRESSURE PROCESSING TREATMENT

Products	Process condition	Hazard	Reduction	References
		Mycotoxin		
Apple juice	Without pulses	Patulin	0.51%	Avsaroglu et al. (2015)
	300–500 MPa, 2 pulses × 150 s at 30–50 °C		0–45%	
	300–500 MPa, 6 pulses × 50 s at 30–50°C		0–62%	
Black table olives	250 MPa, 5 min at 35°C	Citrinin	64–100%	Tokuşoğlu et al. (2010)
Fruit and vegetable juice blends	600 MPa, 300 s at 11°C	Patulin	62 ppb	Hao et al. (2016)
Grape juice	500 MPa, 5 min at room temperature, for 100 µg/L aflatoxin spike	Aflatoxin B_1	17%	Pallarés et al. (2021)
		Aflatoxin B_2	14%	
		Aflatoxin G_1	19%	
		Aflatoxin G_2	29%	
		Pesticide		
Brussels sprouts	200 MPa, 30 min at 5°C + 10% (v/v) ethanol solution	Chlorpyrifos	89%	Iizuka and Shimizu (2014)
Cherry tomatoes	75 MPa, 30 min at 5°C	Chlorpyrifos	75%	Iizuka et al. (2013)

[1]Detection limit (< 1.00 log cfu/mL or < 1.00 log cfu/g)

HPP inhibits enzymes/proteins and inactivates spoilage and pathogenic microorganisms while maintaining nutritional and bioactive components, which are significant quality characteristics in fruits and vegetables (Fernández et al., 2020; Picart-Palmade et al., 2019). In this sense, the acidic nature (< pH 4.6) of fruits and their products can limit the growth and germination of most vegetative spoilage (lactic acid bacteria) and pathogenic bacteria (*Salmonella*, *Escherichia coli*). In contrast, molds and yeasts can easily survive at lower oxygen levels (< 0.5%) and temperatures (-10°C). In general, resistance to HPP treatment may vary as follows: yeast and molds < vegetative bacteria < Gram (+) bacteria < Gram (–) bacteria < bacterial spores (Picart-Palmade et al., 2019). In this regard, thermal-resistant spore-forming bacteria (*Alicyclobacillus acidoterrestris* and *Bacillus coagulans*), and mycotoxin-producing molds (*Byssochlamys fulva*, *Byssochlamys nivea*, *Byssochlamys spectabilis*, and *Neosartorya fischeri*) may germinate under improper refrigeration conditions (Silva & Evelyn, 2020). For low acid foods (e.g. coconut water, melon, watermelon, or vegetable products), bacterial spores of the genus *Clostridium* (*Clostridium botulinum*) and *Bacillus* (*Bacillus cereus*) need to be considered (Augusto et al., 2018). For fruit and vegetable smoothies, mesophilic aerobic bacteria, yeasts and molds, *Enterobacteriaceae*, and lactic acid bacteria counts remained below the detection limit (< 1.0 log cfu/g) after 26 days of storage at 25°C (Fernandez et al., 2019). A recent study conducted by Bulut and Karatzas (2021) studied the application of HPP in combination with freezing in orange juice for *Escherichia coli* K12 inactivation. Then, *E. coli* containing samples were subjected to freezing at three temperatures; 4, –24, and –80°C. Following HPP treatment (400MPa/9 min), *Escherichia coli* log reductions were 1.83, 5.63, and 6.83 at 4, –24, and –80°C, respectively. Another study from Buerman et al. (2020) investigated the use of HPP (450–600 MPa, 1.5–3 min) in a diluted apple juice concentrate at different water activity and pH levels for spoilage fungi (*Penicillium* spp., *Aspergillus niger*, *Byssochlamys spectabilis*, *Rhodotorula mucilaginosa*, *Candida parapsilosis*, *Torulaspora delbrueckii*, or *Brettanomyces bruxellensis*) inactivation. As a result, *Candida parapsilosis* was found to be the most resistant fungus, and HPP-treated food can be stored under refrigeration at lower pH (≤ 4.6) and higher water activity (≥ 0.98) levels.

Small-molecule components (e.g. volatile compounds, pigments, and vitamins) remained intact in HPP-treated fruit and vegetable-based products, but high-molecular-weight (e.g. pectin, starch, and protein) substances changed their native structure. This could be explained by the higher compressibility of non-covalent bonds (ionic bond, hydrogen bond, and hydrophobic interaction) than covalent bonds (Balasubramaniam et al., 2015). In addition, HPP offers fresh-like fruit products with the retention of heat-sensitive compounds (vitamins and aroma) and sensorial attributes in juices by extending shelf life (Pallarés et al., 2021; Silva & Evelyn, 2020). In comparison to high-temperature short-time pasteurization, HPP processing was found to be better at maintaining phenolic compounds of red raspberry juices (Zhang et al., 2021b). Yildiz et al. (2021) also compared different non-thermal processes on strawberry juice, such as ultrasound, high pressure, and pulsed electric fields. They found that HPP was the best option for shelf life extension and phytochemical preservation.

When compared to thermal processing (95°C, 3 min), HPP-treated (500 MPa, 10 min) pineapple fruit juice retained higher amounts of bioactive components, color, flavor, volatile aroma, and antioxidant capacity, along with spoilage microorganism inactivation (Wu et al., 2021). In another study, aqueous pectin solutions were treated at 200, 400, and 600 MPa for 30 min to investigate pectin's microstructure and chemical conversion. The findings showed that HPP increased the pectin solubility in water by improving the hydrophobic hydration of the methyl group. As a result, the apparent viscosity and stability of the aqueous pectin solution increased, which plays a crucial role in determining the physical, chemical, and nutritional properties of fruit- and vegetable-based products (Zhong et al., 2021).

In carbon dioxide-added beverages, packaging may damage during the depressurization step in HPP (Silva & Evelyn, 2020). Air-containing fruits (e.g. strawberry and lettuce) may lose their original shape and become fragile following the HPP treatment (Rastogi et al., 2007).

Recently, HPP-treated (400 MPa/15 min) broccoli contained the highest amounts of isothiocyanates and total phenolic content while vitamin C and flavonoids content remained constant. Furthermore, HPP inhibits oxidase enzymes like polyphenol oxidase and peroxidase, preserving food quality. Unlike traditional thermal processing, HPP prevented the loss of heat-sensitive components while improving nutritional quality (Ke et al., 2021). Accordingly, HPP treatment (600 MPa/3 min) improved consumer acceptance for cold-pressed lemonade, citrus, and green juices (de Souza et al., 2020).

HPP is a non-thermal process that reduces microbial load, mycotoxins, and protein degradation inmicrobial cells while also inactivating enzymes, with little or no effect on the organoleptic and nutritional properties of fruits and vegetables. More studies need to be carried out to remove both hydrophilic and hydrophobic pesticides. More research is required to validate the HPP process on various food models under varying process conditions.

11.5 APPLICATION OF OZONE

The three-atomic form of oxygen, ozone, is a powerful oxidant gas approved as a GRAS (generally recognized as safe) designation by the US Food and Drug Administration in 1997 and allowed for use as a disinfectant or sanitizer in food processing (García-Martín et al., 2018). Generally, ozone sterilization can be achieved as a gaseous or aqueous form (Sung et al., 2014). Ozone is also widely used to control pathogenic bacteria and fungi in the food industry as indicated in Table 11.4. Furthermore, ozone has several advantages over conventional chemical techniques, including a cost-effective and commercially viable aqueous and gaseous state approach (Pandiselvam et al., 2020). Another advantage is that ozone rapidly decomposes to a non-toxic product, oxygen, leaving no residues in foods (Sung et al., 2014).

Ozone can kill many insects, bacteria, fungi, and stored goods insects, and it may convert non-biodegradable organic materials into biodegradable forms in the fruit and vegetable industry (Brodowska et al., 2015; Cravero et al., 2018; Sadło et al., 2017). The inactivation mechanism of ozone has been identified as oxidation of cellular components. Ozone generation efficiency has been influenced by many parameters such as intake gas oxygen content, inlet gas purity, humidity, and electrical parameters (Niveditha et al., 2021). Ozone also affects several cellular components like proteins, peptidoglycans in cell envelopes, enzymes, and nucleic acids in the cytoplasm (Sung et al., 2014). Recently, several researchers have investigated the usage of ozone to control pathogenic diseases and extend the shelf life of fruits and vegetables such as black mulberry (Tabakoglu & Karaca, 2018; Han et al., 2017), lettuce, spinach, and parsley (Karaca & Velioğlu, 2014), berry fruit (Lone et al., 2019), and potatoes. The lag phase duration of oranges stored in an ozone atmosphere (1.6 mg/kg, 12 h/d) was increased three times, and the exponential phase was also increased. Additionally, decay incidence in oranges decreased with the ozone-enriched atmosphere (60 mg/kg) storage conditions at 5°C. Valencia oranges showed lower decay incidence than Lanelate oranges (García-Martín et al., 2018). Similarly, Han et al. (2017) treated mulberry fruits with ozone (2 ppm and 30 min) and precooling. There were no decay symptoms shown on the first third day of storage. In a study on apricot exposed to 0.5, 1.0, and 1.5 mg/L for 60 min, fungal decay development was decreased to 33%, 38%, and 41%, respectively during the storage period. Ozone dose has been significantly important for the control of fungal decay development on fruits. Furthermore, the ozonation process protects apricot sensory characteristics over a 10-day storage period, whereas untreated apricots only last 7 days (Panou et al., 2018). De Santis et al. (2021) found that gaseous ozone treatment effectively controlled post-harvest decay by *Fusarium prolifera-tum* on garlic. The gaseous ozone treatment slightly affected the sensory profile of garlic (2.14 $\mu g/m^3$ during 4 days, 20 h a day).

Concerning other fruit, Brodowska et al. (2015), mesophilic bacteria count significantly decreased with 30 min of treatment. But they also reported that more than 60 min treatment could cause activation of bacteria. Similarly, in a grape tomato fruit smooth surface, *Salmonella* populations were decreased 2.21 and 2.02 log cfu/fruit by 2 and 4 hours 6 mg/L. However, gaseous ozone treatment did not affect the molds and yeast counts (Wang et al., 2019). Similarly, Alexopoulos et al. (2013) also found that yeast and molds require greater ozone concentration and longer exposure time to inactivation than bacteria.

TABLE 11.4   STUDIES CONDUCTED WITH OZONE TREATMENT AGAINST MICROORGANISMS

Products	Process condition	Hazard	Reduction (Log)	References
Apple juice	2.8 mg/L, 40 min	*Alicyclobacillus acidoterrestris*	1.8	Torlak (2014)
	5.3 mg/L, 40 min		2.4	
Bell peppers	0.5 mg/L, 15 min	Aerobic mesophiles	2.07	Alexopoulos et al. (2013)
		Coliforms	1.5	
		Yeast and mold	2.9	
Fresh produce (tomatoes, strawberries, cilantro)	10 ppm, 10 min	*Escherichia coli* O157:H7	1.2–1.5	Gibson et al. (2019)
Grape berries	5 mg/L, 24 h	*Brettanomyces bruxellensis*	2.1	Cravero et al. (2018)
Grape tomato	6.85 mg/L, 2h	*Salmonella*	2.21	Wang et al. (2019)
Juniper berries	130 g O$_3$/m^3, 30 min	Total mesophilic count	1.4	Brodowska et al. (2015)
		Total count of fungi	1	
Juniper berries	130 g O$_3$/m^3, 90 min	Total mesophilic count.	1.6	Brodowska et al. (2015)
		Total count of fungi	1.6	
Lettuce	0.5 mg/L, 15 min	Aerobic mesophiles	2.04	Alexopoulos et al. (2013)
		Coliforms	1.09	
		Yeast and mold	2.5	
Peach juice	10 ppm, 12 min	Coliform	4.3	Loredo et al. (2015)
		Listeria innocua	3.9	
Red bell peppers	0.3 ppm, 2 min	*Listeria innocua*	1.8	Alexandre et al. (2011)
	2 ppm, 2 min		2.4	
Spinach	1 ppm, 10 min	*Listeria innocua, Listeria seeligeri, Escherichia coli*	1	Wani et al. (2015)
Sugarcane juice	0.45 mg/min mL, 30 min	*Leuconostoc mesenteroides*	2.4	Panigrahi et al. (2021)
		Saccharomyces cerevisiae	1.82	
		Total coliforms	2.7	
Tomato	800 ppm, 30 min	*Salmonella*	0.6	Fan et al. (2020)

The application of aqueous ozone to four types of fresh produce (tomatoes, strawberries, cilantro, romaine head lettuce) for 30 minutes, nearly 4 log microbial reduction was achieved. Furthermore, after 30 minutes of treatment, ozone is more effective at inhibiting *Listeria innocua* than *Escherichia coli*. They also claimed that ozonated water acts as a protective barrier against cross-contamination (Gibson et al., 2019). Lettuce, tomato, and carrot disinfection with ozone treatment have been studied. After 15 minutes of treatment, a 2.2 log reduction in tomatoes was detected. Carrots and lettuce, on the other hand, only showed lower inactivation rates after 15 minutes. Also, when the ozone concentration was greater than 2.5 ppm, the overall visual quality of lettuce decreased (Bermúdez-Aguirre & Barbosa-Cánovas, 2013).

Saccharomyces cerevisiae was more susceptible to the application of ozone than *Listeria innocua* in pitaya juice, according to del Rosario García-Mateos et al. (2019). On the other hand, the combination of HPP and ozone in pitaya juice results in a >5 log cfu/mL reduction in both *L. innocua* and *Saccharomyces cerevisiae* at 10 days of storage. These combinations provide a synergistic effect over *Listeria innocua* inactivation because, in contrast, *Listeria innocua* is resistant to HPP. However, there was no need to combine ozone with HPP to control *S. cerevisiae*. Papachristodoulou et al. (2018) found that the application of ozonated water only reduces the population of Gram-negative and *Enterobacteriaceae* sp. in the first days of fresh-cut spinach storage. While the ozone dosage increased, the quality of products decreased. A similar result was revealed by Tabakoğlu and Karaca (2018). They found that the total aerobic mesophilic bacteria and total mold and yeast counts were decreased by ozone application at the beginning of storage in black mulberry fruits, but these effects disappeared over time. Panigrahi et al. (2021) discovered that ozone dose had a significant impact on sugarcane juice inactivation. At the same dosage, a longer treatment time resulted in a lower microbial population. As mentioned previously, ozone treatment efficiency is affected by several factors such as ozone dosage, exposure time, and type of products.

Ozone has also been used to promote mycotoxin degradation of food products (Inan et al., 2007; Diao et al., 2019; Zorlugenç et al., 2008), and efficiency is highlighted in Table 11.5. According to Zorlugenç et al. (2008), gaseous ozone was more effective than ozonated water in degrading aflatoxin B_1 (AFB$_1$) in dried figs. The differences between the two ozone treatments were significant (p<0.05). But ozonated water was found to be effective in reducing microbial counts. Kamber et al. (2016) stated that ozone application in red peppers reduced AFB$_1$ levels by 6.1 to 74.1%, with efficiency varying depending on the O$_3$ dose and application period. The highest degradation rate was obtained with 80 mg/L, 40 min with 25 ppb ozone treatment. Similar results have been reported by Diao et al. (2019). The degradation efficiency of patulin depended on ozonation time in apple juice. After 15 minutes of treatment, patulin was decreased to 49.54±2.60 µg/L from 201.06±0.70 µg/L. However, phenolic compounds and organic acids were significantly destroyed by the ozonation process in apple juice, so they should be optimized for conditions to protect phenols and organic acids.

Additionally, ozone efficiency to remove pesticide residues has been extensively studied in fruits and vegetable products such as bell peppers (Rodrigues et al., 2021), strawberries (Lozowicka et al., 2016), cherry tomatoes (Ikeura et al., 2013), pak choi (Wang et al., 2021), and carrots (de Souza et al., 2018). The optimization of difenoconazole and linuron degradation in carrot products with gaseous ozone and ozonated water have been studied. The efficiency of pesticide removal was influenced by ozone concentration and treatment time but not by temperature. Pesticide removal was higher than 98% for difenoconazole and 95% for linuron (de Souza et al., 2018). In another study, green peppers were treated with immersion in ozonated water (2 ppm) and tap water for 20 minutes, exposure to 2 ppm ozone gas in air and air to control. Ozonated water (2 ppm) was the most effective method for degrading pesticide residues, preserving the green color, and achieving the highest score at the end of the storage (Özen et al., 2021). Rodrigues et al. (2019) used two conventional procedures (immersion in pure water and detergent solution at 0.25 and 1%) to degrade residues of the fungicides azoxystrobin, chlorothalonil, and difenoconazole in tomatoes. Two ozone treatments (immersion in water with bubbling O$_3$ and immersion in ozonated water, both at 1 and 3 mg/L) and reduction are presented in Table 11.5.

TABLE 11.5   STUDIES CONDUCTED WITH OZONE TREATMENT AGAINST MYCOTOXIN AND PESTICIDE

Products	Process condition	Hazard	Reduction (%)	References
		Mycotoxin		
Apple juice	12 mg/L, 15 min	Patulin	75.36	Diao et al. (2019)
Dried fig	Gaseous ozone 13.8 mg/L, 30 min	Aflatoxin B$_1$	48.77	Zorlugenç et al. (2008)
	Gaseous ozone 13.8 mg/L, 60 min		72.99	
	Ozonated water, 13.8 mg/L, 30 min		0.76	
	Ozonated water, 13.8 mg/L, 60 min		83.25	
Red pepper	33 mg/L, 60 min	Aflatoxin B$_1$	80	Inan et al. (2007)
	66 mg/L, 60 min		93	
Red pepper	40–80 mg/L, 20–40 min	Aflatoxin B$_1$	6.1–74.1	Kamber et al. (2016)
		Pesticide		
Apple	Ozonated water, 2 ppm	Captan	81	Sadło et al. (2017)
		Boscalid	40	
		Pyraclostrobin	20	
	Gaseous ozone 10 ppm, 30 min	Captan	-26	
		Boscalid	42	
		Pyraclostrobin	32	
Carrots	Gaseous ozone treatment, 2.5 mg/L, 60 min	Difenoconazole	95.7	De Souza et al. (2018)
		Linuron	74.2	
Dried chillies	Gaseous ozone, 5.5 g/h, 30 min	Malathion	46	Sintuya et al. (2018)
		Chlorpyrifos	39	
		Profenofos	42	
		Ethion	28	
Green pepper	Ozonated water, 2 ppm, 10 min	Acetamiprid	70.08	Özen et al. (2021)
		Malathion	84.8	
		Emamectin benzoate	100	
Tomatoes	Ozonated water, 3 mg/L	Azoxystrobin	70	Rodrigues et al. (2019)
		Chlorothalonil	90	
		Difenoconazole	77	

On the other hand, the most effective method to remove fungicides was reported with the application of water bubbled with ozone. Additionally, ozone microbubbles have been more effective in removing residual pesticides in cherry tomatoes, strawberry, and lettuce than ozone millibubbles due to the combination of ozone's oxidative power with the generation of hydroxyl radicals (Ikeura et al., 2013). Gaseous ozone fumigation was also found to be an efficient method to degrade organophosphate pesticides in dried chilies without significant change in the water activity, color, and capsaicin content (Sintuya et al., 2018). To conclude, future studies should be focused on the ozone efficiency to degrade microorganisms, mycotoxin, and pesticide residues without any significant change with optimum conditions.

11.6 APPLICATION OF PULSED ELECTRIC FIELDS

The pulsed electrical field (PEF) process is a non-thermal technology generally applied on plant cell tissues, leading to pore formation in cell membranes with less damage to nutritional compounds (Janositz et al., 2011). The pulsed electrical field may lead to reversible or irreversible permeabilization in the plant, animal, and microbial cells with the application of very short electric pulses mainly of 1–100 µs with the electric field intensities in the range of 0.1 kV/cm and 40 kV/cm (Misra, 2015). However, for irreversible impacts, strong electric fields (5–50 kV/cm) are applied to lead to cell death (Timmermans et al., 2014). PEF technology has been extensively used to decrease the microbial population of pathogenic microorganisms and extend the shelf life of liquid foods and fresh products while preserving any significant impact on sensory and nutritional properties (Wu et al., 2021). Additionally, the PEF process has been widely used to reduce pathogenic and spoilage microorganisms; however, inactivation mechanism has not been explained briefly, but it could be described with the combination of the electroporation were defined as the formation of pores in cell membrane when electric fields are applied (Timmermans et al., 2014; Arshad et al., 2021; Misra, 2015). An external electric field increases the transmembrane potential of the cell membranes, causing structural changes and formation of pores in the membranes that may be permanent or temporary, depending on the intensity of the treatment (Moens et al., 2020; Jaeger et al., 2010). The PEF treatment's efficiency is affected by the process, such as the type and design of the equipment, pulse shape and intensity, field strength, energy, and the number of pulses. In addition to process parameters, it depends on the type of food matrix and the existence of the preservatives (Moens et al., 2020). Also, inactivation efficiency is affected from the growth stage of the microorganism and initial microbial concentration (Martín-Belloso & Elez-Martínez, 2005).

There have been numerous studies on the effects of PEF on bioactive compounds, enzymes, microbial inactivation, and shelf life extension (Bansal et al., 2015; Mosqueda-Melgar et al., 2012; Rybak et al., 2020). Nevertheless, there have only been a few studies about the formation of undesirable compounds such as furfural and hydroxymethylfurfural (HMF) (Bansal et al., 2015; Agcam et al., 2016). Timmermans et al. (2014) researched the inactivation of *Saccharomyces cerevisiae*, *Salmonella* Panama, *Escherichia coli*, and *Listeria monocytogenes* in apple, orange, and watermelon juice. All of these species were susceptible to PEF treatment, and their responses were non-linear. PEF sensitivity is *Saccharomyces cerevisiae* > *Salmonella* Panama > *Escherichia coli* > *Listeria monocytogenes*, which shows that pathogenic bacteria require more energy to inactivate than yeast. Yildiz et al. (2021) reported that thermal pasteurization (72°C, 15 s), ultrasonication (55°C, 3 min), and HPP (300 MPa, 1 min) extended the shelf life of fresh strawberry juice by at least 42 days while keeping microbial counts around 2 log cfu/mL. However, due to significant microbial growth at the 35th storage, PEF treatment (35 kV/cm, 27 μs) only extended shelf life by at least 28 days. Groot et al. (2019) observed that *Penicillium expansum* conidia are more resistant to PEF treatment than *Penicillium bialowiezense* conidia in orange juice. Researchers proposed an alternative technique for pasteurization of juice combination of PEF and mild heat treatment. In all juice technologies (high pressure, mild heat, PEF) a reduction over 5 logs in *Penicillium* species could have resulted. Additionally, Akdemir Evrendilek et al. (2008) revealed that PEF is effective in inactivating *Penicillium expansum* in sour cherry juice, peach, and apricot nectar. By increasing the electric field strength and exposure time, germination tube elongation and spore germination rate were completely inhibited.

Numerous factors affect the PEF inactivation of microorganisms. These parameters can be grouped as process parameters, microbial characteristics, and product parameters (Martín-Belloso & Elez-Martínez, 2005). Process parameters' effects in the fruits and vegetable industry have been widely studied (Huang et al., 2014a; Agcam et al., 2016; Katiyo et al., 2017; Liang et al., 2006). In red apple juice, Katiyo et al. (2017) found that increasing the electric field intensity from 25 to 30 kV promoted microbial inactivation more than increasing it from 30 to 35 kV. In all treatment conditions, the maximum cell reduction was found with 35 kV/cm for 258 μs and 50°C. As a result, the electrical field strength, treatment time, and post-treatment temperature affect cell inactivation. Similarly, Huang et al. (2014a) studied the two control parameters as treatment time and specific energy against *Staphylococcus aureus, Escherichia coli* DH5 and *Saccharomyces cerevisiae* in grape juice. In another study, Huang et al. (2014b) applied PEF treatments (9–27 kV/cm for 34–275 ms) at an initial treatment temperature of 40°C to inactivate three different strains of microorganisms (*Staphylococcus aureus, Escherichia coli* and *Saccharomyces cerevisiae*) in grape juice (Table 11.6). Due to microbiological characteristics such as microbial species and cell size, they discovered that PEF treatment had varying efficacy against different microorganisms in the same food product. However, few investigations are available about the effect of PEF treatments on mycotoxin reduction in the fruits and vegetable process. In grape juice, AFs levels were reduced 14–29% after high-pressure processing (HPP), whereas AFs levels were significantly reduced 24–84% after PEF treatment. Heat treatment temperatures should be at least 150°C for 30 minutes to achieve a higher reduction in aflatoxins (Pallarés et al., 2021). As an alternative to heat treatment, PEF process efficiency to degrade pesticide has been studied by several researchers. They reported that PEF treatment efficiency against pesticide was strongly influenced by electric field strength (Zhang et al., 2012; Chen et al., 2009), pulse number (Chen et al., 2009), and treatment time (Zhang et al., 2012). In apple juice, diazinon and dimethoate were significantly degraded after being treated with PEF. The maximum degradation of diazinon (47.6%) and dimethoate (34.7%) was observed with 20 kV/cm for 260 μs (Zhang et al., 2012). Similarly, Chen et al. (2009) found that methamidophos and chlorpyrifos residues were degraded with the PEF process in apple juice. Methamidophos was more stable to PEF treatment than chlorpyrifos. As previously stated, most studies in the fruit and vegetable industry focused on juices and beverages.

On the other hand, there is limited information on the efficiency of whole fruits' inactivation on PEF treatment. Jin et al. (2017) studied the combination of PEF and sanitizer solution (peracetic acid) on blueberries. PEF treatments were applied to natural microflora, and total bacterial counts were reduced by 2.6 log cfu/g. On the other hand, the combination was able to reduce *Escherichia coli* and *Listeria* by up to 2.9 and 3 log cfu/g, respectively, as well as native microorganisms by 2 log cfu/g.

During the Maillard reaction, contaminants such as 5-(hydroxy-methyl) furfural (HMF), potentially carcinogenic to humans, and acrylamide are produced. Among the proposed

TABLE 11.6   STUDIES CONDUCTED WITH PULSED ELECTRIC FIELD TREATMENT AGAINST MICRORGANISMS

Products	Process condition	Hazard	Reduction (Log)	References
Apple cider	27–33 kV/cm, 200 pulses/s	Total plate counts	3.1	Liang et al. (2006)
Emblica juice	26 kV/cm, 500 µs	*Zygosaccharomyces bailii*	5.1	Bansal et al. (2015)
Grape juice	27 kV/cm, 45 µs	*Saccharomyces cerevisiae*	>6	Huang et al. (2014a)
Grape juice	24 kV/cm, 180 µs	*Escherichia coli*	3.06	Huang et al. (2014b)
		Staphylococcus aureus	2.69	
		Saccharomyces cerevisiae	6.01	
Grape juice	27 kV/cm, 275 µs	*Escherichia coli*	2.27	Huang et al. (2014a)
		Staphylococcus aureus	3.36	
Orange juice	40 kV/cm, 100 µs	*Listeria innocua*	3.7	McNamee et al. (2010)
		Escherichia coli	6.3	
		Pichia fermentans	4.8	
Orange juice	17 kV/cm, 12.5 µs	*Penicillium bialowiezense*	4.3	Groot et al. (2019)
		Penicillium expansum	0.4	
Red apple juice	25–35 kV/cm, 86–258 µs	*Escherichia coli*	5.21	Katiyo et al. (2017)
		Salmonella enteritidis	6.02	
		Saccharomyces cerevisiae	5.49	
Strawberry juice	18.6 kV/cm, 150 µs	*Escherichia coli* O157:H7	3.09–4.71	Gurtler et al. (2011)
		Escherichia coli	2.86–3.79	

techniques to minimize the undesirable MR, pulsed electric field (PEF) technology is probably one of the most attractive methods for researchers due to its ability to process the food while minimizing the HMF formation (Khaneghah et al., 2020). HMF formation is used as the evaluation criteria for color and flavor, a key player in the heat application. In PEF-treated emblica juices, the formation of HMF was reduced (1.49 ± 0.1 mg/L), and it always remained below the maximum values allowed (5 mg/L) at 4°C for 40 days. However, HMF formation was significantly increased in heat-treated juices (7.95 ± 0.6 mg/L) with the same storage conditions (Bansal et al., 2015). Similar results were reported after the PEF processing of an orange juice, and there were no significant changes in the HMF content of the product (Agcam et al., 2016). According to Akdemir Evrendilek et al. (2017), the formation of furfural and HMF was affected by PEF process parameters in apple juice, such as moderate level of electric field strength and high energy level. There was no considerable formation of furfural or HMF in PEF-treated juices under these conditions, although significant formation was reported after thermal treatment. Recently, PEF treatment also has been used as a pretreatment in the potato industry. Furthermore, Genovese et al. (2019) observed that PEF treatment reduced acrylamide levels compared to conventional blanching. Acrylamide has been identified as a potential contaminant in a variety of fried and oven-cooked meals. To conclude, PEF treatment has been considered a potential alternative to remove or reduce Maillard reaction products and acrylamide content in the fruit and vegetable industry.

11.7 APPLICATION OF PULSED LIGHT

Pulsed light (PL) uses electromagnetic radiation with a very broad wavelength spectrum (100–1100 nm), also defined as high intensity of pulses (Varalakshmi, 2021). Pulsed light is a novel non-thermal decontamination technology. Many researchers have proven it to be effective against various bacteria, fungi, and viruses in the fruit and vegetable industry under appropriate conditions for the extension of shelf life (Huang et al., 2017; Bernal et al., 2019; Preetha et al., 2021). Pulsed light's inactivation process is based on the generation of reactive oxygen species (ROS) due to oxidative damage to microorganisms' lipids and proteins in cell membranes (Varalakshmi, 2021).

However, the efficiency of pulsed light sterilization diminished due to 'shadow' impact, which is produced by inadequate penetrating capacity. In addition, pulsed light has an adverse effect, including the higher cost of energy usage with a shorter process time (Cao et al., 2017). Two forms of PL treatments have been used, including dry and water-assisted, and their decontamination efficiency was changed. For example, *Salmonella* was more effectively inactivated in water-assisted PL treatment than dry PL in blueberries. However, Xu et al. (2013) observed wet PL treatment was not effective as dry PL treatment in green onions. Moreover, higher reductions were obtained with spot inoculation than dip inoculation. Also, dry PL was not effective in inhibiting yeast and molds on the subsurface of blueberries (Cao et al., 2017). Rybak et al. (2021) reported that total aerobic bacteria counts were significantly decreased with the range of 8 to 32 J/cm² PL doses in red bell peppers. While pulse light doses as low as 12 J/cm² reduced total fungi count by 1.2 log cycles, PL doses of 16 and 32 J/cm² reduced total fungi count to the limit of detection. Also, after the PL treatment, the number of *Salmonella* Typhimurium and *Listeria monocytogenes* was reduced to 0.75–2.35 and 0.81–2.59 log cycles, respectively. In another study, PL has been reported to inactivate *Salmonella* and *Escherichia coli* O157:H7 in raspberries. Moreover, fresh raspberries' safety and quality have been maintained with the fluence of 5.0 J/cm² (Xu & Wu, 2016). Additionally, microorganisms showed different susceptibility to PL treatments (Tao et al., 2019; Rybak et al., 2021). According to Tao et al. (2019), PL treatment effectively disinfects microbial flora, preserving nutritional quality, and delaying the browning of fresh-cut lettuce. In this study, the most resistant strain was *Listeria monocytogenes*, while *Staphylococcus aureus* was the most susceptible to PL treatments on fresh-cut lettuce. *Escherichia coli* and *Salmonella* Enteritidis were also moderately susceptible (Tao et al., 2019). Duarte-Molina et al. (2016) observed that the incidence of post-harvest molds decreased 16–42% with different PL doses in strawberries. Generally, PL treatments mainly deal with solid products, but nowadays, researchers investigate liquid foods for inactivation efficiency like orange juice (Preetha et al., 2021), red grape juice (Xu et al., 2019), and apple juice (Ferrario et al., 2015). Due to differences in transparency, Preetha et al. (2021) found that PL treatment was more efficient in tender coconut water than orange and pineapple juice (Table 11.7). The inactivation mechanism is based on photothermal and photochemical effects in this study. Similarly, researchers reported that the presence of pulp-reduced light transmission due to reflection and scattering, which influenced the PL treatment effectiveness (Ferrario et al., 2015).

Xu et al. (2019) reported that the inactivation of *Escherichia coli* in red grape juice with PL treatment significantly depends on the intensity and pulse number. The combination of PL with

TABLE 11.7   STUDIES CONDUCTED WITH PULSED LIGHT TREATMENT

Products	Process condition	Hazard	Reduction (Log)	References
		Microorganism		
Apple juice	Total fluence: 1.27 J/cm2; Distance from the lamp: 0. 1 m	*Saccharomyces cerevisiae*	4.4	Ferrario et al. (2015)
Blueberries	Dry PL; Peak power: 0.066 J/cm²/pulse; Treatment time: 30 s; Total fluence: 6 J/cm²	*Salmonella*	0.9	Cao et al. (2017)
Blueberries	Water assisted PL; Peak power: 0.066 J/cm²/pulse; Treatment time: 45 s; Total fluence: 9 J/cm²	*Salmonella*	4	Cao et al. (2017)
		Yeast and mold	0.9	
Grape juice	Peak power: 0.66 J/cm²/pulse; Flow rate: 30 mL/min; Pulse: 100	*Escherichia coli*	3.7	Xu et al. (2019)
Green onions	Dry PL; Total fluence: 5 J/cm²; Treatment time: 5 s; Pulse-repetition-rate (pulses/s): 3	*Escherichia coli* O157:H7	4.8 cfu/g	Xu et al. (2013)
Green onions	Wet PL; Total fluence: 5 J/cm²; Treatment time: 5 s; Pulse-repetition-rate (pulses/s): 3	*Escherichia coli* O157:H7	0.5 cfu/g	Xu et al. (2013)
Lettuce	Total fluence: 16.8 J/cm²; Treatment time: 5 s; Distance from the lamp: 9 cm; Pulse width (µs): 300	*Staphylococcus aureus*	6.56 cfu/g	Tao et al. (2019)
		Listeria monocytegenes	4 cfu/g	
		Escherichia coli O157:H7	5.08 cfu/g	
		Salmonella enteritidis	5.4 cfu/g	
Orange juice	Total fluence: 95.2 J/cm²; Pulse width (µs): 300	*Escherichia coli*	4.0	Preetha et al. (2021)
Pineapple juice			5.33	
Raspberry	Total fluence: 28.2 J/cm², Treatment time: 30 s; Distance from the lamp: 13 cm	*Salmonella*	4.5 cfu/g	Xu and Wu (2016)
		Escherichia coli O157:H7	3.9 cfu/g	
		Total yeast and mold	1.6 cfu/g	
Red bell peppers	Total fluence: 16 J/cm² Distance from the lamp: 7.58 cm; Voltage: 2.1 kV	Total aerobic bacteria	1.81	Rybak et al. (2021)
	Total fluence: 32 J/cm², distance from the lamp: 7.58 cm; Voltage: 2.1 kV		2.01	
Strawberry	Total fluence: 11.9 J/cm², 1 Hz, distance from the lamp: 0.1 cm; Treatment time: 10 s	*Botrytis cinerea*	2	Bernal et al. (2019)
		Parasite		
Cilantro	Total fluence: 0.0675 J/cm², 1 Hz, Distance from the lamp: 40 cm; Treatment time: 10 s; Pulse width (µs): 360	*Cryptosporidium parvum*	1.47	Craighead et al. (2021)
Mesclun lettuce			1.69	
Spinach			1.14	
Tomato			0.28	

other non-thermal technologies is used, including ultrasound (Ferrario et al., 2015), manothermosonication (Palgan et al., 2011) and pulsed electric fields (Caminiti et al., 2011). The most effective combination was 30 minutes of ultrasound followed by 60 seconds of PL treatment, which resulted in *Saccharomyces cerevisiae* reductions of 6.4 and 5.8 logs in commercial and natural apple juice, respectively. On the other hand, PL treatment was more effective than the combinations for inactivating *Alicyclobacillus acidoterrestris* (Ferrario et al., 2015). The microbial load affects inactivation efficiency; in a study with apple juice, a higher *Penicillium expansum* load resulted in lower inactivation, defined with shadow impact (Maftei et al., 2014). Additionally, the usage of organic acids was another strategy to ensure the food safety of fruit and vegetables. In fresh-cut mango, the combination of malic acid and the PL treatment resulted in a 4.5 log reduction in *Listeria innocua* counts. (Salinas-Roca et al., 2016). *Cryptosporidium parvum*, a coccidian protozoan parasite, was inactivated with PL treatment in green leafy vegetables in a recent study conducted by Craighead et al. (2021). Patulin was degraded in apple purée after PL treatment. Patulin levels were reduced by 51% after 10 seconds when 11.9 J/cm^2 was applied, and patulin levels were found to be below the detection limit after 20 seconds (Funes et al., 2013). Huang et al. (2017) studied the application of berry surfaces to inactivate murine norovirus, Tulane virus, *Escherichia coli* O157:H7, and *Salmonella*. Bacteria were more susceptible to the effects of PL than viruses. A PL application of 5.9 J/cm^2 resulted in a 0.7 and 3.1 log reduction on strawberry and blueberry, respectively.

11.8 APPLICATION OF SUPERCRITICAL CARBON DIOXIDE

Supercritical carbon dioxide (SCCD) has emerged as anon-thermal food processing method, also known as high pressure carbon dioxide (HPCD) or dense phase carbon dioxide (DPCD) (Ferrentino & Spilimbergo, 2017; Plazzotta & Manzocco, 2019; Tang et al., 2021). SCCD has been used to extract bioactive components, inactivate pathogenic/spoilage microorganisms, inhibit endogenous enzymes, and remove mycotoxins and pesticides. For microbial inactivation and enzymatic inhibition in plant-based juices, selected temperature and pressure ranges should be 35–55°C and 10–60 MPa, respectively, while maintaining the original taste of fresh juices (Silva et al., 2020).

SCCD leads to a decrease in the intracellular pH of cell membrane with CO_2 solubilization. Then, disruption and modification of cell membrane with the extraction of cellular components, and also inhibition of enzymes can cause microbial inactivation (Fabroni et al., 2010). SCCD has some applications for ensuring food safety; for example, the combination of high carbon dioxide and low oxygen was investigated in mycotoxins:, aflatoxin by *Aspergillus flavus*, patulin by *Byssochlamys nivea*. As a result, a significant reduction was observed in all mycotoxins at the tested conditions (20, 40, and 60% CO_2 and < 0.5% O_2) (Taniwaki et al., 2009).

The mechanism of action of SCCD for mycotoxin removal can act in three ways: firstly, direct removal of mycotoxins from food; secondly, modification of mycotoxin structure; and thirdly, inactivation of mycotoxin-producing microorganisms (Park & Kim, 2013). Carbon dioxide can extract non-polar pesticides efficiently due to its low viscosity and high diffusivity. In other words, increasing pressure and density are proportional to process efficacy and yield (Sartori et al., 2017). In addition, there are some experimental design and method optimizations in the literature (Saito-Shida et al., 2014; Sartori et al., 2017; Tolcha et al., 2020).

SCCD treatment (10–60 MPa, 45°C, 30 min) increased cloudy apple juices' physicochemical, nutritional, and phytochemical quality for 10 weeks of storage at 4°C (Marszałek et al., 2018).

TABLE 11.8   STUDIES CONDUCTED WITH SUPERCRITICAL CARBON DIOXIDE TREATMENT

Products	Process condition	Target	Reduction	References
Apples in syrup	12 MPa, 55°C, 15 min	Mesophilic microorganisms	3.4 log to > D.L.	Ferrentino and Spilimbergo (2017)
		Total coliforms	2.02 log to > D.L.[1]	
		Yeast and mold	2.10 log to > D.L.	
Apple juice	10–20 MPa, 35–55°C, 10–30 min	*Lactobacillus casei*	0–6.93 log cfu/mL	Silva et al. (2018)
Blood orange juice	23 MPa, 3.91 L/h CO_2 flow rate 5.08 L/h juice flow rate, at 36°C	Yeast and mold	8.60×10^3 cfu/mL to > 1.54 log cfu/mL	Fabroni et al. (2010)
		Orange juice spoilage microorganisms	1.09×10^4 cfu/mL to > 3.77 log cfu/mL	
Elderberry juice	8–18 mPa, 35–45°C, 15–150 min	Mesophilic bacteria	5.23 log cfu/mL	Torabian et al. (2018)
		Yeast and mold	5.38 log cfu/mL	
Mango in syrup	20 MPa, 60°C, 30 min	*Staphylococcus aureus*	8.22 log to > D.L.	Tang et al. (2021)
		Escherichia coli	3.12 log to > D.L.	

[1]Detection limit (< 1.00 log cfu/mL or < 1.00 log cfu/g)

Browning can be prevented using SCCD (Kong et al., 2021; Zhang et al., 2021c). SCCD can be an alternative to high pressure processing on an industrial scale (Marszałek et al., 2017). Following the SCCD treatment, the shelf life of fruit juices can be extended to 20 days (Silva et al., 2020). Studies regarding the microbial inactivation by SCCD treatment were summarized in Table 11.8. SCCD is a promising non-thermal technique in fruit- and vegetable-based products which needs more investigation for process optimization in different types of juice.

11.9 APPLICATION OF ULTRASOUND

Ultrasound (US), which is among the non-thermal methods, has been applied to food products for various purposes combined with a single or another method. In addition to studies carried out with the idea of replacing conventional processes, the effectiveness of the US to ensure food safety has become an issue for scientists (Firouz, 2021). US, including three major main elements (generator, transducer, and emitter), may mainly be applied at two different frequency ranges. While higher than 10 MHz is known as the high-frequency US, the range of 20 kHz–10 MHz is called low-frequency US, commonly used in food processing (Radhakrishnan & Tiwari, 2021). Many studies cover US applications to fruits and vegetables, especially in the form of juice (Charoux et al., 2017; Khandpur & Gogate, 2015; Khandpur & Gogate, 2016). For the application of the US, various factors classified under two main classes: intrinsic and extrinsic (Figure 11.2), should be considered (Rojas et al., 2017).

Alongside microbial decontamination, the usage of US has many purposes, including enzyme inactivation, several physicochemical changes, and being an alternative to conventional processes (extraction, drying, homogenization etc.). Each purpose is achieved by cavitation resulting in chemical, physical, and thermal impacts, which change the outcome of the process (Radhakrishnan & Tiwari, 2021). While the US affects microorganisms with different

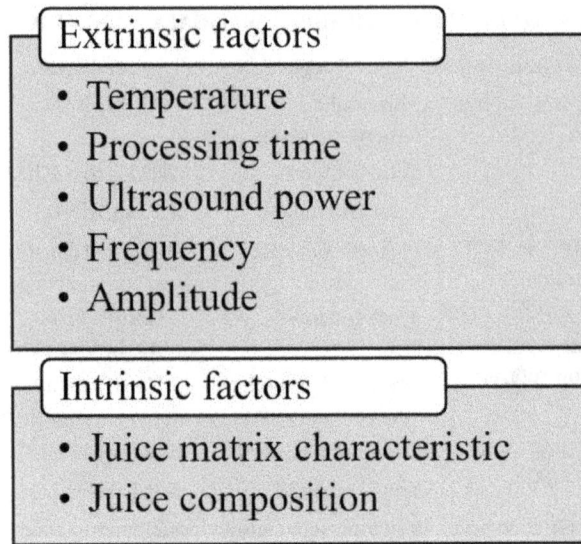

Figure 11.2 The effective factors of ultrasound treatment for juices.

mechanisms, it may ensure that a microbiologically safe food product reaches the consumer. These mentioned mechanisms include physical effects such as cavitation and a chemical effect with the formation of free radicals. As can be seen, due to the different effects of the US, it should be viewed from a food safety perspective.

Several studies have been performed to understand the effect of the US in terms of microbiology. For instance, US was conducted to control the growth of several microorganisms in fruits and vegetables (Table 11.9). For the effectiveness of US in food preservation, fruit and vegetable juices such as Chinese bayberry (*Myrica rubra*) juice (Li et al., 2019), mango juice (Wang et al., 2020), and apple juice (Tremarin et al., 2017) are the main focus for food products, the direct treatment of US to fruit and vegetables such as cherry tomatoes (He et al., 2021) and lettuce (Takundwa et al., 2021) have become the subject of studies. There is a wide range of microbial risks for fruits and vegetables and their processed products. In addition to yeast and mold, bacteria such as *Alicyclobacillus acidoterrestris* must be inactivated to provide microbiological safety. In this context, although different parameters have to be taken into consideration, it has been observed that the US may eliminate these risks in optimized conditions (Jambrak et al., 2018). Ding et al. (2015) conducted studies with both US and slightly acidic electrolyzed water to treat cherry tomatoes and strawberries. Though there was a slight reduction in microbial load, the treatment with both methods showed a remarkable reduction. The cactus pear juices were also treated by US (amplitude level: 60–90%, duration: 1–5 min) to observe inactivation of *Escherichia coli*, and duration of 5 min at 60–80% amplitude was found effective for total inactivation (Cruz-Cansino et al., 2016).

It was shown that the effect of the US was as effective as thermal processes. In the study of Khandpur and Gogate (2016), US has resulted in almost the same microbial reduction in fruit (orange and sweet lime) and vegetable (carrot and spinach) juices when compared with thermal treatments. As a result, the products subjected to the US preserved their microbiologically safe conditions during 10 weeks of storage. Although there were attempts to eliminate the

TABLE 11.9 STUDIES CONDUCTED WITH ULTRASOUND TREATMENT

Products	Process condition	Hazard	Reduction	References
		Microorganism		
Apple juice (100%) and nectar (50%)	20 kHz, 600 W, 3–9 min, 20–60°C, 60–120 µm amplitude	*Alicyclobacillus acidoterrestris*	0–0.067 log cfu/mL	Jambrak et al. (2018)
		Aspergillus ochraceus, Penicillium expansum, Rhodotorula sp., *Saccharomyces cerevisiae*	0–5.934 log cfu/mL	
Blueberry juice (100%) and nectar (40%)	20 kHz, 600 W, 3–9 min, 20–60°C, 60-120 µm amplitude	*Alicyclobacillus acidoterrestris*	0–0.155 log cfu/mL	Jambrak et al. (2018)
		Aspergillus ochraceus, Penicillium expansum, Rhodotorula sp., *Saccharomyces cerevisiae*	0–5.765 log cfu/mL	
Carrot	40 kHz, 30 W/L, 5 min (with Tween 20 0.1%)	*Bacillus cereus* spores	2.22 log cfu/g	Sagong et al. (2013)
Cherry tomatoes	20 kHz, 167–500 W/L, 3–9 min (with thyme essential oil)	*Escherichia coli* O157:H7	4.49–6.72 log cfu/g	He et al. (2021)
Cherry tomatoes	45 kHz, 10–30 min, 20°C	*Salmonella* Typhimurium	0.83–1.73 log cfu/g	São José and Vanetti (2012)
	45 kHz, 10 min, 20°C (with peracetic acid 40 mg/L)		3.88 log cfu/g	
Cherry tomatoes	40 kHz, 240 W, 10 min	Total aerobic bacteria	0.71 log cfu/g	Ding et al. (2015)
		Yeast and mold	0.62 log cfu/g	
Chinese bayberry (*Myrica rubra*) juice	20 kHz, 200–400 W, 9.6-36.7 min, 55–63°C	*Bacillus subtilis*	5.00 log cfu/mL	Li et al. (2019)

(*Continued*)

TABLE 11.9 (CONTINUED) STUDIES CONDUCTED WITH ULTRASOUND TREATMENT

Products	Process condition	Hazard	Reduction	References
Cranberry juice (100%) and nectar (30%)	20 kHz, 600 W, 3–9 min, 20–60°C, 60–120 μm amplitude	Alicyclobacillus acidoterrestris	0–0.104 log cfu/mL	Jambrak et al. (2018)
		Aspergillus ochraceus, Penicillium expansum, Rhodotorula sp., Saccharomyces cerevisiae	0–5.869 log cfu/mL	
Lettuce	50 kHz, 600 W, 14.65 min (with 771.2 IU/g nisin + 0.185% v/v oregano)	Escherichia coli O157:H7	3.43 log cfu/mL	Takundwa et al. (2021)
		Listeria monocytogenes	9.20 log cfu/mL	
Lettuce	37 kHz, 30 W/L	Escherichia coli, Listeria innocua, Salmonella Enteritidis, Staphylococcus aureus	1.71–5.72 log cfu/g	Birmpa et al. (2013)
Lettuce	40 kHz, 30 W/L, 5 min (with lactic acid 0.3–2.0%)	Escherichia coli O157:H7, Listeria monocytogenes, Salmonella Typhimurium	1.23–2.75 log cfu/g	Sagong et al. (2011)
	40 kHz, 30 W/L, 5 min (with citric acid 0.3–2.0%)		1.25–3.18 log cfu/g	
Lettuce	40 kHz, 30 W/L, 5 min (with Tween 20 0.1%)	Bacillus cereus spores	2.49 log cfu/g	Sagong et al. (2013)
Parsley	45 kHz, 20 min, 25°C (with peracetic acid)	Aerobic mesophiles	5.10 log cfu/g	de São José and Vanetti (2015)
Purple cabbage	40 kHz, 500 W (with benzalkonium chloride or sodium dichloroisocyanurate)	Salmonella Typhimurium	4.00 log cfu/g (approximately)	Duarte et al. (2018)
Strawberry	37 kHz, 30 W/L	Escherichia coli, Listeria innocua, Salmonella Enteritidis, Staphylococcus aureus	2.41–6.12 log cfu/g	Birmpa et al. (2013)

(Continued)

TABLE 11.9 (CONTINUED) STUDIES CONDUCTED WITH ULTRASOUND TREATMENT

Products	Process condition	Hazard	Reduction	References
Strawberry	40 kHz, 240 W, 10 min	Total aerobic bacteria	0.52 log cfu/g	Ding et al. (2015)
		Yeast and mold	0.30 log cfu/g	
Strawberry	45 kHz, 20 min, 25°C (with 50 mg/L sodium dichloroisocyanurate)	Yeast and mold	1.00 log cfu/g	de São José and Vanetti (2015)
		Aerobic mesophiles	1.20 log cfu/g	
	45 kHz, 20 min, 25°C (with hydrogen peroxide)	Aerobic mesophiles	3.40 log cfu/g	
Watercress	45 kHz, 20 min, 25°C (with chlorine dioxide)	Aerobic mesophiles	1.70 log cfu/g	de São José and Vanetti (2015)
		Pesticide		
Lettuce	24 kHz, 10 min, 25°C (with low-intensity electrical currents, 1400 mA)	Captan	92.57%	Cengiz et al. (2021)
		Thiamethoxam	81.99%	
		Metalaxyl	93.09%	
Lettuce	Triple-frequency in sequential mode (20, 40, and 60 kHz)	Abamectin b1	92.31%	Azam et al. (2021)
		Alphamethrin	89.36%	
		Emamectin benzoate	95.25%	
Strawberry	40 kHz, 1–5 min	16 different pesticide residues	Up to 91.2%	Lozowicka et al. (2016)
Tomato	US (with low-intensity electrical currents)	Captan	94.4%	Cengiz et al. (2018)
		Thiamethoxam	69.8%	
		Metalaxyl	95.1%	

disadvantages of heat in food quality with non-thermal processes, studies were also carried out by combining non-thermal processes with mild thermal processes as an intermediate transition. It was shown that the combined US with mild heat provided microbial safety in various juices such as Chinese bayberry (*Myrica rubra*) juice (Li et al., 2019), mango juice (Kiang et al., 2013), and orange juice (Walkling-Ribeiro et al., 2009).

In addition to single usage of the US, it may be combined with other methods to deal with consumers' concerns about green technology. However, their combinations should be optimized for better reductions. The combined effect of nisin, oregano, and US treatment was examined against *Escherichia coli* O157:H7 and *Listeria monocytogenes* on lettuce (Takundwa et al., 2021). In addition to an effective reduction of microorganisms, the studied combination was suggested as an alternative to conventional systems by means of a slight change in physical properties. Besides, the synergistic effect of essential oil with US treatment for lettuce was shown by Millan-Sango et al. (2016). While the treatment method was essential to compare reductions in microbial levels, the type of essential oil was another factor causing variations. Among tested essential oils (oregano and thyme), the highest reduction was observed at 0.022% concentration of oregano essential oil (3.73 ± 0.61 CFU/cm^2) which was higher than single US treatment (2.23 ± 0.29 cfu/cm^2). In addition to the decrease in the microbial population, the observation of a significant total inactivation and the increase of the applied essential oil emulsion's stability were also mentioned in He et al. (2021) *Escherichia coli* O157:H7 and cherry tomatoes as the advantage of US. As a positive aspect of the US, it had no serious impact on color and firmness, which are important quality parameters. In particular, properties such as color and shelf life may change in relation to the enzymes whose activity is affected by US application.

For this reason, the effects of US application on this issue should be examined in order to keep these quality criteria of the product within acceptable safe limits. For instance, the success of the US in inactivating enzymes responsible for enzymatic browning in bayberry juice was shown by Cao et al. (2018). Like Yildiz et al. (2021), who studied the effect of the US on strawberry juice, prolonged shelf life was observed. Moreover, the US may be assisted by other non-thermal processes. With the combination of ultraviolet light (UV), US-UV successfully sterilized mango juice, and there was no remarkable change in bioactive compounds. On the other hand, it was stated that degradation of some compounds depending on the processing conditions might be seen, which should be optimized by processing conditions to provide safety from both physicochemical and microbial aspects (Wang et al., 2020). Another example of the process combined with the US was seen in the study of Pérez-Grijalva et al. (2018), and the US was combined with the microwave, one of the thermal processes, unlike the previous study. Microwave treatment helped for juice extraction by increasing the content of bioactive compounds, US-treated blackberry juice reduced mold, yeast, and aerobic mesophiles. However, the efficiency of treatment may change according to target microbial risk. For instance, *Alicyclobacillus acidoterrestris* spores showed high resistance against US treatment (35 kHz, 120–480 W, maximum 30 min.) (Tremarin et al., 2017).

Green technologies are being studied to be an alternative to the sanitizers used today. Even if the usage of sanitizers such as sodium hypochlorite, sodium dichloroisocyanurate, and benzalkonium chloride, which are frequently used in the food industry, may not be completely removed, it has been shown by the study by Duarte et al. (2018) that efficiency may be increased by US. Alongside the studied microbial effect, it is observed that US also provides advantages when used either for replacement of any conventional processing step or as an assistant to process. For instance, ultrasound-assisted dehydration of carrot and strawberry showed better quality as opposed to non-assisted ones (Villamiel et al., 2015).

It is known that apart from microorganisms, some contaminants cause contamination in food and threaten food safety. Although there are many studies about the US effect on contaminants, few are about removing pesticides (Azam et al., 2021; Lozowicka et al., 2016) as indicated in Table 11.9. Cengiz et al. (2018) studied US to reduce pesticide residues on tomatoes, and they observed a reduction in the content of captan, thiamethoxam, and metalaxyl, which are commonly found pesticides, after application of US combined with low-density electrical current. In another study with the same treatment combination, the pesticide residues on lettuce were noticeably reduced (Cengiz et al., 2021).

In the light of these studies, it is understood that the parameters of the US should be optimized at a level to ensure food safety in terms of microbial and food quality. In addition, the effects of technology should be observed by ensuring the diversity of vegetables and fruits used in research, and its usability should be increased.

11.10 CONCLUSION

The rate of foodborne outbreaks originating from fruits and vegetables is high. At the same time, alternative methods have become the focus of researchers due to consumers' preference toward natural products and an increasing awareness of the chemicals used to ensure food safety. Fruits and vegetables carry the risk of pathogens and threaten human health in terms of other contaminants such as pesticides and toxins. Therefore, nowadays, non-thermal processes that may eliminate the negative impacts of thermal treatment in fruit and vegetables are emphasized. While ensuring food safety by removing various contaminants, food consumption should also be safe in terms of parameters such as food quality. As a result, it has been observed that the effects of non-thermal treatments (cold plasma, irradiation, high hydrostatic pressure, ozone, pulsed electric field, pulsed light, supercritical carbon dioxide, and ultrasound etc.) vary with changing parameters and application conditions. For this reason, further studies should be carried out with varied parameters for different fresh fruit-vegetable and processed product groups.

REFERENCES

Aaliya, B., Valiyapeediyekkal Sunooj, K., Navaf, M., Parambil Akhila, P., Sudheesh, C., Ahmed Mir, S., Sabu, S., Sasidharan, A., Theingi Hlaing, M., & George, J. (2021). Recent trends in bacterial decontamination of food products by hurdle technology: A synergistic approach using thermal and non-thermal processing techniques. *Food Research International, 147*, 110514.

Aday, S., Pala, Ç. U., Çam, B. A., & Bulut, S. (2018). Storage quality and microbiological safety of high pressure pasteurized liquorice root sherbet. *LWT, 90*, 613–619.

Agcam, E., Akyildiz, A., & Evrendilek, G. A. (2016). A comparative assessment of long-term storage stability and quality attributes of orange juice in response to pulsed electric fields and heat treatments. *Food and Bioproducts Processing, 99*, 90–98.

Akdemir Evrendilek, G., Celik, P., Agcam, E., & Akyildiz, A. (2017). Assessing impacts of pulsed electric fields on quality attributes and furfural and hydroxymethylfurfural formations in apple juice. *Journal of Food Process Engineering, 40*(5), e12524.

Akdemir Evrendilek, G. A., Tok, F. M., Soylu, E. M., & Soylu, S. (2008). Inactivation of *Penicillum expansum* in sour cherry juice, peach and apricot nectars by pulsed electric fields. *Food Microbiology, 25*(5), 662–667.

Alexandre, E. M., Santos-Pedro, D. M., Brandão, T. R., & Silva, C. L. (2011). Influence of aqueous ozone, blanching and combined treatments on microbial load of red bell peppers, strawberries and watercress. *Journal of Food Engineering, 105*(2), 277–282.

Alexopoulos, A., Plessas, S., Ceciu, S., Lazar, V., Mantzourani, I., Voidarou, C., Stavropoulou, E., & Bezirtzoglou, E. (2013). Evaluation of ozone efficacy on the reduction of microbial population of fresh cut lettuce (*Lactuca sativa*) and green bell pepper (*Capsicum annuum*). *Food Control*, *30*(2), 491–496.

Alighourchi, H., Barzegar, M., & Abbasi, S. (2008). Effect of gamma irradiation on the stability of anthocyanins and shelf-life of various pomegranate juices. *Food Chemistry*, *110*(4), 1036–1040.

Almeida, F. D. L., Cavalcante, R. S., Cullen, P. J., Frias, J. M., Bourke, P., Fernandes, F. A., & Rodrigues, S. (2015). Effects of atmospheric cold plasma and ozone on prebiotic orange juice. *Innovative Food Science and Emerging Technologies*, *32*, 127–135.

Arjeh, E., Barzegar, M., & Sahari, M. A. (2015). Effects of gamma irradiation on physicochemical properties, antioxidant and microbial activities of sour cherry juice. *Radiation Physics and Chemistry*, *114*, 18–24.

Arshad, R. N., Abdul-Malek, Z., Roobab, U., Munir, M. A., Naderipour, A., Qureshi, M. I., Bekhit, A. E. D., Liu, Z. W., & Aadil, R. M. (2021). Pulsed electric field: A potential alternative towards a sustainable food processing. *Trends in Food Science & Technology*, *111*, 43–54.

Augusto, P. E., Tribst, A. A., & Cristianini, M. (2018). High hydrostatic pressure and high-pressure homogenization processing of fruit juices. In G. Rajauria & B. K. Tiwari (Eds), *Fruit juices* (pp. 393–421). Academic Press.

Avsaroglu, M. D., Bozoglu, F., Alpas, H., Largeteau, A., & Demazeau, G. (2015). Use of pulsed-high hydrostatic pressure treatment to decrease patulin in apple juice. *High Pressure Research*, *35*(2), 214–222.

Azam, S. R., Ma, H., Xu, B., Devi, S., Stanley, S. L., Siddique, M. A. B., Mujumdar, A. S., & Zhu, J. (2021). Multi-frequency multi-mode ultrasound treatment for removing pesticides from lettuce (*Lactuca sativa* L.) and effects on product quality. *LWT*, *143*, 111147.

Balasubramaniam, V. M., Martinez-Monteagudo, S. I., & Gupta, R. (2015). Principles and application of high pressure–based technologies in the food industry. *Annual Review of Food Science and Technology*, *6*, 435–462.

Bang, I. H., Lee, E. S., Lee, H. S., & Min, S. C. (2020). Microbial decontamination system combining antimicrobial solution washing and atmospheric dielectric barrier discharge cold plasma treatment for preservation of mandarins. *Postharvest Biology and Technology*, *162*, 111102.

Bansal, V., Sharma, A., Ghanshyam, C., Singla, M. L., & Kim, K. H. (2015). Influence of pulsed electric field and heat treatment on *Emblica officinalis juice* inoculated with *Zygosaccharomyces bailii*. *Food and Bioproducts Processing*, *95*, 146–154.

Bassey, E. J., Cheng, J. H., & Sun, D. W. (2021). Novel nonthermal and thermal pretreatments for enhancing drying performance and improving quality of fruits and vegetables. *Trends in Food Science and Technology*, *112*, 137–148.

Bermudez-Aguirre, D. (Ed.). (2019). *Advances in cold plasma applications for food safety and preservation*. Academic Press.

Bermudez-Aguirre, D. (2020). Disinfection of high-moisture food using cold plasma. In D. Bermudez-Aguirre (Ed.), *Advances in cold plasma applications for food safety and preservation* (pp. 147–183). Academic Press.

Bermúdez-Aguirre, D., & Barbosa-Cánovas, G. V. (2013). Disinfection of selected vegetables under nonthermal treatments: Chlorine, acid citric, ultraviolet light and ozone. *Food Control*, *29*(1), 82–90.

Bernal, A. R. R., Contigiani, E. V., González, H. H., Alzamora, S. M., Gómez, P. L., & Raffellini, S. (2019). *Botrytis cinerea* response to pulsed light: Cultivability, physiological state, ultrastructure and growth ability on strawberry fruit. *International Journal of Food Microbiology*, *309*, 108311.

Bidawid, S., Farber, J. M., & Sattar, S. A. (2000). Inactivation of hepatitis A virus (HAV) in fruits and vegetables by gamma irradiation. *International Journal of Food Microbiology*, *57*(1–2), 91–97.

Birmpa, A., Sfika, V., & Vantarakis, A. (2013). Ultraviolet light and ultrasound as non-thermal treatments for the inactivation of microorganisms in fresh ready-to-eat foods. *International Journal of Food Microbiology*, *167*(1), 96–102.

Brodowska, A. J., Śmigielski, K., Nowak, A., Czyżowska, A., & Otlewska, A. (2015). The impact of ozone treatment in dynamic bed parameters on changes in biologically active substances of juniper berries. *PloS One*, *10*(12), e0144855.

Buerman, E. C., Worobo, R. W., & Padilla-Zakour, O. I. (2020). High pressure processing of spoilage fungi as affected by water activity in a diluted apple juice concentrate. *Food Control*, *107*, 106779.

Bulut, S., & Karatzas, K. A. (2021). Inactivation of *Escherichia coli* K12 in phosphate buffer saline and orange juice by high hydrostatic pressure processing combined with freezing. *LWT, 136*, 110313.

Caminiti, I. M., Palgan, I., Noci, F., Muñoz, A., Whyte, P., Cronin, D. A., Morgan, J. D. & Lyng, J. G. (2011). The effect of pulsed electric fields (PEF) in combination with high intensity light pulses (HILP) on *Escherichia coli* inactivation and quality attributes in apple juice. *Innovative Food Science and Emerging Technologies, 12*(2), 118–123.

Cao, X., Cai, C., Wang, Y., & Zheng, X. (2018). The inactivation kinetics of polyphenol oxidase and peroxidase in bayberry juice during thermal and ultrasound treatments. *Innovative Food Science and Emerging Technologies, 45*, 169–178.

Cao, X., Huang, R., & Chen, H. (2017). Evaluation of pulsed light treatments on inactivation of *Salmonella* on blueberries and its impact on shelf-life and quality attributes. *International Journal of Food Microbiology, 260*, 17–26.

Cengiz, M. F., Basançelebi, O., Başlar, M., & Certel, M. (2021). A novel technique for the reduction of pesticide residues by a combination of low-intensity electrical current and ultrasound applications: A study on lettuce samples. *Food Chemistry, 354*, 129360.

Cengiz, M. F., Başlar, M., Basançelebi, O., & Kılıçlı, M. (2018). Reduction of pesticide residues from tomatoes by low intensity electrical current and ultrasound applications. *Food Chemistry, 267*, 60–66.

Chang, Y. H., Wu, S. J., Chen, B. Y., Huang, H. W., & Wang, C. Y. (2017). Effect of high-pressure processing and thermal pasteurization on overall quality parameters of white grape juice. *Journal of the Science of Food and Agriculture, 97*(10), 3166–3172.

Charoux, C. M., O'Donnell, C. P., & Tiwari, B. K. (2017). Ultrasound processing and food quality. In D. Bermudez-Aguirre (Ed.), *Ultrasound: Advances for food processing and preservation* (pp. 215–235). Academic Press.

Chaudry, M. A., Bibi, N., Khan, M., Khan, M., Badshah, A., & Qureshi, M. J. (2004). Irradiation treatment of minimally processed carrots for ensuring microbiological safety. *Radiation Physics and Chemistry, 71*(1–2), 171–175.

Chen, F., Zeng, L., Zhang, Y., Liao, X., Ge, Y., Hu, X., & Jiang, L. (2009). Degradation behaviour of methamidophos and chlorpyrifos in apple juice treated with pulsed electric fields. *Food Chemistry, 112*(4), 956–961.

Craighead, S., Huang, R., Chen, H., & Kniel, K. E. (2021). The use of pulsed light to inactivate *Cryptosporidium parvum* oocysts on high-risk commodities (cilantro, mesclun lettuce, spinach, and tomatoes). *Food Control, 126*, 107965.

Cravero, F., Englezos, V., Rantsiou, K., Torchio, F., Giacosa, S., Segade, S. R., Gerbi, V., Rolle, L., & Cocolin, L. (2018). Control of *Brettanomyces bruxellensis* on wine grapes by post-harvest treatments with electrolyzed water, ozonated water and gaseous ozone. *Innovative Food Science and Emerging Technologies, 47*, 309–316.

Cruz-Cansino, N. D. S., Reyes-Hernández, I., Delgado-Olivares, L., Jaramillo-Bustos, D. P., Ariza-Ortega, J. A., & Ramírez-Moreno, E. (2016). Effect of ultrasound on survival and growth of *Escherichia coli* in cactus pear juice during storage. *Brazilian Journal of Microbiology, 47*(2), 431–437.

Dasan, B. G., & Boyaci, I. H. (2018). Effect of cold atmospheric plasma on inactivation of *Escherichia coli* and physicochemical properties of apple, orange, tomato juices, and sour cherry nectar. *Food and Bioprocess Technology, 11*(2), 334–343.

de Jesus Filho, M., Scolforo, C. Z., Saraiva, S. H., Pinheiro, C. J. G., Silva, P. I., & Della Lucia, S. M. (2018). Physicochemical, microbiological and sensory acceptance alterations of strawberries caused by gamma radiation and storage time. *Scientia Horticulturae, 238*, 187–194.

De Santis, D., Garzoli, S., & Vettraino, A. M. (2021). Effect of gaseous ozone treatment on the aroma and clove rot by *Fusarium proliferatum* during garlic postharvest storage. *Heliyon, 7*(4), e06634.

de São José, J. F. B., & Vanetti, M. C. D. (2015). Application of ultrasound and chemical sanitizers to watercress, parsley and strawberry: Microbiological and physicochemical quality. *LWT - Food Science and Technology, 63*(2), 946–952.

de Souza, L. P., Faroni, L. R. D. A., Heleno, F. F., Pinto, F. G., de Queiroz, M. E. L. R., & Prates, L. H. F. (2018). Ozone treatment for pesticide removal from carrots: Optimization by response surface methodology. *Food Chemistry, 243*, 435–441.

de Souza, V. R., Popović, V., Bissonnette, S., Ros, I., Mats, L., Duizer, L., Warriner, K., & Koutchma, T. (2020). Quality changes in cold pressed juices after processing by high hydrostatic pressure, ultraviolet-c light and thermal treatment at commercial regimes. *Innovative Food Science and Emerging Technologies, 64*, 102398.

del Rosario García-Mateos, M., Quiroz-González, B., Corrales-García, J., Ybarra-Moncada, M. C., & Leyva-Ruelas, G. (2019). Ozone-high hydrostatic pressure synergy for the stabilization of refrigerated pitaya (*Stenocereus pruinosus*) juice. *Innovative Food Science and Emerging Technologies, 56*, 102187.

Denoya, G. I., Colletti, A. C., Vaudagna, S. R., & Polenta, G. A. (2021). Application of non-thermal technologies as a stress factor to increase the content of health-promoting compounds of minimally processed fruits and vegetables. *Current Opinion in Food Science, 42*, 224–236.

Diao, E., Wang, J., Li, X., Wang, X., Song, H., & Gao, D. (2019). Effects of ozone processing on patulin, phenolic compounds and organic acids in apple juice. *Journal of Food Science and Technology, 56*(2), 957–965.

Ding, T., Ge, Z., Shi, J., Xu, Y. T., Jones, C. L., & Liu, D. H. (2015). Impact of slightly acidic electrolyzed water (SAEW) and ultrasound on microbial loads and quality of fresh fruits. *LWT - Food Science and Technology, 60*(2), 1195–1199.

Duarte, A. L. A., do Rosário, D. K. A., Oliveira, S. B. S., de Souza, H. L. S., de Carvalho, R. V., Carneiro, J. C. S., Silva, P. I., & Bernardes, P. C. (2018). Ultrasound improves antimicrobial effect of sodium dichloroisocyanurate to reduce *Salmonella* typhimurium on purple cabbage. *International Journal of Food Microbiology, 269*, 12–18.

Duarte-Molina, F., Gómez, P. L., Castro, M. A., & Alzamora, S. M. (2016). Storage quality of strawberry fruit treated by pulsed light: Fungal decay, water loss and mechanical properties. *Innovative Food Science and Emerging Technologies, 34*, 267–274.

Elias, M. I., Madureira, J., Santos, P. M. P., Carolino, M. M., Margaça, F. M. A., & Verde, S. C. (2020). Preservation treatment of fresh raspberries by e-beam irradiation. *Innovative Food Science and Emerging Technologies, 66*, 102487.

Espinosa, A. C., Jesudhasan, P., Arredondo, R., Cepeda, M., Mazari-Hiriart, M., Mena, K. D., & Pillai, S. D. (2012). Quantifying the reduction in potential health risks by determining the sensitivity of poliovirus type 1 chat strain and rotavirus SA-11 to electron beam irradiation of iceberg lettuce and spinach. *Applied and Environmental Microbiology, 78*(4), 988–993.

Evelyn, Kim, H. J., & Silva, F. V. M. (2016). Modeling the inactivation of *Neosartorya fischeri* ascospores in apple juice by high pressure, power ultrasound and thermal processing. *Food Control, 59*, 530–537.

Evrendilek, G. A., Tok, F. M., Soylu, E. M., & Soylu, S. (2008). Inactivation of *Penicillum expansum* in sour cherry juice, peach and apricot nectars by pulsed electric fields. *Food Microbiology, 25*(5), 662–667.

Fabroni, S., Amenta, M., Timpanaro, N., & Rapisarda, P. (2010). Supercritical carbon dioxide-treated blood orange juice as a new product in the fresh fruit juice market. *Innovative Food Science and Emerging Technologies, 11*(3), 477–484.

Fan, X., Sokorai, K. J., & Gurtler, J. B. (2020). Advanced oxidation process for the inactivation of *Salmonella typhimurium* on tomatoes by combination of gaseous ozone and aerosolized hydrogen peroxide. *International Journal of Food Microbiology, 312*, 108387.

Fernandez, M. V., Denoya, G. I., Agüero, M. V., Jagus, R. J., & Vaudagna, S. R. (2018). Optimization of high pressure processing parameters to preserve quality attributes of a mixed fruit and vegetable smoothie. *Innovative Food Science and Emerging Technologies, 47*, 170–179.

Fernández, M. V., Denoya, G. I., Agüero, M. V., Vaudagna, S. R., & Jagus, R. J. (2020). Quality preservation and safety ensurement of a vegetable smoothie by high-pressure processing. *Journal of Food Processing and Preservation, 44*(2), 1–9.

Fernandez, M. V., Denoya, G. I., Jagus, R. J., Vaudagna, S. R., & Agüero, M. V. (2019). Microbiological, antioxidant and physicochemical stability of a fruit and vegetable smoothie treated by high pressure processing and stored at room temperature. *LWT, 105*, 206–210.

Ferrario, M., Alzamora, S. M., & Guerrero, S. (2015). Study of the inactivation of spoilage microorganisms in apple juice by pulsed light and ultrasound. *Food Microbiology, 46*, 635–642.

Ferrentino, G., & Spilimbergo, S. (2017). Non-thermal pasteurization of apples in syrup with dense phase carbon dioxide. *Journal of Food Engineering, 207*, 18–23.

Firouz, M. S. (2021). Application of high-intensity ultrasound in food processing for improvement of food quality. In F. J. Barba, G. Cravotto & P. E. Sichetti Munekata (Eds), *Design and optimization of innovative food processing techniques assisted by ultrasound* (pp. 143–167). Academic Press.

Funes, G. J., Gómez, P. L., Resnik, S. L., & Alzamora, S. M. (2013). Application of pulsed light to patulin reduction in McIlvaine buffer and apple products. *Food Control, 30*(2), 405–410.

García-Martín, J. F., Olmo, M., & García, J. M. (2018). Effect of ozone treatment on postharvest disease and quality of different citrus varieties at laboratory and at industrial facility. *Postharvest Biology and Technology, 137*, 77–85.

Gavahian, M., Pallares, N., Al Khawli, F., Ferrer, E., & Barba, F. J. (2020). Recent advances in the application of innovative food processing technologies for mycotoxins and pesticide reduction in foods. *Trends in Food Science and Technology, 106*, 209–218.

Genovese, J., Tappi, S., Luo, W., Tylewicz, U., Marzocchi, S., Marziali, S., Romani, S., Ragnş, L. & Rocculi, P. (2019). Important factors to consider for acrylamide mitigation in potato crisps using pulsed electric fields. *Innovative Food Science and Emerging Technologies, 55*, 18–26.

Gibson, K. E., Almeida, G., Jones, S. L., Wright, K., & Lee, J. A. (2019). Inactivation of bacteria on fresh produce by batch wash ozone sanitation. *Food Control, 106*, 106747.

Gomes, C., Moreira, R. G., & Castell-Perez, E. (2011). Radiosensitization of *Salmonella* spp. and *Listeria* spp. in ready-to-eat baby spinach leaves. *Journal of Food Science, 76*(1), E141–E148.

Grasso, E. M., Uribe-Rendon, R. M., & Lee, K. (2011). Inactivation of *Escherichia coli* inoculated onto fresh-cut chopped cabbage using electron-beam processing. *Journal of Food Protection, 74*(1), 115–118.

Groot, M. N., Abee, T., & van Bokhorst-van de Veen, H. (2019). Inactivation of conidia from three *Penicillium* spp. isolated from fruit juices by conventional and alternative mild preservation technologies and disinfection treatments. *Food Microbiology, 81*, 108–114.

Guimarães, I. C., Menezes, E. G. T., Abreu, P. S. D., Rodrigues, A. C., Borges, P. R. S., Batista, L. R., Cirilo, M. A., & Lima, L. C. D. O. (2013). Physicochemical and microbiological quality of raspberries (*Rubus idaeus*) treated with different doses of gamma irradiation. *Food Science and Technology, 33*(2), 316–322.

Gurtler, J. B., Bailey, R. B., Geveke, D. J., & Zhang, H. Q. (2011). Pulsed electric field inactivation of *E. coli* O157: H7 and non-pathogenic surrogate *E. coli* in strawberry juice as influenced by sodium benzoate, potassium sorbate, and citric acid. *Food Control, 22*(10), 1689–1694.

Hajare, S. N., Dhokane, V. S., Shashidhar, R., Saroj, S., Sharma, A., & Bandekar, J. R. (2006). Radiation processing of minimally processed pineapple (*Ananas comosus* Merr.): Effect on nutritional and sensory quality. *Journal of Food Science, 71*(6), S501–S505.

Han, Q., Gao, H., Chen, H., Fang, X., & Wu, W. (2017). Precooling and ozone treatments affects postharvest quality of black mulberry (Morus nigra) fruits. *Food Chemistry, 221*, 1947–1953.

Hao, H., Zhou, T., Koutchma, T., Wu, F., & Warriner, K. (2016). High hydrostatic pressure assisted degradation of patulin in fruit and vegetable juice blends. *Food Control, 62*, 237–242.

He, Q., Guo, M., Jin, T. Z., Arabi, S. A., & Liu, D. (2021). Ultrasound improves the decontamination effect of thyme essential oil nanoemulsions against *Escherichia coli* O157: H7 on cherry tomatoes. *International Journal of Food Microbiology, 337*, 108936.

Hu, X., Sun, H., Yang, X., Cui, D., Wang, Y., Zhuang, J., Wang, X., Ma, R., & Jiao, Z. (2021). Potential use of atmospheric cold plasma for postharvest preservation of blueberries. *Postharvest Biology and Technology, 179*, 111564.

Huang, K., Jiang, T., Wang, W., Gai, L., & Wang, J. (2014b). A comparison of pulsed electric field resistance for three microorganisms with different biological factors in grape juice via numerical simulation. *Food and Bioprocess Technology, 7*(7), 1981–1995.

Huang, K., Yu, L., Wang, W., Gai, L., & Wang, J. (2014a). Comparing the pulsed electric field resistance of the microorganisms in grape juice: Application of the Weibull model. *Food Control, 35*(1), 241–251.

Huang, Y., Ye, M., Cao, X., & Chen, H. (2017). Pulsed light inactivation of murine norovirus, Tulane virus, *Escherichia coli* O157: H7 and *Salmonella* in suspension and on berry surfaces. *Food Microbiology, 61*, 1–4.

Hussain, P. R., Omeera, A., Suradkar, P. P., & Dar, M. A. (2014). Effect of combination treatment of gamma irradiation and ascorbic acid on physicochemical and microbial quality of minimally processed eggplant (*Solanum melongena* L.). *Radiation Physics and Chemistry, 103*, 131–141.

Iizuka, T., Maeda, S., & Shimizu, A. (2013). Removal of pesticide residue in cherry tomato by hydrostatic pressure. *Journal of Food Engineering, 116*(4), 796–800.

Iizuka, T., & Shimizu, A. (2014). Removal of pesticide residue from Brussels sprouts by hydrostatic pressure. *Innovative Food Science and Emerging Technologies, 22*, 70–75.

Ikeura, H., Kobayashi, F., & Tamaki, M. (2013). Ozone microbubble treatment at various water temperatures for the removal of residual pesticides with negligible effects on the physical properties of lettuce and cherry tomatoes. *Journal of Food Science, 78*(2), T350–T355.

Inan, F., Pala, M., & Doymaz, I. (2007). Use of ozone in detoxification of aflatoxin B1 in red pepper. *Journal of Stored Products Research, 43*(4), 425–429.

Ishihara, J., Umesawa, M., Okada, C., Kokubo, Y., & Iso, H. (2018). Relationship between vegetables and fruits (antioxidant vitamins, minerals, and fiber) intake and risk of cardiovascular disease. *Encyclopedia of Cardiovascular Research and Medicine, 1*, 249–283.

Jaeger, H., Janositz, A., & Knorr, D. (2010). The Maillard reaction and its control during food processing: The potential of emerging technologies. *Pathologie Biologie, 58*(3), 207–213.

Jambrak, A. R., Šimunek, M., Evačić, S., Markov, K., Smoljanić, G., & Frece, J. (2018). Influence of high power ultrasound on selected moulds, yeasts and *Alicyclobacillus acidoterrestris* in apple, cranberry and blueberry juice and nectar. *Ultrasonics, 83*, 3–17.

Janositz, A., Noack, A. K., & Knorr, D. (2011). Pulsed electric fields and their impact on the diffusion characteristics of potato slices. *LWT - Food Science and Technology, 44*(9), 1939–1945.

Jin, T. Z., Yu, Y., & Gurtler, J. B. (2017). Effects of pulsed electric field processing on microbial survival, quality change and nutritional characteristics of blueberries. *LWT, 77*, 517–524.

Jo, C., Ahn, D. U., & Lee, K. H. (2012). Effect of gamma irradiation on microbiological, chemical, and sensory properties of fresh ashitaba and kale juices. *Radiation Physics and Chemistry, 81*(8), 1076–1078.

Joshi, B., Moreira, R. G., Omac, B., & Castell-Perez, M. E. (2018). A process to decontaminate sliced fresh cucumber (*Cucumis sativus*) using electron beam irradiation. *LWT, 91*, 95–101.

Kamber, U., Gülbaz, G., Aksu, P., & Doğan, A. (2016). Detoxification of aflatoxin B_1 in red pepper (*Capsicum annuum* L.) by ozone treatment and its effect on microbiological and sensory quality. *Journal of Food Processing and Preservation, 41*(5), e13102.

Karaca, H., & Velioglu, Y. S. (2014). Effects of ozone treatments on microbial quality and some chemical properties of lettuce, spinach, and parsley. *Postharvest Biology and Technology, 88*, 46–53.

Katiyo, W., Yang, R., & Zhao, W. (2017). Effects of combined pulsed electric fields and mild temperature pasteurization on microbial inactivation and physicochemical properties of cloudy red apple juice (Malus pumila Niedzwetzkyana (Dieck)). *Journal of Food Safety, 37*(4), e12369.

Ke, Y. Y., Shyu, Y. T., & Wu, S. J. (2021). Evaluating the anti-inflammatory and antioxidant effects of broccoli treated with high hydrostatic pressure in cell models. *Foods, 10*(1), 167.

Khandpur, P., & Gogate, P. R. (2015). Effect of novel ultrasound based processing on the nutrition quality of different fruit and vegetable juices. *Ultrasonics Sonochemistry, 27*, 125–136.

Khandpur, P., & Gogate, P. R. (2016). Evaluation of ultrasound based sterilization approaches in terms of shelf life and quality parameters of fruit and vegetable juices. *Ultrasonics Sonochemistry, 29*, 337–353.

Khaneghah, A. M., Gavahian, M., Xia, Q., Denoya, G. I., Roselló-Soto, E., & Barba, F. J. (2020). Effect of pulsed electric field on Maillard reaction and hydroxymethylfurfural production. In F. J. Barba, O. Parniakov & A. Wiktor (Eds.), *Pulsed electric fields to obtain healthier and sustainable food for tomorrow* (pp. 129–140). Academic Press.

Kiang, W. S., Bhat, R., Rosma, A., & Cheng, L. H. (2013). Effects of thermosonication on the fate of *Escherichia coli* O157: H7 and *Salmonella* enteritidis in mango juice. *Letters in Applied Microbiology, 56*(4), 251–257.

Kim, D., Song, H., Lim, S., Yun, H., & Chung, J. (2007). Effects of gamma irradiation on the radiation-resistant bacteria and polyphenol oxidase activity in fresh kale juice. *Radiation Physics and Chemistry, 76*(7), 1213–1217.

Kim, H. Y., Ahn, J. J., Shahbaz, H. M., Park, K. H., & Kwon, J. H. (2014). Physical-, chemical-, and microbiological-based identification of electron beam-and γ-irradiated frozen crushed garlic. *Journal of Agricultural and Food Chemistry, 62*(31), 7920–7926.

Kim, J. H., & Min, S. C. (2017). Microwave-powered cold plasma treatment for improving microbiological safety of cherry tomato against *Salmonella. Postharvest Biology and Technology, 127*, 21–26.

Kim, J. H., & Min, S. C. (2018). Moisture vaporization-combined helium dielectric barrier discharge-cold plasma treatment for microbial decontamination of onion flakes. *Food Control, 84*, 321–329.

Kong, M., Murtaza, A., Hu, X., Iqbal, A., Zhu, L., Ali, S. W., Xu, X., Pan, S., & Hu, W. (2021). Effect of high-pressure carbon dioxide treatment on browning inhibition of fresh-cut Chinese water chestnut (Eleocharis tuberosa): Based on the comparison of damaged tissue and non-damaged tissue. *Postharvest Biology and Technology, 179*, 111557.

Kovačević, D. B., Putnik, P., Dragović-Uzelac, V., Pedisić, S., Jambrak, A. R., & Herceg, Z. (2016). Effects of cold atmospheric gas phase plasma on anthocyanins and color in pomegranate juice. *Food Chemistry, 190*, 317–323.

Kuete, V., Karaosmanoğlu, O., & Sivas, H. (2017). Anticancer activities of African medicinal spices and vegetables. In V. Kuete (Ed.), *Medicinal spices and vegetables from Africa* (pp. 271–297). Academic Press.

Lacombe, A., Niemira, B. A., Gurtler, J. B., Fan, X., Sites, J., Boyd, G., & Chen, H. (2015). Atmospheric cold plasma inactivation of aerobic microorganisms on blueberries and effects on quality attributes. *Food Microbiology, 46*, 479–484.

Lacombe, A., Niemira, B. A., Gurtler, J. B., Sites, J., Boyd, G., Kingsley, D. H., Li, X., & Chen, H. (2017). Nonthermal inactivation of norovirus surrogates on blueberries using atmospheric cold plasma. *Food Microbiology, 63*, 1–5.

Lee, H., Kim, J. E., Chung, M. S., & Min, S. C. (2015). Cold plasma treatment for the microbiological safety of cabbage, lettuce, and dried figs. *Food Microbiology, 51*, 74–80.

Lee, N. Y., Jo, C., Shin, D. H., Kim, W. G., & Byun, M. W. (2006). Effect of γ-irradiation on pathogens inoculated into ready-to-use vegetables. *Food Microbiology, 23*(7), 649–656.

Lee, T., Puligundla, P., & Mok, C. (2018). Intermittent corona discharge plasma jet for improving tomato quality. *Journal of Food Engineering, 223*, 168–174.

Li, J., Cheng, H., Liao, X., Liu, D., Xiang, Q., Wang, J., Chen, S., Ye, X., & Ding, T. (2019). Inactivation of *Bacillus subtilis* and quality assurance in Chinese bayberry (*Myrica rubra*) juice with ultrasound and mild heat. *LWT, 108*, 113–119.

Liang, Z., Cheng, Z., & Mittal, G. S. (2006). Inactivation of spoilage microorganisms in apple cider using a continuous flow pulsed electric field system. *LWT - Food Science and Technology, 39*(4), 351–357.

Limnaios, A., Pathak, N., Bovi, G. G., Fröhling, A., Valdramidis, V. P., Taoukis, P. S., & Schlüter, O. (2021). Effect of cold atmospheric pressure plasma processing on quality and shelf life of red currants. *LWT*, 112213.

Liu, F., Zhang, X., Zhao, L., Wang, Y., & Liao, X. (2016). Potential of high-pressure processing and high-temperature/short-time thermal processing on microbial, physicochemical and sensory assurance of clear cucumber juice. *Innovative Food Science and Emerging Technologies, 34*, 51–58.

Lone, S. A., Raghunathan, S., Davoodbasha, M., Srinivasan, H., & Lee, S. Y. (2019). An investigation on the sterilization of berry fruit using ozone: An option to preservation and long-term storage. *Biocatalysis and Agricultural Biotechnology, 20*, 101212.

Lopez, M. E., Gontijo, M. T., Boggione, D. M., Albino, L. A., Batalha, L. S., & Mendonça, R. C. (2018). Microbiological contamination in foods and beverages: Consequences and alternatives in the era of microbial resistance. In A. M. Holban & A. M. Grumezescu (Eds), *Microbial contamination and food degradation* (pp. 49–84). Academic Press.

Loredo, A. B. G., Guerrero, S. N., & Alzamora, S. M. (2015). Inactivation kinetics and growth dynamics during cold storage of *Escherichia coli* ATCC 11229, *Listeria innocua* ATCC 33090 and *Saccharomyces cerevisiae* KE162 in peach juice using aqueous ozone. *Innovative Food Science and Emerging Technologies, 29*, 271–279.

Lozowicka, B., Jankowska, M., Hrynko, I., & Kaczynski, P. (2016). Removal of 16 pesticide residues from strawberries by washing with tap and ozone water, ultrasonic cleaning and boiling. *Environmental Monitoring and Assessment, 188*(1), 51.

Lu, Z., Yu, Z., Gao, X., Lu, F., & Zhang, L. (2005). Preservation effects of gamma irradiation on fresh-cut celery. *Journal of Food Engineering, 67*(3), 347–351.

Maftei, N. A., Ramos-Villarroel, A. Y., Nicolau, A. I., Martín-Belloso, O., & Soliva-Fortuny, R. (2014). Influence of processing parameters on the pulsed-light inactivation of *Penicillium expansum* in apple juice. *Food Control, 41*, 27–31.

Majeed, A., Muhammad, Z., Majid, A., Shah, A. H., & Hussain, M. (2014). Impact of low doses of gamma irradiation on shelf life and chemical quality of strawberry (*Fragaria* x *ananassa*) cv.'Corona'. *Journal of Animal and Plant Sciences, 24*(5), 1531–1536.

Marszałek, K., Krzyżanowska, J., Woźniak, Ł., & Skąpska, S. (2017). Kinetic modelling of polyphenol oxidase, peroxidase, pectin esterase, polygalacturonase, degradation of the main pigments and polyphenols in beetroot juice during high pressure carbon dioxide treatment. *LWT - Food Science and Technology, 85*, 412–417.

Marszałek, K., Woźniak, Ł., Barba, F. J., Skąpska, S., Lorenzo, J. M., Zambon, A., & Spilimbergo, S. (2018). Enzymatic, physicochemical, nutritional and phytochemical profile changes of apple (Golden Delicious L.) juice under supercritical carbon dioxide and long-term cold storage. *Food Chemistry, 268*, 279–286.

Martín-Belloso, O., & Elez-Martínez, P. (2005). Food safety aspects of pulsed electric fields. In Da-Wen Sun (Ed.), *Emerging technologies for food processing* (pp. 183–217). Academic Press.

Martins, C. G., Behrens, J. H., Destro, M. T., Franco, B. D. G. M., Vizeu, D. M., Hutzler, B., & Landgraf, M. (2004). Gamma radiation in the reduction of *Salmonella* spp. inoculated on minimally processed watercress (*Nasturtium officinalis*). *Radiation Physics and Chemistry, 71*(1–2), 89–93.

McNamee, C., Noci, F., Cronin, D. A., Lyng, J. G., Morgan, D. J., & Scannell, A. G. M. (2010). PEF based hurdle strategy to control *Pichia fermentans, Listeria innocua* and *Escherichia coli* K12 in orange juice. *International Journal of Food Microbiology, 138*(1–2), 13–18.

Merkulow, N., Eicher, R., & Ludwig, H. (2000). Pressure inactivation of fungal spores in aqueous model solutions and in real food systems. *International Journal of High Pressure Research, 19*(1–6), 253–262.

Millan-Sango, D., Garroni, E., Farrugia, C., Van Impe, J. F., & Valdramidis, V. P. (2016). Determination of the efficacy of ultrasound combined with essential oils on the decontamination of *Salmonella* inoculated lettuce leaves. *LWT, 73*, 80–87.

Min, S. C., Roh, S. H., Niemira, B. A., Sites, J. E., Boyd, G., & Lacombe, A. (2016). Dielectric barrier discharge atmospheric cold plasma inhibits *Escherichia coli* O157: H7, *Salmonella, Listeria monocytogenes*, and Tulane virus in Romaine lettuce. *International Journal of Food Microbiology, 237*, 114–120.

Mirza Alizadeh, A., Hashempour-Baltork, F., Mousavi Khaneghah, A., & Hosseini, H. (2021). New perspective approaches in controlling fungi and mycotoxins in food using emerging and green technologies. *Current Opinion in Food Science, 39*, 7–15.

Misra, N. N. (2015). The contribution of non-thermal and advanced oxidation technologies towards dissipation of pesticide residues. *Trends in Food Science and Technology, 45*(2), 229–244.

Moens, L. G., De Laet, E., Van Wambeke, J., Van Loey, A. M., & Hendrickx, M. E. (2020). Pulsed electric field and mild thermal processing affect the cooking behaviour of carrot tissues (*Daucus carota*) and the degree of methylesterification of carrot pectin. *Innovative Food Science and Emerging Technologies, 66*, 102483.

Mohácsi-Farkas, C., Nyirő-Fekete, B., Daood, H., Dalmadi, I., & Kiskó, G. (2014). Improving microbiological safety and maintaining sensory and nutritional quality of pre-cut tomato and carrot by gamma irradiation. *Radiation Physics and Chemistry, 99*, 79–85.

Mosqueda-Melgar, J., Raybaudi-Massilia, R. M., & Martín-Belloso, O. (2012). Microbiological shelf life and sensory evaluation of fruit juices treated by high-intensity pulsed electric fields and antimicrobials. *Food and Bioproducts Processing, 90*(2), 205–214.

Mostafidi, M., Sanjabi, M. R., Shirkhan, F., & Zahedi, M. T. (2020). A review of recent trends in the development of the microbial safety of fruits and vegetables. *Trends in Food Science and Technology, 103*, 321–332.

Mutlu-Ingok, A., Devecioglu, D., Dikmetas, D. N., & Karbancioglu-Guler, F. (2021). Safety issues for non-thermal food processing applications in grain industries. In M. Selvamuthukumaran (Ed.), *Non-thermal processing technologies for the grain industry* (pp. 1–233). CRC Press.

Nam, H. A., Ramakrishnan, S. R., & Kwon, J. H. (2019). Effects of electron-beam irradiation on the quality characteristics of mandarin oranges (*Citrus unshiu* (Swingle) Marcov) during storage. *Food Chemistry, 286*, 338–345.

Naresh, K., Varakumar, S., Variyar, P. S., Sharma, A., & Reddy, O. V. S. (2015a). Effect of γ-irradiation on physico-chemical and microbiological properties of mango (*Mangifera indica* L.) juice from eight Indian cultivars. *Food Bioscience, 12*, 1–9.

Naresh, K., Varakumar, S., Variyar, P. S., Sharma, A., & Reddy, O. V. S. (2015b). Enhancing antioxidant activity, microbial and sensory quality of mango (*Mangifera indica* L.) juice by γ-irradiation and its in vitro radioprotective potential. *Journal of Food Science and Technology*, *52*(7), 4054–4065.

Niveditha, A., Pandiselvam, R., Prasath, V. A., Singh, S. K., Gul, K., & Kothakota, A. (2021). Application of cold plasma and ozone technology for decontamination of *Escherichia coli* in foods-A review. *Food Control*, *130*, 108338.

Oh, Y. J., Song, A. Y., & Min, S. C. (2017). Inhibition of *Salmonella typhimurium* on radish sprouts using nitrogen-cold plasma. *International Journal of Food Microbiology*, *249*, 66–71.

Özen, T., Koyuncu, M. A., & Erbaş, D. (2021). Effect of ozone treatments on the removal of pesticide residues and postharvest quality in green pepper. *Journal of Food Science and Technology*, *58*(6), 2186–2196.

Palgan, I., Caminiti, I. M., Muñoz, A., Noci, F., Whyte, P., Morgan, D. J., Cronin, D. & Lyng, J. G. (2011). Combined effect of selected non-thermal technologies on *Escherichia coli* and *Pichia fermentans* inactivation in an apple and cranberry juice blend and on product shelf life. *International Journal of Food Microbiology*, *151*(1), 1–6.

Pallarés, N., Berrada, H., Tolosa, J., & Ferrer, E. (2021). Effect of high hydrostatic pressure (HPP) and pulsed electric field (PEF) technologies on reduction of aflatoxins in fruit juices. *LWT*, *142*, 111000.

Pandiselvam, R., Kaavya, R., Jayanath, Y., Veenuttranon, K., Lueprasitsakul, P., Divya, V., Kothakota, A. & Ramesh, S. V. (2020). Ozone as a novel emerging technology for the dissipation of pesticide residues in foods–a review. *Trends in Food Science and Technology*, *97*, 38–54.

Panigrahi, C., Mishra, H. N., & De, S. (2021). Modelling the inactivation kinetics of *Leuconostoc mesenteroides*, *Saccharomyces cerevisiae* and total coliforms during ozone treatment of sugarcane juice. *LWT*, *144*, 111218.

Pankaj, S. K., Misra, N. N., & Cullen, P. J. (2013). Kinetics of tomato peroxidase inactivation by atmospheric pressure cold plasma based on dielectric barrier discharge. *Innovative Food Science and Emerging Technologies*, *19*, 153–157.

Panou, A. A., Karabagias, I. K., & Riganakos, K. A. (2018). The effect of different gaseous ozone treatments on physicochemical characteristics and shelf life of apricots stored under refrigeration. *Journal of Food Processing and Preservation*, *42*(5), e13614.

Papachristodoulou, M., Koukounaras, A., Siomos, A. S., Liakou, A., & Gerasopoulos, D. (2018). The effects of ozonated water on the microbial counts and the shelf life attributes of fresh-cut spinach. *Journal of Food Processing and Preservation*, *42*(1), e13404.

Park, H. S., & Kim, K. H. (2013). Enhancement of supercritical CO_2 inactivation of spores of *Penicillium oxalicum* by ethanol cosolvent. *Journal of Microbiology and Biotechnology*, *23*(6), 833–836.

Pasquali, F., Stratakos, A. C., Koidis, A., Berardinelli, A., Cevoli, C., Ragni, L., Mancusi, R., Manfreda, G., & Trevisani, M. (2016). Atmospheric cold plasma process for vegetable leaf decontamination: A feasibility study on radicchio (red chicory, *Cichorium intybus* L.). *Food Control*, *60*, 552–559.

Pérez-Grijalva, B., Herrera-Sotero, M., Mora-Escobedo, R., Zebadúa-García, J. C., Silva-Hernández, E., Oliart-Ros, R., Pérez-Cruz, C., & Guzmán-Gerónimo, R. (2018). Effect of microwaves and ultrasound on bioactive compounds and microbiological quality of blackberry juice. *LWT*, *87*, 47–53.

Phan, K. T. K., Phan, H. T., Boonyawan, D., Intipunya, P., Brennan, C. S., Regenstein, J. M., & Phimolsiripol, Y. (2018). Non-thermal plasma for elimination of pesticide residues in mango. *Innovative Food Science and Emerging Technologies*, *48*, 164–171.

Picart-Palmade, L., Cunault, C., Chevalier-Lucia, D., Belleville, M. P., & Marchesseau, S. (2019). Potentialities and limits of some non-thermal technologies to improve sustainability of food processing. *Frontiers in Nutrition*, *5*, 130.

Plazzotta, S., & Manzocco, L. (2019). High-pressure carbon dioxide treatment of fresh fruit juices. *Value-Added Ingredients and Enrichments of Beverages*, 429–463.

Preetha, P., Pandiselvam, R., Varadharaju, N., Kennedy, Z. J., Balakrishnan, M., & Kothakota, A. (2021). Effect of pulsed light treatment on inactivation kinetics of *Escherichia coli* (MTCC 433) in fruit juices. *Food Control*, *121*, 107547.

Puligundla, P., Lee, T., & Mok, C. (2018). Effect of intermittent corona discharge plasma treatment for improving microbial quality and shelf life of kumquat (*Citrus japonica*) fruits. *LWT*, *91*, 8–13.

Radhakrishnan, M., & Tiwari, B. (2021). *Application of ultrasound in fruit and vegetable processing*. Elsevier.

Ramos, B., Miller, F. A., Brandão, T. R. S., Teixeira, P., & Silva, C. L. M. (2013). Fresh fruits and vegetables—An overview on applied methodologies to improve its quality and safety. *Innovative Food Science and Emerging Technologies*, *20*, 1–15.

Rastogi, N. K., Raghavarao, K. S. M. S., Balasubramaniam, V. M., Niranjan, K., & Knorr, D. (2007). Opportunities and challenges in high pressure processing of foods. *Critical Reviews in Food Science and Nutrition*, *47*(1), 69–112.

Rodrigues, A. A. Z., de Queiroz, M. E. L. R., Faroni, L. R. D. A., Prates, L. H. F., Neves, A. A., de Oliveira, A. F., de Freitas, J. F., Heleno, F. F., & Zambolim, L. (2021). The efficacy of washing strategies in the elimination of fungicide residues and the alterations on the quality of bell peppers. *Food Research International*, *147*, 110579.

Rodrigues, A. A. Z., de Queiroz, M. E. L. R., Neves, A. A., de Oliveira, A. F., Prates, L. H. F., de Freitas, J. F., Heleno, F. F. & Faroni, L. R. D. A. (2019). Use of ozone and detergent for removal of pesticides and improving storage quality of tomato. *Food Research International*, *125*, 108626.

Rojas, M. L., Miano, A. C., & Augusto, P. E. (2017). Ultrasound processing of fruit and vegetable juices. In D. Bermudez-Aguirre (Ed.), *Ultrasound: Advances for food processing and preservation* (pp. 181–199). Academic Press.

Roobab, U., Shabbir, M. A., Khan, A. W., Arshad, R. N., Bekhit, A. E. D., Zeng, X. A., Inam-Ur-Raheem, M. & Aadil, R. M. (2021). High-pressure treatments for better quality clean-label juices and beverages: Overview and advances. *LWT*, *149*, 111828.

Rybak, K., Samborska, K., Jedlinska, A., Parniakov, O., Nowacka, M., Witrowa-Rajchert, D., & Wiktor, A. (2020). The impact of pulsed electric field pretreatment of bell pepper on the selected properties of spray dried juice. *Innovative Food Science and Emerging Technologies*, *65*, 102446.

Rybak, K., Wiktor, A., Pobiega, K., Witrowa-Rajchert, D., & Nowacka, M. (2021). Impact of pulsed light treatment on the quality properties and microbiological aspects of red bell pepper fresh-cuts. *LWT*, *149*, 111906.

Sadło, S., Szpyrka, E., Piechowicz, B., Antos, P., Józefczyk, R., & Balawejder, M. (2017). Reduction of captan, Boscalid and pyraclostrobin residues on apples using water only, gaseous ozone, and ozone Aqueous solution. *Ozone: Science and Engineering*, *39*(2), 97–103.

Sagong, H. G., Cheon, H. L., Kim, S. O., Lee, S. Y., Park, K. H., Chung, M. S., Choi, Y. J., & Kang, D. H. (2013). Combined effects of ultrasound and surfactants to reduce *Bacillus cereus* spores on lettuce and carrots. *International Journal of Food Microbiology*, *160*(3), 367–372.

Sagong, H. G., Lee, S. Y., Chang, P. S., Heu, S., Ryu, S., Choi, Y. J., & Kang, D. H. (2011). Combined effect of ultrasound and organic acids to reduce *Escherichia coli* O157: H7, *Salmonella Typhimurium*, and *Listeria monocytogenes* on organic fresh lettuce. *International Journal of Food Microbiology*, *145*(1), 287–292.

Saito-Shida, S., Nemoto, S., & Matsuda, R. (2014). Multiresidue analysis of pesticides in vegetables and fruits by supercritical fluid extraction and liquid chromatography-tandem mass spectrometry. *Shokuhin Eiseigaku Zasshi. Journal of the Food Hygienic Society of Japan*, *55*(3), 142–151.

Salinas-Roca, B., Soliva-Fortuny, R., Welti-Chanes, J., & Martín-Belloso, O. (2016). Combined effect of pulsed light, edible coating and malic acid dipping to improve fresh-cut mango safety and quality. *Food Control*, *66*, 190–197.

São José, J. F. B., & Vanetti, M. C. D. (2012). Effect of ultrasound and commercial sanitizers in removing natural contaminants and *Salmonella enterica* typhimurium on cherry tomatoes. *Food Control*, *24*(1–2), 95–99.

Sarangapani, C., O'Toole, G., Cullen, P. J., & Bourke, P. (2017). Atmospheric cold plasma dissipation efficiency of agrochemicals on blueberries. *Innovative Food Science and Emerging Technologies*, *44*, 235–241.

Sartori, R. B., Higino, M. L., Bastos, L. H. P., & Mendes, M. F. (2017). Supercritical extraction of pesticides from banana: Experimental and modeling. *Journal of Supercritical Fluids*, *128*, 149–158.

Silva, E. K., Alvarenga, V. O., Bargas, M. A., Sant'Ana, A. S., & Meireles, M. A. A. (2018). Non-thermal microbial inactivation by using supercritical carbon dioxide: Synergic effect of process parameters. *Journal of Supercritical Fluids*, *139*, 97–104.

Silva, E. K., Meireles, M. A. A., & Saldaña, M. D. (2020). Supercritical carbon dioxide technology: A promising technique for the non-thermal processing of freshly fruit and vegetable juices. *Trends in Food Science and Technology*, *97*, 381–390.

Silva, F. V. M., & Evelyn. (2020). Resistant moulds as pasteurization target for cold distributed high pressure and heat assisted high pressure processed fruit products. *Journal of Food Engineering, 282*, 109998.

Sintuya, P., Narkprasom, K., Jaturonglumlert, S., Whangchai, N., Peng-Ont, D., & Varith, J. (2018). Effect of gaseous ozone fumigation on organophosphate pesticide degradation of dried chilies. *Ozone: Science and Engineering, 40*(6), 473–481.

Song, H. P., Byun, M. W., Jo, C., Lee, C. H., Kim, K. S., & Kim, D. H. (2007). Effects of gamma irradiation on the microbiological, nutritional, and sensory properties of fresh vegetable juice. *Food Control, 18*(1), 5–10.

Song, W. J., Sung, H. J., Kim, S. Y., Kim, K. P., Ryu, S., & Kang, D. H. (2014). Inactivation of *Escherichia coli* O157: H7 and *Salmonella* typhimurium in black pepper and red pepper by gamma irradiation. *International Journal of Food Microbiology, 172*, 125–129.

Sung, H. J., Song, W. J., Kim, K. P., Ryu, S., & Kang, D. H. (2014). Combination effect of ozone and heat treatments for the inactivation of *Escherichia coli* O157: H7, *Salmonella Typhimurium*, and *Listeria monocytogenes* in apple juice. *International Journal of Food Microbiology, 171*, 147–153.

Surowsky, B., Froehling, A., Gottschalk, N., Schlüter, O., & Knorr, D. (2014). Impact of cold plasma on *Citrobacter freundii* in apple juice: Inactivation kinetics and mechanisms. *International Journal of Food Microbiology, 174*, 63–71.

Szczepańska, J., Barba, F. J., Skąpska, S., & Marszałek, K. (2020). High pressure processing of carrot juice: Effect of static and multi-pulsed pressure on the polyphenolic profile, oxidoreductases activity and colour. *Food Chemistry, 307*, 125549.

Tabakoglu, N., & Karaca, H. (2018). Effects of ozone-enriched storage atmosphere on postharvest quality of black mulberry fruits (*Morus nigra* L.). *LWT, 92*, 276–281.

Takundwa, B. A., Bhagwat, P., Pillai, S., & Ijabadeniyi, O. A. (2021). Antimicrobial efficacy of nisin, oregano and ultrasound against *Escherichia coli* O157: H7 and *Listeria monocytogenes* on lettuce. *LWT, 139*, 110522.

Tang, Y., Jiang, Y., Jing, P., & Jiao, S. (2021). Dense phase carbon dioxide treatment of mango in syrup: Microbial and enzyme inactivation, and associated quality change. *Innovative Food Science and Emerging Technologies, 70*, 102688.

Taniwaki, M. H., Hocking, A. D., Pitt, J. I., & Fleet, G. H. (2009). Growth and mycotoxin production by food spoilage fungi under high carbon dioxide and low oxygen atmospheres. *International Journal of Food Microbiology, 132*(2–3), 100–108.

Tao, T., Ding, C., Han, N., Cui, Y., Liu, X., & Zhang, C. (2019). Evaluation of pulsed light for inactivation of foodborne pathogens on fresh-cut lettuce: Effects on quality attributes during storage. *Food Packaging and Shelf Life, 21*, 100358.

Tappi, S., Berardinelli, A., Ragni, L., Dalla Rosa, M., Guarnieri, A., & Rocculi, P. (2014). Atmospheric gas plasma treatment of fresh-cut apples. *Innovative Food Science and Emerging Technologies, 21*, 114–122.

Tappi, S., Gozzi, G., Vannini, L., Berardinelli, A., Romani, S., Ragni, L., & Rocculi, P. (2016). Cold plasma treatment for fresh-cut melon stabilization. *Innovative Food Science and Emerging Technologies, 33*, 225–233.

Tawema, P., Han, J., Vu, K. D., Salmieri, S., & Lacroix, M. (2016). Antimicrobial effects of combined UV-C or gamma radiation with natural antimicrobial formulations against *Listeria monocytogenes*, *Escherichia coli* O157: H7, and total yeasts/molds in fresh cut cauliflower. *LWT - Food Science and Technology, 65*, 451–456.

Timmermans, R. A. H., Groot, M. N., Nederhoff, A. L., Van Boekel, M. A. J. S., Matser, A. M., & Mastwijk, H. C. (2014). Pulsed electric field processing of different fruit juices: Impact of pH and temperature on inactivation of spoilage and pathogenic micro-organisms. *International Journal of Food Microbiology, 173*, 105–111.

Tokuşoğlu, Ö., Alpas, H., & Bozoğlu, F. (2010). High hydrostatic pressure effects on mold flora, citrinin mycotoxin, hydroxytyrosol, oleuropein phenolics and antioxidant activity of black table olives. *Innovative Food Science and Emerging Technologies, 11*(2), 250–258.

Tolcha, T., Gemechu, T., Al-Hamimi, S., Megersa, N., & Turner, C. (2020). High density supercritical carbon dioxide for the extraction of pesticide residues in onion with multivariate response surface methodology. *Molecules, 25*(4), 1012.

Torabian, G., Bahramian, B., Zambon, A., Spilimbergo, S., Adil, Q., Schindeler, A., Valtchev, P., & Dehghani, F. (2018). A hybrid process for increasing the shelf life of elderberry juice. *Journal of Supercritical Fluids*, *140*, 406–414.

Torlak, E. (2014). Efficacy of ozone against *Alicyclobacillus acidoterrestris* spores in apple juice. *International Journal of Food Microbiology*, *172*, 1–4.

Tremarin, A., Brandão, T. R., & Silva, C. L. (2017). Application of ultraviolet radiation and ultrasound treatments for *Alicyclobacillus acidoterrestris* spores inactivation in apple juice. *LWT*, *78*, 138–142.

Trinetta, V., Vaidya, N., Linton, R., & Morgan, M. (2011). A comparative study on the effectiveness of chlorine dioxide gas, ozone gas and e-beam irradiation treatments for inactivation of pathogens inoculated onto tomato, cantaloupe and lettuce seeds. *International Journal of Food Microbiology*, *146*(2), 203–206.

Ukuku, D. O., Niemira, B. A., & Ukanalis, J. (2019). Nisin-based antimicrobial combination with cold plasma treatment inactivate *Listeria monocytogenes* on Granny Smith apples. *LWT*, *104*, 120–127.

Varalakshmi, S. (2021). A review on the application and safety of non-thermal techniques on fresh produce and their products. *LWT*, *149*, 111849.

Verde, S. C., Trigo, M. J., Sousa, M. B., Ferreira, A., Ramos, A. C., Nunes, I., Janqueira, C., Melo, R., Santos, P. M. P., & Botelho, M. L. (2013). Effects of gamma radiation on raspberries: Safety and quality issues. *Journal of Toxicology and Environmental Health, Part A*, *76*(4–5), 291–303.

Villamiel, M., Gamboa, J., Soria, A. C., Riera, E., García-Pérez, J. V., & Montilla, A. (2015). Impact of power ultrasound on the quality of fruits and vegetables during dehydration. *Physics Procedia*, *70*, 828–832.

Waghmare, R. (2021). Cold plasma technology for fruit based beverages: A review. *Trends in Food Science and Technology*, *114*, 60–69.

Walkling-Ribeiro, M., Noci, F., Riener, J., Cronin, D. A., Lyng, J. G., & Morgan, D. J. (2009). The impact of thermosonication and pulsed electric fields on *Staphylococcus aureus* inactivation and selected quality parameters in orange juice. *Food and Bioprocess Technology*, *2*(4), 422–430.

Wang, J., Liu, Q., Xie, B., & Sun, Z. (2020). Effect of ultrasound combined with ultraviolet treatment on microbial inactivation and quality properties of mango juice. *Ultrasonics Sonochemistry*, *64*, 105000.

Wang, L., Fan, X., Sokorai, K., & Sites, J. (2019). Quality deterioration of grape tomato fruit during storage after treatments with gaseous ozone at conditions that significantly reduced populations of *Salmonella* on stem scar and smooth surface. *Food Control*, *103*, 9–20.

Wang, S., Wang, J., Li, C., Xu, Y., & Wu, Z. (2021). Ozone treatment pak choi for the removal of malathion and carbosulfan pesticide residues. *Food Chemistry*, *337*, 127755.

Wani, S., Maker, J. K., Thompson, J. R., Barnes, J., & Singleton, I. (2015). Effect of ozone treatment on inactivation of *Escherichia coli* and *Listeria* sp. on spinach. *Agriculture*, *5*(2), 155–169.

Woldemariam, H. W., & Emire, S. A. (2019). High pressure processing of foods for microbial and mycotoxins control: Current trends and future prospects. *Cogent Food and Agriculture*, *5*(1), 1622184.

Won, M. Y., Lee, S. J., & Min, S. C. (2017). Mandarin preservation by microwave-powered cold plasma treatment. *Innovative Food Science and Emerging Technologies*, *39*, 25–32.

Wu, W., Xiao, G., Yu, Y., Xu, Y., Wu, J., Peng, J., & Li, L. (2021). Effects of high pressure and thermal processing on quality properties and volatile compounds of pineapple fruit juice. *Food Control*, *130*, 108293.

Wu, X., Wang, C., & Guo, Y. (2020). Effects of the high-pulsed electric field pretreatment on the mechanical properties of fruits and vegetables. *Journal of Food Engineering*, *274*, 109837.

Xu, F., Wang, B., Hong, C., Telebielaigen, S., Nsor-Atindana, J., Duan, Y., & Zhong, F. (2019). Optimization of spiral continuous flow-through pulse light sterilization for *Escherichia coli* in red grape juice by response surface methodology. *Food Control*, *105*, 8–12.

Xu, W., Chen, H., Huang, Y., & Wu, C. (2013). Decontamination of *Escherichia coli* O157: H7 on green onions using pulsed light (PL) and PL–surfactant–sanitizer combinations. *International Journal of Food Microbiology*, *166*(1), 102–108.

Xu, W., & Wu, C. (2016). The impact of pulsed light on decontamination, quality, and bacterial attachment of fresh raspberries. *Food Microbiology*, *57*, 135–143.

Yildiz, S., Pokhrel, P. R., Unluturk, S., & Barbosa-Cánovas, G. V. (2021). Shelf life extension of strawberry juice by equivalent ultrasound, high pressure, and pulsed electric fields processes. *Food Research International*, *140*, 110040.

Yousuf, B., Deshi, V., Ozturk, B., & Siddiqui, M. W. (2020). Fresh-cut fruits and vegetables: Quality issues and safety concerns. In M. W. Siddiqui (Ed.), *Fresh-cut fruits and vegetables* (pp. 1–15). Academic Press.

Zarbakhsh, S., & Rastegar, S. (2019). Influence of postharvest gamma irradiation on the antioxidant system, microbial and shelf life quality of three cultivars of date fruits (*Phoenix dactylifera* L.). *Scientia Horticulturae, 247*, 275–286.

Zeng, L. F., Yang, W. Y., Liang, G. H., Luo, M. H., Cao, Y., Chen, H. Y., Pan, J. K., Huang, H. T., Han, Y. H., Zhao, D., Lin, J. T., Hou, S. R., Ou, A. H., Guan, Z. T., Wang, Q., & Liu, J. (2019). Can increasing the prevalence of vegetable-based diets lower the risk of osteoporosis in postmenopausal subjects? A systematic review with meta-analysis of the literature. *Complementary Therapies in Medicine, 42*, 302–311.

Zhang, J., Iqbal, A., Murtaza, A., Zhou, X., Xu, X., Pan, S., & Hu, W. (2021c). Effect of high pressure carbon dioxide on the browning inhibition of sugar-preserved orange peel. *Journal of CO2 Utilization, 46*, 101467.

Zhang, L., Lu, Z., Lu, F., & Bie, X. (2006). Effect of γ irradiation on quality-maintaining of fresh-cut lettuce. *Food Control, 17*(3), 225–228.

Zhang, W., Liang, L., Pan, X., Lao, F., Liao, X., & Wu, J. (2021b). Alterations of phenolic compounds in red raspberry juice induced by high-hydrostatic-pressure and high-temperature short-time processing. *Innovative Food Science and Emerging Technologies, 67*, 102569.

Zhang, Y., Hou, Y., Zhang, Y., Chen, J., Chen, F., Liao, X., & Hu, X. (2012). Reduction of diazinon and dimethoate in apple juice by pulsed electric field treatment. *Journal of the Science of Food and Agriculture, 92*(4), 743–750.

Zhang, Y., Zhang, J., Zhang, Y., Hu, H., Luo, S., Zhang, L., Zhou, H., & Li, P. (2021a). Effects of in-package atmospheric cold plasma treatment on the qualitative, metabolic and microbial stability of fresh-cut pears. *Journal of the Science of Food and Agriculture, 101*(11), 4473–4480.

Zhao, B., Hu, S., Wang, D., Chen, H., & Huang, M. (2020). Inhibitory effect of gamma irradiation on *Penicillium digitatum* and its application in the preservation of Ponkan fruit. *Scientia Horticulturae, 272*, 109598.

Zhong, L., Li, X., Duan, M., Song, Y., He, N., & Che, L. (2021). Impacts of high hydrostatic pressure processing on the structure and properties of pectin. *LWT, 148*, 111793.

Zhou, Y. H., Vidyarthi, S. K., Zhong, C. S., Zheng, Z. A., An, Y., Wang, J., Wei, Q., & Xiao, H. W. (2020). Cold plasma enhances drying and color, rehydration ratio and polyphenols of wolfberry via microstructure and ultrastructure alteration. *LWT, 134*, 110173.

Ziuzina, D., Patil, S., Cullen, P. J., Keener, K. M., & Bourke, P. (2014). Atmospheric cold plasma inactivation of *Escherichia coli, Salmonella enterica* serovar Typhimurium and *Listeria monocytogenes* inoculated on fresh produce. *Food Microbiology, 42*, 109–116.

Zook, C. D., Parish, M. E., Braddock, R. J., & Balaban, M. O. (1999). High pressure inactivation kinetics of Saccharomyces cerevisiae ascospores in orange and apple juices. *Journal of Food Science, 64*(3), 533–535.

Zorlugenç, B., Zorlugenç, F. K., Öztekin, S., & Evliya, I. B. (2008). The influence of gaseous ozone and ozonated water on microbial flora and degradation of aflatoxin B1 in dried figs. *Food and Chemical Toxicology, 46*(12), 3593–3597.

Index

For Product Safety Concerns and Information please contact our EU
representative GPSR@taylorandfrancis.com
Taylor & Francis Verlag GmbH, Kaufingerstraße 24, 80331 München, Germany

www.ingramcontent.com/pod-product-compliance
Lightning Source LLC
Chambersburg PA
CBHW061349210326
41598CB00035B/5925

* 9 7 8 1 0 3 2 1 1 9 2 7 4 *